INORGANIC
CRYSTAL
STRUCTURES

INORGANIC CRYSTAL STRUCTURES

B.G. Hyde

Research School of Chemistry
The Australian National University
Canberra, A.C.T.
Australia

Sten Andersson

Kemicentrum
Oorganisk Kemi 2
Lund University
Lund, Sweden

WILEY

A WILEY-INTERSCIENCE PUBLICATION

JOHN WILEY & SONS

New York   Chichester   Brisbane   Toronto   Singapore

Library of Congress Cataloging in Publication Data:

Hyde, Bruce G.
    Inorganic crystal structures.

    "A Wiley-Interscience publication."
    Includes bibliographical references.
    1. Crystallography.   2. Inorganic compounds.

I. Andersson, S. (Sten)   II. Title.
QD921.H93   1988   548   88-5492
ISBN 0-471-62897-2

Printed in the United States of America

10   9   8   7   6   5   4   3   2   1

To David Wadsley, who led the way

It is a basic principle of mathematics to build more and
more complicated structures out of simple ones.
S. Hildebrandt and A. Troma,
Mathematics and Optional Form

. . . ; überall finden sich Querverbindungen, Anknüpfungspunkte
und allerlei Mystifikationen.
Karl Schumann (on the sleeve of
a record of Robert Schumann's
Kreisleriana and Kinderszenen)

Preface

With crystal structures, as (we are told) with gambling, one needs a system; otherwise the number of possibilities is too great. One purpose of this book is to provide such a system (for crystal structures): a system of classifying and relating structures that introduces some order into the vast and burgeoning mass of factual information that constitutes the body of crystal structure data. It is this which distinguishes the present book from previous structure texts.

While our treatment is by no means exhaustive (in the sense that we do not consider *every* structure that fits into this pattern), it is sufficiently comprehensive that the reader will be able to acquire the ability to deal with many more structures in a similar way.

The structures are invariably depicted by projecting them along a principal axis of a unit cell, atom heights being given as a fraction of the projection axis (usually hundredths). Each type of polyhedron is depicted in a uniform way—octahedra by line-shading (usually) one face, tetrahedra by dotting, and so on. They are almost invariably cation-centered polyhedra of coordinating anions and include trigonal prisms, bisdisphenoids, square antiprisms, and so forth, in addition to the more familiar tetrahedra and octahedra. "Reading" such drawings entails expending some effort to learn the "language"; but we feel that such a uniform method has strong merit (and is usually preferable to the more "three-dimensional" drawings such as clinographic projections). After a little perseverance we think the reader will agree. Our experience with graduate students shows that a facility with this "language" is readily acquired, particularly if the process is accompanied by the construction, or at least the handling, of (three-dimensional) models composed of appropriate articulated polyhedra.

There is no one "best way" of depicting *all* structure types (although this is not always appreciated). This is apparent from time to time in the text, where some structures are described in more than one way. While we usually emphasize the articulation of cation-centered polyhedra or anions, the reverse can also be useful; for example, A-type La_2O_3 is more elegantly described as articulated *anion*-centered polyhedra of *cations* (octahedra and tetrahedra) or as an "anion-stuffed" ("hcp") array of cations. The most appropriate mode of description may depend on the circumstances, and will sometimes certainly depend on the structure.

[The reader may like to try to describe, in the conventional terms of cation-centered polyhedra, the structure of such simple compounds as β-K_2SO_4, Na_2CO_3 (not dealt with here), and some of the complex sulfosalts of, e.g. Sb and Sn or Pb. In his book,* Wyckoff comments on the problems involved in comprehending the structure of these and $BaSO_4$, etc. It transpires that there is a convenient method, which has been dealt with elsewhere.†]

While we have not tried to be completely comprehensive, we have tried to range widely enough to cover a large number of important structure types—because they are those of important compounds, or because they are significant in defining our systematic approach, or because at first sight they appear to be impossibly complex but yield readily to the system used here. In the course of the book we start with simple descriptions (e.g., those based on cubic close-packing/eutaxy or hexagonal close-packing/eutaxy) and proceed to less familiar and finally more complex descriptions. Each chapter starts with quite simple structures—often rather familiar ones—and proceeds to less simple ones and, finally, to those more complex structures that are generated by applying the standard symmetry operations to elementary units ("fundamental building blocks") of the simple structure types. A great deal of crystal chemistry is thereby reduced from an inchoate mass of fact to a logical and relatively simple whole—which aids the memory and the organization and manipulation of the facts (the separate structures).

We have ignored the artificial barrier between inorganic and mineral structures on the one hand and metallurgical structures (intermetallic compounds, borides, carbides, etc.) on the other. This does not greatly increase the number of structure types to be considered, which emphasizes the fact that the two groups have a great deal in common and that knowledge of each impinges usefully on the other.

Most of the drawings we have done ourselves (starting before computer graphics became useful for the purpose), but some were produced by our colleagues Ingrid Mellqvist, Harry Nyman, and Carlos Otero-Diaz, to whom we are grateful. Clearly, we also owe a great debt to those who produced high quality photographic reproductions of these many drawings, especially Byam Wight, Lilo Hennig, and Jane Sutton (in Canberra) and Ingrid Mellqvist (in Lund). Their help was essential and generous, as was the dedicated typing and retyping and word processor production of the text by Ingrid Mellqvist (Lund) and Jane Sutton and Caroline Twang (Canberra).

---

* R.W.G. Wyckoff, *Crystal Structures*, 2nd ed., Vol. 4, Wiley, New York, 1968, p. 545.
† M. O'Keeffe and B.G. Hyde, *Structure and Bonding* (Berlin) **61** (1985) 77–144.

The development of these ideas started longer ago than either of us cares to remember (and fragments were published as papers from time to time). During the intervening years we have, of course, benefited from discussions, criticisms, and even heated arguments (!) with many colleagues, far too many to mention individually. These have included our (post-) graduate students in Lund and Perth and Leiden and Canberra. But we must take a special acknowledgment of Michael O'Keeffe, who has been mentor and penetrating critic for most of this time. We also happily acknowledge our great debt to the eminent authors of previous structure books, especially Wyckoff, Wells, Schubert, and Bragg and Claringbull. They would, we think, be satisfied to see how tattered and scribbled and drawn upon are our copies of their important works. Not least, we thank our families for years of patient tolerance of crystal psychiatrists.

We hope that the reader will have at least half as much fun and excitement in this subject as we have.

B.G. HYDE
STEN ANDERSSON

Canberra, Australia
Lund, Sweden
August 1988

Contents

CHAPTER I

Introduction

In this short chapter we set out the essence of the system used to describe and relate crystal structures in subsequent chapters, in which the system is used without further description or explanation. Some very simple, two-dimensional examples are here used as illustrations.

Our approach is simply a geometrical one. Starting with relatively few basic structures, we apply to segments of such structures one or more of a few geometrical operations that are essentially those symmetry operators long used by crystallographers to generate structures from a set of point (atom) positions. The main difference is that in generating complex structures we operate on a building unit (element of a basic parent structure) instead of on a point. This approach involves the well-known "hierarchy of structures."

The aim is to facilitate the *visualization* of structures (in geometrical terms, usually cation-centered coordination polyhedra of anions), so we do not rely on space groups, especially space group hierarchies such as sub- and supergroup relations. These are very useful and very powerful and are used by other authors but are not used explicitly here. Our goal is an intuitive "feel" for structures, rather than formal symmetry relations.

We follow Hilbert's approach to geometry: "Let us consider three distinct systems of things. The things composing the first system we call points, those of the second system we will call straight lines, and those of the third system we will call planes" (1). Working in three dimensions, the analogues of these three things may need to be expanded to blocks (or building units or clusters; bounded in three dimensions), rods or columns (bounded in two dimensions, infinite in the third), and slabs or lamellae, sheets, or layers (bounded in only one dimension, infinite in the other two). That is, we will also be concerned with packets of points or groups of planes and/or lines.

As is frequently the case in analyzing the concepts of science, any attempt at a strict definition of point, line, or plane encounters logical difficulties

1

Figure 1. A simple example of a structure.

Figure 2. A structure formed by two interpenetrating identical systems of points.

[see, for example, Synge (2)]. In our consideration of crystal structures we will use geometry but leave these simple concepts undefined. In this way (with a point representing the position of an atom or a complex of atoms) we will arrive at useful and comprehensible descriptions of crystal structures (and crystal "defects"). A general, formal description is formulated in the following way (3).

I A system of points in which each point is repeated by symmetry operations (a crystal structure, Figure 1).

II *N* identical systems of points in which each point is repeated by symmetry operations, and therefore each system also. [Identical structures interpenetrate to form a new structure (Figures 2 and 3).]

III *N* different systems of points, each system repeated by symmetry operations as in II. [Different structures interpenetrate to form a new structure (Figure 4).]

IV A part of one or several systems of points of I, II, or III repeated by symmetry operations (Figures 5–7). Of the new system thus obtained, a part may also be repeated, et seq. (Complex structures leading to more complex ones: a hierarchy of structures.)

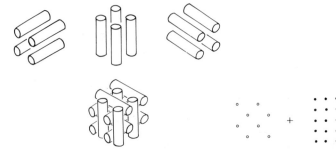

Figure 3. A three-dimensional structure formed by four interpenetrating identical systems of rods.

Figure 4. Two different systems of points that interpenetrate to give a structure (cf. Figure 2).

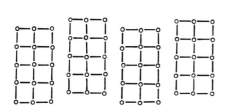

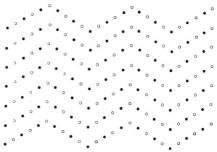

Figure 5. A part of a square array of points repeated (by translation) to form a structure.

Figure 6. A part of another array of points repeated (by reflection) to form a structure.

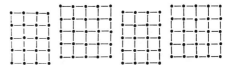

Figure 7. Part of an array of points repeated by reflection cyclically.

Figure 8. A structure produced by the repetition of two parts, of different size, of a system of points.

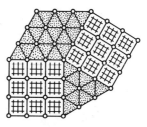

Figure 9. A structure produced by the repetition of two parts as in Figure 8, but the two parts are now of different systems of points.

Figure 10. Cyclic repetition of a part of each of two different systems of points.

Figure 11. A topological transformation of a triangular array of points (on the left) to a square array (on the right).

    V   Two or more parts, either of different sizes of the same system of points of I, II, or III (Figure 8) or of different systems of points of I, II, or III (Figures 9 and 10), repeated by symmetry operations. (Complex structures; intergrowths.)

   VI   A topological transformation changes one system of points continuously into another (Figure 11). (Note that an ordinary symmetry operation is discontinuous.)

  VII   A part of a system of points is topologically transformed and is repeated by a symmetry operation. (Modulated structures with solitons; vernier structures with antiphase boundaries.)

 VIII   Two or several different parts of different systems are repeated by topological transformations. (Modulated structures; vernier structures.)

Portions of a system of points correspond to groups of planes, or lines, or packets of points. The first is commonly called a slab, sheet, layer, or lamella; the second a rod or column; the third a block, building unit, or cluster.* Structures may be constructed from such portions in two ways:

---

* But note that the term "block structure" is also used to describe structures in which the building unit is really a column, for example, those double-crystallographic shear structures derived from the ReO_3 type.

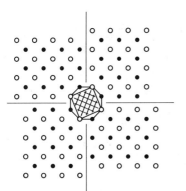

Figure 12. A fourling structure produced by reflection repeated in a cyclic way. (Note the different structures at each mirror line and at the intersection of the two mirrors.)

1. By (discontinuous) symmetry operations, like translation or reflection or their combinations, repeated in a *parallel* way (Figures 5, 6, 8, and 9). Such structures include those derived by crystallographic shear (or, better, slip) and chemical or unit-cell twinned structures.
2. By similar symmetry operations repeated in a cyclic way (involving rotation) (Figures 7 and 10). Macroscopically twinned structures of this sort are called trillings, fourlings (Figure 12), sixlings, and so on (with, respectively 3, 4, 6, etc., twins).

The foregoing description summarizes and consolidates the various steps that emerged, piecemeal, over a number of years, but its brevity may make it difficult to comprehend fully. What follows is less brief and less difficult.

References

1. D. Hilbert, *The Foundations of Geometry,* 10th ed. Open Court, La Salle, IL, 1971.
2. J. L. Synge, *Science: Sense and Nonsense.* Books for Libraries Press, New York, 1972.
3. S. Andersson and B. G. Hyde, *Z. Kristallogr.* **158,** 119 (1982).

CHAPTER II

CCP I
Cubic-Close-Packed Arrays (Usually of Anions) Projected along the Diagonal of an Octahedron—A Four-fold Axis

The following main structure types are described in this chapter: NaCl, zinc blende (cubic ZnS), NbO, $MnMg_6O_8$, high-temperature TiO and VO, low-temperature TiO and Ti_4O_5, Mg_3NF_3, Mg_2NF, Cu_2O, CuO, Cu_4O_3, anatase (TiO_2), α-$ZnCl_2$, MoO_3, V_2O_5, R-Nb_2O_5, SnF_4, SnI_4, $NbOCl_3$, UF_5, $VMoO_5$, SiS_2, ReO_3, Cu_3N, $SrTiO_3$ (cubic perovskite), $ZnCMn_3$, Fe_4N, Nb_3O_7F, $(W_{0.2}V_{0.8})_3O_7$, M-Nb_2O_5, $TiNb_2O_7$, N-Nb_2O_5, $AlNbO_4$, V_6O_{13}, PNb_9O_{25}, H-Nb_2O_5, $Ti_2Nb_{10}O_{29}$, $WO_{3-\delta}$, $(Mo,W)O_{3-\delta}$, K_2MgF_4, the new high-T_c superconductors $La_{2-x}Ba_xCuO_4$ and $Ba_2YCu_3O_7$ (and also $Ba_2YCu_3O_6$), $Na_2Ti_3O_7$, $Na_2Ti_6O_{13}$, $Ag_xV_2O_5$, $Na_xTi_4O_8$, Al_2MgO_4 (normal spinel), Fe_2MgO_4 (inverse spinel), PrI_2–V, $Cu_2(OH)_3Cl$ (atacamite), Ti_2C, $BaFe_{12}O_{19}$, β-alumina, UCl_5, $(NH_4)_3FeF_6$, K_2PtCl_6, elpasoite (K_2NaAlF_6), $Ba_3TiNb_4O_{15}$ (a "tetragonal tungsten bronze" = TTB type), $BaNb_2O_6$, $BaTi_4O_9 = KTiNb_3O_9$.

Introduction

This projection is simple to visualize and is one of the most common for depicting solid oxides. It is represented by the net given in Figure 1, with filled and unfilled circles representing anions at heights of 0 and $\frac{1}{2}$, respectively. For oxides and fluorides of the transition metals, and for the metals themselves, the projection axis is about 3.8–4.0 Å long; for chlorides, about 5 Å. The tetrahedra and octahedra drawn in the figure are obvious; the two octahedra on different levels share an edge. The centers of the octahedra are

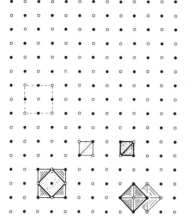

Figure 1. Cubic close packing, ccp, projected along an octahedral diagonal. The largest polyhedron is a cuboctahedron; tetrahedra and two edge-sharing octahedra are also shown. The circles represent atoms at heights of 0 and $\frac{1}{2}$ (open/closed circles). Atoms at the centers of octahedra will also be at $0,\frac{1}{2}$; those at the centers of the tetrahedra, at $\frac{1}{4}$, $\frac{3}{4}$. One face-centered-cubic unit cell (in projection) is outlined.

at 0 or $\frac{1}{2}$, and those of the tetrahedra at $\frac{1}{4}$ or $\frac{3}{4}$, respectively. The more complicated polyhedron drawn in the left lower part of the figure is a cuboctahedron, with an anion at its center. This central position is often taken by a large alkali or alkaline earth metal; an example is Sr in the cubic perovskite type of $SrTiO_3$ (see below), giving it 12 oxygen neighbors.

Many of the structures described in this chapter have anions only partly occupying the positions in cubic close-packing (ccp). In the ReO_3 structure type the 000 position in the standard face-centered-cubic description of ccp is vacant. It is characteristic of the so-called crystallographic shear structures derived from ReO_3 that the crystallographic shear (CS) planes have a complete ccp structure (see below).

Many layer structures that are easily described in this projection show small deviations from ccp. These will be discussed below.

Several important oxide and fluoride structures have an axis of ~8 Å corresponding to two octahedral diagonals. This is often due to the ordering of two cations. In such cases we have found it convenient to use so-called bounded projections, in each of which half of the unit cell is drawn in projection. In order to demonstrate the structure, at least two pictures are then necessary, and if the reader transfers one to transparent paper and then superimposes it on the other, the structure can be studied in that way. We recommend model building, which is easy using these bounded projection drawings.

The column structures with double CS, of which some simple examples are described here, were not easily solved; and models of the structures were normally derived by means of reciprocal lattice algebra. In such an ideal model, distortions had to be introduced before atomic positions could be refined by standard methods. Some simple rules exist for such distortions and we will describe them below.

In this chapter we introduce the important structural principles of crystallographic shear* and intergrowth (in both of which the symmetry operation is translation). Many mixed metal oxide and oxide-fluoride compounds with complex compositions† and large unit cells have been discovered that have structures that follow these principles.

It was mainly on the basis of these structures that the now well-known structure and lattice-image techniques of electron microscopy were developed. Pictures with a resolution of one octahedron were obtained, and the complete structure was directly studied in the projection used in this chapter. Planar defects were observed as random crystallographic shear planes—or variations of slab or column sizes—and named Wadsley defects after the man who forecast them, A. D. Wadsley (1).

At the end of this chapter we also give examples of how to describe some complex structures by using other symmetry operations such as reflection and cyclic translation.

Distortion in Octahedral Structures

While we have hardly any understanding at all of why a certain compound has a particular structure, we do have a picture of some of the reasons behind the observed variations in interatomic distances. In order to give this picture so that it can be used in day-to-day work with structures, it is convenient to simplify. The starting point is an idealized model with regular polyhedra, which we discuss, draw, or build.

Distortions of polyhedra may be due to the manner in which they are arranged, to differences in charges and ionic sizes, and/or to the electronic structures of the central atom and the ligands. Special cases are the distortions due to the presence of d electrons (ligand field distortions), distortions due to magnetic ordering, or distortions associated with ferroelectric effects. Many of the structures we discuss in this chapter contain transition metal ions, and of these Mo^{6+} and V^{5+} and sometimes W^{6+} are too small for octahedral but too big for tetrahedral configurations. Two cases can then be distinguished. One is when the anions form a regular packing, with regular octahedra. In such arrangements the small cations are off-center in the octahedra. They "move" in the polyhedra to achieve lower coordination number, examples being MoO_3, $MoVO_5$, and V_2O_5. The other case is when anions form nets which have little or no resemblance to ideal close-packed arrays. Anions may then be arranged so as to give trigonal bipyramid or

---

* The term *crystallographic shear* or CS was introduced to differentiate between the kind of "shear" that occurs in inorganic compounds and the (mechanical) shear in, for example, metals.

† In Wadsley's phrase, "grotesque stoichiometries" (as distinct from nonstoichiometry).

Figure 2. Distortions that occur when octahedra share an edge. The magnitudes of the distortion vectors are relative only and are exaggerated in size.

square pyramid polyhedra; examples are $K_2Ti_2O_5$ (2) and many vanadates (3).

When octahedra share edges or faces, the metal atoms repel each other from their centers. Such repulsion forces may be represented by vectors (as also can the attractive forces between anions and cations), and we can then deduce the distortion of the octahedra by simple geometry.

The simplest case is when two octahedra share an edge. As depicted in Figure 2, we obtain (for each cation) two longer anion–cation bonds and four shorter, which, for example, is the case with UCl_5 (see below). Note also that the anion–anion distance is shorter in the shared edge. If we have a string of octahedra sharing opposite edges, we just extend the discussion for a pair and obtain four longer and two shorter bonds.*

Face-sharing of pairs of octahedra, when three oxygens are shared between the two metal atoms, consequently means three longer (to the shared face) and three shorter anion–cation distances in each octahedron, as is observed in corundum.

The Nb_3O_7F structure, an excellent example of the crystallographic shear principle and described below, is easily constructed by taking blocks of the ReO_3 or NbO_2F structure type three octahedra wide and joining them together by edge-sharing. From the structure determination we know the distortions. Figure 3 illustrates that part of the structure where edge-sharing occurs and indicates the attraction and repulsion vectors. The magnitudes of the vectors are again somewhat exaggerated. In Figure 4 we show the net resultant vectors. Finally, in Figure 5 we have drawn the derived forms of the octahedra, which agree with those obtained from the structure determination of Nb_3O_7F (4).

Following these rules for distortions, it is very easy to explain complicated distortions such as those that are observed in compounds like $Na_xTi_4O_8$, $Ag_xV_2O_5$, $NaNb_{13}O_{33}$, all the many forms of Nb_2O_5, and the very great number of complex structures that exist for the compounds found in such systems as $Nb_2O_5 + TiO_2$ and $Nb_2O_5 + WO_3$ (5).

---

* That things are not always quite so simple as this is easily shown by examining the simple structure of rutile (cf. Chapters III and V). It contains strings of edge-sharing octahedra, but the relative bond lengths are now reversed—four short and two long!

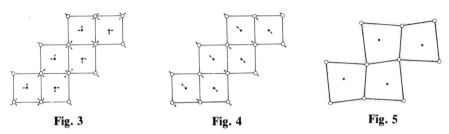

Fig. 3   Fig. 4   Fig. 5

Figure 3. The vector pattern to derive the distortions in Nb_3O_7F.

Figure 4. Resultant vectors in Nb_3O_7F. Compare Figure 3.

Figure 5. Part of the Nb_3O_7F structure as derived from Figure 4. Broken lines represent regular octahedra.

Some Simple Structures with Tetrahedra and Octahedra, and Their Relations

The structures of zinc blende and NaCl are shown in Figure 6, using the net of Figure 1. (Note that the cubic form of diamond is isostructural with zinc blende, with C in place of both Zn and S.*) Cations are not marked but are in the centers of the tetrahedra (zinc blende) or the octahedra (NaCl). A path for the cations in the (high-pressure) structural transformation from the former to the latter (cations go from tetrahedra to octahedra) is shown with an arrow in the middle of the figure. In zinc blende, tetrahedra share corners; in NaCl, octahedra share edges.

Together with the wurtzite structure, these structures are the most common for AB compounds. Alloys, oxides, halides, nitrides, carbides, and mixtures of these are examples. Many metals in fcc (face-centered cubic) packing (the same as ccp) are also represented by Figure 6, and, for example, TiC and TiN can be regarded as titanium metal with carbon and nitrogen occupying the octahedral interstices.

NaCl and zinc blende are both cubic, with $a = 5.64$ Å for NaCl and $a = 5.41$ Å for zinc blende. The space groups are $Fm3m$ and $F\bar{4}3m$, respectively, with $Z = 4$. Many structures are related to these two. Considering first the NaCl type, NbO is one such derivative, with one fourth of each of the cation and anion sites unoccupied, in an ordered manner so that each atom is in square-planar coordination. $MnMg_6O_8$ is another, with one-eighth of the cations missing in an ordered way. The higher-temperature forms of TiO and VO occur over rather large ranges of composition. (The cubic lattice parameter varies with composition, but with different slopes for the two oxides!) Both oxides have atoms missing from some of the NaCl sites. In $TiO_{0.64}$, ~36% of the oxygens are missing; in $TiO_{1.26}$, ~23% of the titaniums are

---

* It is perhaps worth mentioning the remarkable fact that at room temperature the thermal conductivity of diamond exceeds that of even the best of the metals: Ag, Au, and Cu.

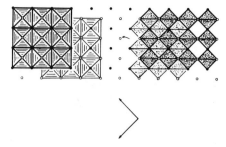

Figure 6. Sodium chloride to the left and zinc blende to the right. The hatched octahedra and dotted tetrahedra are those occupied by cations. (Heavier shading at the greater height—closer to the viewer.) Note that, in ZnS, the tetrahedra share only corners (not edges).

missing; and in the one-to-one composition, TiO, both kinds of atoms are missing, ~15% of each. Why is this so? In Figure 7 we see how the density varies for the various titanium oxides. Full occupancy of sites in TiO would indeed give a remarkably high density as shown in the figure. This is not a full explanation, but it is an interesting observation. [At lower temperatures there is an ordered TiO, with five-sixths of the NaCl cation and anion sites filled, and Ti_4O_5, in which four-fifths of the cation sites and all the anion sites are occupied (6).]

A simple example of an ordered compound related to TiO is Mg_3NF_3, which is illustrated in Figure 8. Larger circles are nitrogens; smaller, fluorines. The structure is indeed very similar to that of MgO (NaCl). The anion

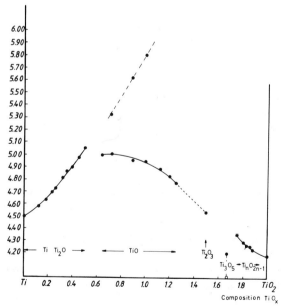

Figure 7. Observed densities of titanium oxides in grams per cubic centimeter. The broken line gives a hypothetical TiO with 100% occupancy of cation sites in the NaCl structure type.

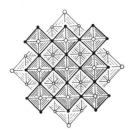

Figure 8. The structure of Mg_3NF_3. Larger circles are nitrogens; smaller circles are fluorines. Mg is in the centers of the octahedra.

Figure 9. The structure of Cu_2O: oxygens in the center of corner-connected Cu_4 tetrahedra. (There are, in fact, two such arrays of corner-connected Cu_4 tetrahedra—and they appear not to be connected! We leave their detection as an exercise for the reader.)

arrangement is intact, although the two different anions are ordered. The difference between the two structures, which have nearly the same cell dimensions, is that the position (000) containing Mg in MgO is empty (in an ordered way) in Mg_3NF_3 (Data Table 1).

Data Table 1 Mg_3NF_3 (7)

Cubic, space group $Pm3m$, No. 221; a = 4.216 Å; Z = 1, V = 74.98 Å^3

Atomic Positions
Mg in 3(*c*): $0,\frac{1}{2},\frac{1}{2}$; $\frac{1}{2},0,\frac{1}{2}$; $\frac{1}{2},\frac{1}{2},0$;
F in 3(*d*): $\frac{1}{2},0,0$; $0,\frac{1}{2},0$; $0,0,\frac{1}{2}$;
N in 1(*b*): $\frac{1}{2},\frac{1}{2},\frac{1}{2}$.
Atomic Distances
4 Mg–F = 2.108, 2 Mg–N = 2.108, N–F = 2.980, F–F = 2.980, N–N = 4.216 Å

The L-Mg_2NF structure (7) is intermediate between the zinc blende and rock salt structures. Nitrogen and fluorine are approximately in a cubic-close-packed arrangement with magnesium in square pyramidal (fivefold) coordination, "on the way" to an octahedral configuration. Elevated pressure transforms L-Mg_2NF into H-Mg_2NF, which has the NaCl structure.

Cu_2O is an example of an interstitial structure. It can be described as a ccp array of copper atoms with oxygen in interstitial positions in copper tetrahedra (Figure 9). The space group is $Pn3$, with a = 4.27 Å, so that Cu–Cu distances are 3.01 Å (cf. 2.56 Å in Cu metal) and O–O are 3.69 Å. Cu has two collinear bonds (1.87 Å) with oxygen. In CuO (~PdO structure), copper (~ ccp) is in the center of a rectangle with 1.92 and 1.97 Å distances to oxygen. The oxide Cu_4O_3 is intermediate between these two in structure and composition, containing Cu^+ as well as Cu^{2+} (8). Ag_2O is isostructural with Cu_2O, and $Cd(CN)_2$ and $Zn(CN)_2$ are antiforms.

Anatase, a white pigment and one of the forms of TiO_2, has its oxygens in ccp, and its structure is compared with the α-$ZnCl_2$ structure in Figure 10. The two structures are similar to NaCl and zinc blende, respectively, and both can be derived from these basic structures by the ordered subtraction of cations or by simple translations of sheets of the basic structures. There is again a simple transformation path between the cation positions in the two structures (cf. Figure 6). At elevated pressures, α-$ZnCl_2$ would probably transform to a structure of the anatase type. In α-$ZnCl_2$ the tetrahedra are joined by corner-sharing; in anatase the octahedra are joined by edge-sharing. Crystal data for the two compounds are given in Data Tables 2 and 3. The anatase structure has a zigzag arrangement of edge-sharing octahedra and is related to the structure of MoO_3 (which has similar zigzags) by a simple CS mechanism (compare Figures 10 and 12b). R-Nb_2O_5 has a CS structure (see below, Figure 24) intermediate between these two.

Data Table 2 α-ZnCl$_2$ (9)

Tetragonal, space group $I\bar{4}2d$, No. 122; $a = 5.398$, $c = 10.33$ Å; $Z = 4$, $V = 300.99$ Å^3

Atomic Positions
Zn in 4(a): $0,0,0$; $0,\frac{1}{2},\frac{1}{4}$; bc*
Cl in 8(d): $x,\frac{1}{4},\frac{1}{8}$; $\bar{x},\frac{3}{4},\frac{1}{8}$; $\frac{3}{4},x,\frac{7}{8}$; $\frac{1}{4},\bar{x},\frac{7}{8}$; bc; $x = 0.25$

* bc indicates body-centered; i.e., to the given coordinates also add $\frac{1}{2},\frac{1}{2},\frac{1}{2}$.

The structure of α-$ZnCl_2$ is very close to ideal ccp but is, in fact, similar to and topologically identical with α- and β-cristobalite (10) (cf. Chapter XV). It has the same space group as the latter but a difference in bond angle: $\theta(ZnClZn) \approx 110°$ and $\theta(SiOSi) \approx 147°$ (and, of course, in the unit cell

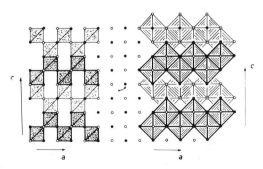

Figure 10. The α-$ZnCl_2$ structure (left) compared with the anatase (TiO_2) structure (right). Both structures idealized.

parameters, anion parameters [$x(O) = 0.09$], and bond lengths). The transformation between the two structures is readily achieved by tilting the tetrahedra about [001] (10). [Interestingly, the much better established polymorph, orthorhombic ($Pna2_1$) $ZnCl_2$ (11), is also topologically identical, but with approximately hcp anions (cf. Ref. 10).]

Data Table 3   Anatase, TiO_2 (12)

Tetragonal, space group $I4_1/amd$, No. 141; $a = 3.785$, $c = 9.515$ Å; $Z = 4$, $V = 136.31$ Å^3

Atomic Positions
Ti in 4(a): $0,0,0; 0,\frac{1}{2},\frac{1}{4}$; bc
O in 8(e): $0,0,z; 0,0,\bar{z}; 0,\frac{1}{2},z + \frac{1}{4}; 0,\frac{1}{2},\frac{1}{4} - z$; bc; $z = 0.2066$
Atomic Distances
Ti–O = 1.89_3 Å (4×), 1.96_6 Å (2×); mean = 1.91_7 Å

The structure of MoO_3 is given in Figure 11, with crystal data in Data Table 4. It is a layer structure in which zigzag rows of edge-sharing octahedra (parallel to **c**) are joined by sharing corners (along **a**). The distortion of the octahedron is simple (cf. Figure 2), though extreme. The small Mo^{6+} ions in each octahedron move off-center so that tetrahedral configuration is almost attained.

Data Table 4 MoO_3 (13)

Orthorhombic, space group $Pbnm$, No. 62; $a = 3.9628$, $b = 13.855$, $c = 3.6964$ Å; $Z = 4$, $V = 202.95$ Å^3

Atomic Positions
All atoms in 4(c): $\pm(x,y,\frac{1}{4}; \frac{1}{2} - x,\frac{1}{2} + y,\frac{1}{4})$

	x	y
Mo	0.08669	0.10164
O(1)	0.4494	0.4351
O(2)	0.5212	0.0866
O(3)	0.0373	0.2214

Atomic Distances
Mo–O: 2.33, 1.95 (2×), 1.73, 2.25, and 1.67 Å

In Figures 12a and b the MoO_3 structure is idealized to perfect cubic close-packing of anions in two different projections. Joining the same layers by corner-sharing gives the structure of V_2O_5 or R-Nb_2O_5, and joining them by edge-sharing gives the anatase structure.

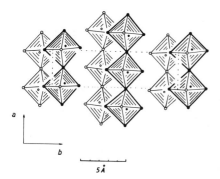

a

b

_____
5 Å

Figure 11. The structure of MoO_3. Small circles are molybdenum.

Crystallographic data for V_2O_5 are given in Data Table 5, and Figures 13*a* and *b* show the structure. In Figure 13*a* the octahedra are drawn. This is close to the real R-Nb_2O_5 structure, and the similarities with the structures of MoO_3 and anatase are easily seen.

Data Table 5 V_2O_5 (14)

Orthorhombic, space group P*mmn*, No. 59; $a = 11.51$, $b = 3.56$, $c = 4.37$ Å; $Z = 2$, $V = 179$ Å^3

Atomic Positions
V in 4(*f*): $x,0,z; \bar{x},0,z; \frac{1}{2} - x,\frac{1}{2},\bar{z}; \frac{1}{2} + x,\frac{1}{2},\bar{z}; x = 0.1487, z = 0.1086$
O(1) in 4(*f*): $x = 0.1460, z = 0.4713$
O(2) in 4(*f*): $x = 0.3191, z = -0.0026$
O(3) in 2(*a*): $0,0,z; \frac{1}{2},\frac{1}{2},\bar{z}; z = -0.0031$

However, in the V_2O_5 structure there are five V–O distances between 1.59 and 2.02 Å and a sixth distance of 2.79 Å, which makes it better to draw this structure as in Figure 13*b*, with square pyramids of oxygen around the V atoms. Clearly it is really a layer structure. In this way the resemblance to the structures of vanadates and $K_2Ti_2O_5$ is obvious.

The octahedral description is simply related to the ReO_3 structure by

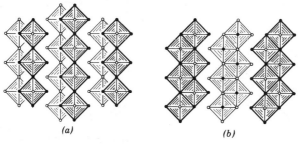

(a)   *(b)*

Figure 12*a,b*. The MoO_3 structure (idealized) is shown in two projections along the two shortest axes. Compare with Figure 11.

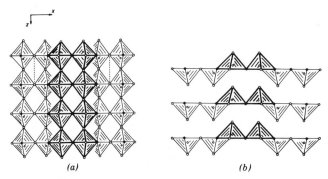

Figure 13. The structure of V_2O_5 shown as (a) octahedral VO_6 and (b) square pyramidal VO_5.

translation (CS). It is, of course, related to many others also, such as V_6O_{13}, Nb_3O_7F, and $Na_xTi_4O_8$.

The idealized structure of SnF_4 is given in Figure 14. It consists of (001) layers of corner-sharing octahedra. The CS relationships with ReO_3, MoO_3, and K_2MgF_4 are obvious. PbF_4 and NbF_4 are isostructural, and one form of Ni_4N is an antitype.

The structure of SnI_4 is molecular; it is shown in two bounded projections in Figure 15. SnI_4 tetrahedra pack together so that the anion array is very nearly perfect ccp. The structure is easily imagined to transform, by Sn shifts, to the SnF_4 structure. High pressure will probably do this.

An arrangement of pairs of edge-sharing octahedra is shown in Figure 16. These pairs, joined by corner-sharing along **c** (the projection direction), form chains in the tetragonal structure of $NbOCl_3$. Large circles are chlorines; small circles are oxygens.

The tetragonal structure of UF_5 (Figure 17) consists of chains of corner-sharing octahedra along **c**, with anions in almost regular ccp. The chains pack together so that tetrahedral holes around each chain are empty; but these holes can be filled, as is the case in the structure of $VMoO_5$, in which Mo ions occupy the tetrahedra (Figure 18a). α-$VOSO_4$, $NbPO_5$, and $MoPO_5$ are isostructural. [In reality, cation–cation repulsion slightly distorts the ccp anion array by a small rotation of the octahedral strings about the **c** axis (see Figure 18b).]

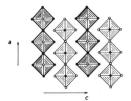

Figure 14. The idealized structure of SnF_4.

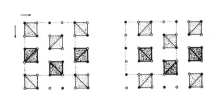

Figure 15. The idealized (but very close to the real) structure of SnI_4 (as two bounded projections).

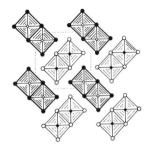

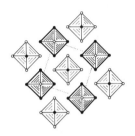

Figure 16. The idealized structure of $NbOCl_3$. The large circles are chlorines. Simple and obvious jumps of half the cations will transform this to the SnF_4 type. (Cf. Figure 14).

Figure 17. The idealized structure of UF_5.

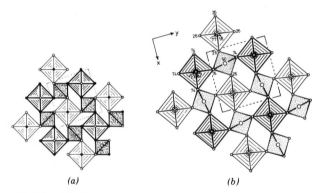

(a)   *(b)*

Figure 18. (*a*) The idealized structure of $VOMoO_4$. (*b*) The real structure of $VOMoO_4$.

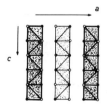

Figure 19. The idealized structure of SiS_2.

In SiS_2 the silicon-centered tetrahedra share edges to form chains along **c** as shown in Figure 19. In the structure of $Ba(FeS_2)_2$ (Chapter XII) and $MFeS_2$ (M = K, Rb, or Cs), there are similar chains of FeS_2, with edge-sharing FeS_4 tetrahedra, but now they are held together by intervening (parallel) rows of the large cations. Crystal data for SnF_4, SnI_4, $NbOCl_3$, UF_5, $VMoO_5$, and SiS_2 are given in Data Tables 6–11.

Data Table 6   SnF$_4$ (15)

Tetragonal, space group $I4/mmm$, No. 139; a = 4.048, c = 7.930 Å; Z = 2, V = 129.94 Å^3

Atomic Positions
Sn in 2(a):   0,0,0; $\frac{1}{2},\frac{1}{2},\frac{1}{2}$
F(1) in 4(c):   0,$\frac{1}{2}$,0; $\frac{1}{2}$,0,0; $\frac{1}{2}$,0,$\frac{1}{2}$; 0,$\frac{1}{2}$,$\frac{1}{2}$
F(2) in 4(c): $\pm$(0,0,z; $\frac{1}{2},\frac{1}{2},z$ + $\frac{1}{2}$); z = 0.245

Isostructural compounds: PbF$_4$, NbF$_4$
Antitype: One form of Ni$_4$N

Data Table 7   SnI$_4$ (16)

Cubic, space group $Pa3$, No. 205; a = 12.273 Å; Z = 8, V = 1848.6 Å^3

Anions are almost perfect ccp
Atomic Positions
See ref. 16.
Isostructural compounds: CBr$_4$, GeI$_4$, SiI$_4$, TiBr$_4$, TiI$_4$, ZrBr$_4$, ZrCl$_4$

Data Table 8   NbOCl$_3$ (17)

Tetragonal, space group $P4_2/mnm$, No. 136; a = 10.87, c = 3.96 Å; Z = 4, V = 467.90 Å^3

Atomic Positions
Nb in 4(f): $\pm$(x,x,0; $\frac{1}{2}$ + $x,\frac{1}{2}$ − $x,\frac{1}{2}$); x = 0.127
Cl in 8(i): $\pm$(x,y,0; y,x,0; $\frac{1}{2}$ + $x,\frac{1}{2}$ − $y,\frac{1}{2}$; $\frac{1}{2}$ + $y,\frac{1}{2}$ − $x,\frac{1}{2}$); x = 0.331, y = 0.105
Cl in 4(g): $\pm$($x,\bar{x}$,0; $\frac{1}{2}$ + $x,\frac{1}{2}$ + $x,\frac{1}{2}$); x = 0.104
O in 4(g): x = 0.385
Atomic Distances
2 Nb–O = 1.99
Nb–Cl = 2.24 Å (2×), 2.53 Å (2×)

Data Table 9   UF$_5$ (18)

Tetragonal, $I4/m$, No. 87; a = 6.525, c = 4.472 Å; Z = 2, V = 190.40 Å

Atomic Positions
U in 2(a):   0,0,0; $\frac{1}{2},\frac{1}{2},\frac{1}{2}$
F(1) in 2(b):   0,0,$\frac{1}{2}$; $\frac{1}{2},\frac{1}{2}$,0
F(2) in 8(h): $\pm$(x,y,0; y,x,0); bc; x = 0.315, y = 0.113
Atomic Distances
U–F(1) = 2.23 Å (2×)
U–F(2) = 2.18 Å (4×)

Data Table 10   VMoO$_5$ (19)

Tetragonal, space group $P4/n$, No. 85; $a = 6.6078$, $c = 4.2646$ Å; $Z = 2$,
$V = 186.20$ Å^3

Atomic Positions
V in 2(c): $\pm(\frac{1}{4},\frac{1}{4},z)$; $z = 0.8395$
Mo in 2(b): $\pm(\frac{1}{4},\frac{3}{4},\frac{1}{2})$
O(1) in 2(c): $z = 0.2327$
O(2) in 8(g): $\pm(x,y,z; \frac{1}{2} - x,\frac{1}{2} - y,z; \frac{1}{2} - y,x,z; y,\frac{1}{2} - x,z)$, $x = 0.7034$,
                  $y = 0.4623$, $z = 0.2597$

Atomic Distances
Within octahedra: V–O = 1.877 Å, 2.588 Å, 1.972 Å (4×)
Within tetrahedra: (Mo–O) = 1.764 Å (4×)

This is a typical off-center distortion. The octahedra are rather regular in the structure. It is close to ideal ccp.

Data Table 11   SiS$_2$ (20,21)

Orthorhombic, space group *Ibam*, No. 72; $a = 9.57$, $b = 5.65$, $c = 5.54$ Å (20);
$a = 9.55$, $b = 5.60$, $c = 5.53$ Å (21); $Z = 4$, $V = 299.6$ Å (20), 295.7 Å^3 (21)

Atomic Positions
Si in 4(a): $0,0,\frac{1}{4}$; $0,0,\frac{3}{4}$; bc
S in 8(j): $\pm(x,y,0; \bar{y},x,\frac{1}{2})$; bc; $x = 0.177$, $y = 0.217$ (20); $x = 0.119$,
                 $y = 0.208$ (21)

Slab and Column Crystallographic Shear and Intergrowth Compounds

The importance of these structures for understanding various aspects of nonstoichiometry was early realized by Wadsley. He introduced the term *crystallographic shear* (CS), and after the structure determination of W$_4$Nb$_{26}$O$_{77}$ (an intergrowth compound, cf. Chapter I) it was obvious that shear planes could occur at random spacings in a crystal. Experimentally this was first shown to be the case by Wadsley and co-workers using lattice images obtained by electron diffraction from single crystals (22). The importance of this work should not be underestimated; it was the starting point for a field in solid state that has now contributed so much to our insight into phenomena such as diffusion, reaction mechanisms, reactivities, and the structure of defects in solids—taken together, a new approach to the crystalline solid state.

The number of so-called (crystallographic) shear compounds is enormous and still growing. In the TiO$_{2-x}$ system the shear planes change their orienta-

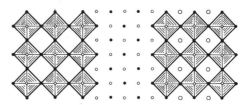

Figure 20. The structures of ReO_3 (left) and $SrTiO_3$. Complete ccp is shown in the center.

tion (relative to the rutile substructure) with composition, that is, they "swing"; and the number of phases in this system seems to be unlimited (23).

Two of the most important parent-structure types in this area are depicted in Figure 20. The ReO_3 structure (on the left) consists of regular octahedra sharing corners. This structure is also taken by such compounds as NbO_2F and TaO_2F and also by some fluorides. There are also many examples of distorted ReO_3, WO_3 being one.* The ReO_3 structure type can be topologically transformed to the hcp PdF_3 structure type (see Chapter XI). Several fluorides like CrF_3, CoF_3, FeF_3, and TiF_3 have the anions in positions that correspond to structures intermediate between PdF_3 and ReO_3.

Cu_3N is an anti-ReO_3 type, with $a = 3.807$ Å.

Only three-fourths of the atoms necessary for ccp are present in ReO_3; the position $\frac{1}{2},\frac{1}{2},\frac{1}{2}$ is empty. If it is occupied by a large cation such as strontium, the structure of (cubic) perovskite is derived (see Figure 20). Many perovskites, like $BaTiO_3$ and $NaNbO_3$, are distorted to other symmetries, and some are ferroelectric (cf. p. 42), but $CaTiO_3$ above 900°C seems to be truly cubic with $a = 3.84$ Å. (Below 900°C it too is distorted; see Chapter XI.)

There are many compounds with an antiperovskite structure, such as $ZnCMn_3$, $SnCFe_3$, and $AlNNi_3$ (with C or N, or sometimes B, in octahedral coordination). Oxide examples, BOA_3, have A = Ca, Sr, or Ba; B = Sn or Pb. Fe_4N ($a = 3.75$ Å) is similar ($FeNFe_3$), with Fe in both Ca and O positions and N in the Ti site. (But it is more simply described as ccp Fe with N in one-fourth of the octahedral sites, those at $\frac{1}{2},\frac{1}{2},\frac{1}{2}$; cf. Ni_4N above.)

The ReO_3 structure serves as the model type for describing crystallographic shear. Similarly, the $SrTiO_3$ structure type is often used to derive and describe the so-called bronze and bronzelike structures, of which the tungsten (or wolfram) bronzes are so well known. The bonding forces in ReO_3 are perpendicular, which obviously allows the structure to deform in response to a mechanical shear. We will see that a shear plane has a complete ccp anion array, which stiffens the structure. Compared to ReO_3, the $SrTiO_3$ structure also has diagonal forces and should be more resistant to shear deformations. We can illustrate this with a simple example: Everyone

* Gentle preparation methods (*chimie douce*) have recently produced an isostructural β'-MoO_3 (24) and, it appears, a slightly differently distorted β-MoO_3 (25).

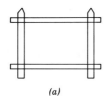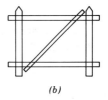

Figure 21. Two gate designs: (*a*) bad; (*b*) good.

who has made a gate knows that whatever number of nails is used in a model of Figure 21*a*, it will shear. The diagonal support, shown in Figure 21*b*, gives the gate stability and serves the same purpose as the large cation strontium in the perovskite structure.

The structure of Nb_3O_7F will serve as one of the simplest compounds to demonstrate CS. It is given in Figure 22. The distortions have been described above. A formal construction of this structure type is given in Figure 23, where, on the left, two slabs of ReO_3 type are on the way to being joined by edge-sharing across the CS plane. This is performed in the right-hand part of the figure in an idealized way; in Figure 24, Nb_3O_7F is again shown,

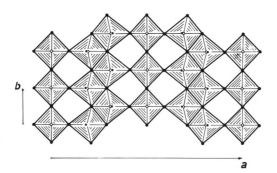

Figure 22. The structure of Nb_3O_7F. The smaller circles are niobium atoms.

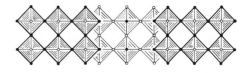

Figure 23. Demonstration of crystallographic shear, with regular octahedra.

Figure 24. The idealized structures of Nb_3O_7F (*above*) and R-Nb_2O_5 (*below*) are compared.

idealized but projected along the other short axis. The structure of $R\text{-Nb}_2O_5$ is also shown for comparison; the ReO_3 slabs in this structure are only two octahedra wide. Crystal data for Nb_3O_7F and $R\text{-Nb}_2O_5$ are given in Data Tables 12 and 13.

Data Table 12 Nb_3O_7F (4)

Orthorhombic, space group *Cmmm*, No. 65; $a = 20.67$, $b = 3.833$, $c = 3.927$ Å; $Z = 2$, $V = 311.13$ Å^3

Atomic Positions*: $(0,0,0)$ and $(\frac{1}{2},\frac{1}{2},0)+$
Nb(1) in 2(a): $0,0,0$
Nb(2) in 4(g): $\pm(x,0,0)$; $x = 0.1836$
O(1) in 2(b): $\frac{1}{2},0,0$
O(2) in 2(d): $0,0,\frac{1}{2}$
O(3) in 4(g): $x = 0.094$
O(4) in 4(g): $x = 0.710$
O(5) in 4(h): $\pm(x,0,\frac{1}{2})$; $x = 0.189$
Atomic Distances
Nb(1)–O(1) = 1.917 Å (2×)
Nb(1)–O(2) = 1.964 Å (2×)
Nb(1)–O(3) = 1.937 Å (2×)
Nb(2)–O(3) = 1.852 Å
Nb(2)–O(4) = 1.993 Å (2×), 2.199 Å
Nb(2)–O(5) = 2.007 Å (2×)

* Oxygens and fluorines are assumed to be randomly distributed over sites O(1) to O(5) in the structure.

Data Table 13 $R\text{-Nb}_2O_5$* (26)

Monoclinic, space group $A2/m$, No. 12; $a = 3.983$, $b = 3.826$, $c = 12.79$ Å, $\beta = 90.75°$; $Z = 2$, $V = 194.89$ Å^3

Atomic Positions: $(0,0,0)$ and $(0,\frac{1}{2},\frac{1}{2})+$
Nb in 4(i): $\pm(x,0,z)$; $x = 0.07$, $z = 0.146$
O(1) in 2(a): $0,0,0$
O(2) in 4(i): $x = 0$, $z = 0.68$
O(3) in 4(i): $x = 0.5$, $z = 0.16$

* The structure was derived from powder data by analogy with that of Nb_3O_7F; the only difference, except the block size, is the slight monoclinic distortion in $R\text{-Nb}_2O_5$.

In Nb_3O_7F the CS operation is a simple translation, one-dimensional in nature. If the resulting structure is now crystallographically sheared perpendicular to the shear planes already present, and if this is done regularly so that columns with a cross section of 3×3 octahedra are formed, then the result is shown in Figure 25. The two translation operations produce a tetragonal structure, and a vanadium tungsten oxide of this composition (M_3O_7, with $M = W_{0.2}V_{0.8}$) and structure has been found (27). Structures

Figure 25. The idealized structure of $(W_{0.2}V_{0.8})_3O_7$ as an example of double translation or shear.

such as this and the following ones were called block structures, but according to the terminology in the Introduction (Chapter I), since they have double CS, or double translation, they are better called *column* structures.

For the compounds that follow, the idealized structures will always be given. The distortions are easy to derive following the rules given above.

By letting the columns grow to 4×4 octahedra in size, the structure of M-Nb$_2$O$_5$ is obtained. It is shown in Figure 26.*

* There is some doubt as to whether M-Nb$_2$O$_5$ has this perfectly ordered structure (28).

Figure 26. The idealized structure of M-Nb$_2$O$_5$.

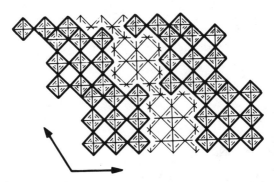

Figure 27. The idealized structure of TiNb$_2$O$_7$.

If parts of these two tetragonal structures slip one octahedral diagonal along the shear planes (there is no change of composition), two structures represented by TiNb$_2$O$_7$ and N-Nb$_2$O$_5$ are formed (Figures 27 and 28). In reality the TiNb$_2$O$_7$ and N-Nb$_2$O$_5$ structures were found first, and the two tetragonal structures were predicted and were afterwards also found. The analogous monoclinic structures with columns of 2×2 and 2×3 octahedra are those of AlNbO$_4$ and V$_6$O$_{13}$, respectively.

It is easy to see that in the structure of PNb$_9$O$_{25}$ shown in Figure 29, columns of ReO$_3$ of size 3×3 octahedra are so arranged that tetrahedral

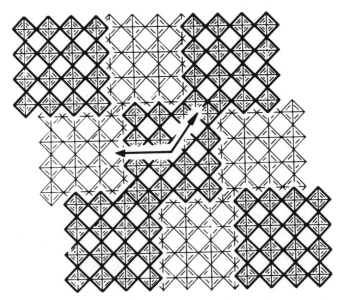

Figure 28. The idealized structure of N-Nb$_2$O$_5$.

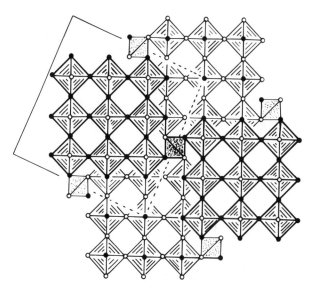

Figure 29. The idealized structure of PNb_9O_{25}; P in tetrahedra.

sites for P are created. One tetrahedron plus the eight neighboring octahedra form a region of cubic-close-packed anions.*

The structure of $H-Nb_2O_5$ was an unsolved problem for many years. The reason became clear when Wadsley and Gatehouse finally presented the solution. The structure was very difficult to solve, and it is easy to invent much simpler structures for the stable high-temperature form of niobium pentoxide. It is shown, idealized, in Figure 30. In the unit cell there are 27 niobium atoms in octahedral positions and one niobium, anchored in the origin, in tetrahedral coordination. The ReO_3 columns are of different sizes on the two levels, 3×5 and 3×4 octahedra in cross section, and they are joined together by the kind of edge-sharing present in $TiNb_2O_7$ and $Ti_2Nb_{10}O_{29}$.† Along the edges of these columns, rows of tetrahedral niobium atoms join four columns together exactly as the phosphorus atoms do in PNb_9O_{25}.

After the solution of the structure of $H-Nb_2O_5$, numerous niobium oxides, niobium oxide fluorides, and niobium tungsten oxides were prepared and studied with X-rays and observed with the electron microscope. Indeed this structure was the key to the structures of all these column compounds. By

---

* Note that $VMoO_5$, discussed earlier, may be described as the smallest possible column structure of this type, with ReO_3 columns only 1×1 octahedron in extent. Compare Figures 18 and 29.

† $Ti_2Nb_{10}O_{29}$ has two forms, monoclinic and orthorhombic. Both are analogues of $TiNb_2O_7$ and $N-Nb_2O_5$, but with intermediate-sized columns of ReO_3, 3×4 octahedra in cross section.

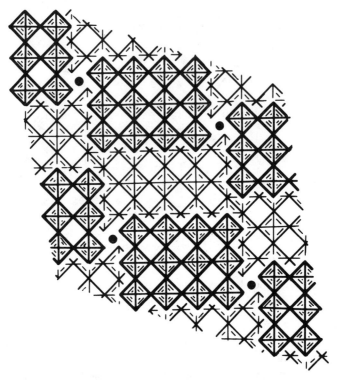

Figure 30. The idealized structure of H-Nb$_2$O$_5$. The tetrahedrally coordinated Nb atoms at the unit cell origin are shown as filled circles.

varying the column sizes and the ways the columns are joined together, the structures and the compositions of these compounds were obtained.

Some data for (V,W)$_3$O$_7$, M-Nb$_2$O$_5$, TiNb$_2$O$_7$, N-Nb$_2$O$_5$, PNb$_9$O$_{25}$, and H-Nb$_2$O$_5$ are given in Data Tables 14–19. Atomic coordinates are to be found in the cited references.

Data Table 14   M-Nb$_2$O$_5$ (29)

Tetragonal, space group $I4/mmm$, No. 139; $a = 20.44$, $c = 3.832$ Å; $Z = 16$, $V = 1600.98$ Å^3

Data Table 15   TiNb$_2$O$_7$ (30)

Monoclinic, space group $A2/m$, No. 12; $a = 11.93$, $b = 3.81$, $c = 20.44$ Å, $\beta = 120.17°$; $Z = 6$, $V = 803.21$ Å^3

Data Table 16   N-Nb$_2$O$_5$ (31)

Monoclinic, space group $C2/m$, No. 12; $a = 28.51$, $b = 3.830$, $c = 17.48$ Å, $\beta = 124.80°$; $Z = 4$, $V = 1567.33$ Å^3

Data Table 17   PNb$_9$O$_{25}$ (32)

Tetragonal, space group $I\bar{4}$, No. 82; $a = 15.60$, $c = 3.828$ Å; $Z = 2$, $V = 931.58$ Å^3

Data Table 18   (W$_{0.2}$V$_{0.8}$)$_3$O$_7$ (27)

Tetragonal, space group $I4/mmm$, No. 139; $a = 14.01$, $c = 3.720$ Å; $Z = 6$, $V = 730.2$ Å^3

Data Table 19   H-Nb$_2$O$_5$ (33)

Monoclinic, space group $P2$, No. 3; $a = 21.16$, $b = 3.822$, $c = 19.35$ Å, $\beta = 119.83°$; $Z = 14$, $V = 1357.56$ Å^3

Ionic radius of 0.59 Å for Nb^{5+} was derived.

Swinging Shear Structures

A slip plane in the ReO$_3$ structure is shown in Figure 31a. This is an anti-phase boundary (APB), with the displacement vector parallel to the plane. Between this APB case (with no change of composition) and the Nb$_3$O$_7$F-type CS plane (Figure 31f),* an infinite number of shear plane types can be derived; some are shown in Figures 31b–e. Analogous CS planes, but with *continuously* changing orientation, have been found in the system TiO$_{2-\delta}$ (23). Those shown in Figure 31 occur in the well-known WO$_{3-\delta}$ and (Mo,W)O$_{3-\delta}$ systems, apparently without intermediate orientations.

Layer Structures Projected along an Octahedral Diagonal, and Their Relation to Certain Tunnel Structures or Three-Dimensional Arrangements of Octahedra and to ccp Anion Arrays

Here we define layer structures as consisting of polyhedra joined in two dimensions by one or several of the common ways to join polyhedra. Such layers may be held together by van der Waals forces (we have already described some simple examples) or, if they are charged, by extra cations inserted between the layers. Invariably it seems that such a layer has a

---

* A CS plane is also an APB but involves a composition change. Its displacement vector is *not* parallel to the plane.

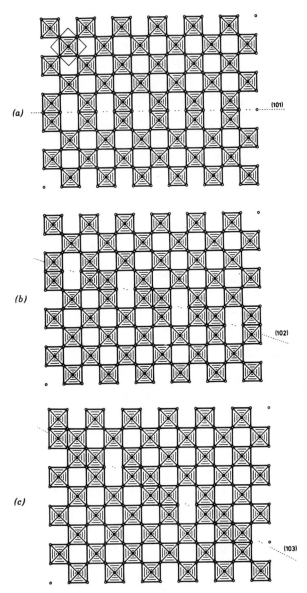

Figure 31. (001) projections of the keO₃ type with a single, isolated fault plane (*hOl*) of various types, all with the same displacement vector, $R = \frac{1}{2}[1\bar{1}0]$. (*a*) (101) (an APB with no oxygen loss); (*b–f*), (10*l*) with l = 2, 3, 4, 5, and ∞ [i.e., (001)]. All these last are CS planes and involve a loss of oxygen.

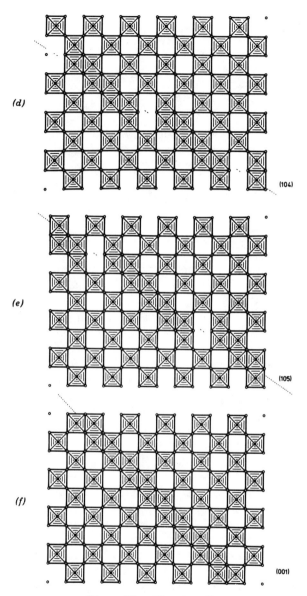

Figure 31. (*Continued*)

simple packing, ccp or hcp, but that this packing is often broken when crossing from one layer to the next in the structure. An important observation is that such layers can often be united (with simultaneous loss of some of the inserted cations together with some anions) by polyhedral corner- or edge-sharing, and then a new structure is obtained. This observation was the starting point for the use of the idea of intergrowth between phases to ex-

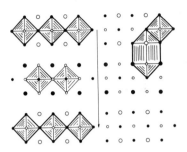

Figure 32. The structure of K_2MgF_4. Larger circles are K. A bicapped cube of potassium and fluorine atoms is shown to the right.

plain nonstoichiometry, diffusion, and reaction mechanisms in certain solids (5).

In K_2MgF_4 there are ReO_3-type layers of composition MgF_4^{2-}, extended in the *ab* plane as shown in Figure 32, which is an exact drawing of the structure (cf. SnF_4, Figure 14). These layers in the structure are held together by potassium ions, which, relative to each layer, are in positions corresponding to those in perovskite. Formally, by subtracting KF from K_2MgF_4 (and collapsing the structure) one obtains the perovskite structure of $KMgF_3$ by corner-sharing between the layers. Crystal data for K_2MgF_4 are given in Data Table 20, where calculated parameters (assuming perfectly regular polyhedra) are also given.

The empty polyhedra of potassiums plus fluorines existing between the layers are cubes. These atoms thus form a layer of primitive cubic packing.* A great number of compounds, oxides or fluorides, have this (K_2NiF_4-type) structure; see Wyckoff (34). The terminal Mg–F distances are longer than the bridging ones (Mg–F–Mg). In the case of Sr_2TiO_4, the result is the opposite (and more normal); the terminal distance is 1.90 Å, while the bridging distance is 1.94 Å.

Sr_2TiO_4 is one of a family of structures sometimes known as the Ruddlesden–Popper phases. Other members are $Sr_3Ti_2O_7$ and $Sr_4Ti_3O_{10}$, in which the perovskite-like layers are respectively two and three octahedra thick (35). Their general formula, $Sr_{n+1}Ti_nO_{3n+1}$, may be written as $SrO \cdot nSrTiO_3$, which is in accord with the description already given; they are intergrowths of NaCl-type SrO with perovskite-type $SrTiO_3$.

---

* Note that such a primitive cubic packing of alternate cations and anions is, of course, a layer of NaCl type, but with its fcc unit cell axis rotated 45° about **c** with respect to that of the MgF_4 part and therefore $\sqrt{2}$ times as long. Compare the ratio of the mean bond lengths K–F/Mg–F = 2.82/2.01 = 1.407, and $\sqrt{2}$ = 1.414. Hence the K_2MgF_4 structure can be quite accurately described as a layer intergrowth of perovskite-like $KMgF_3$ plus NaCl-like KF: K_2MgF_4 = $KMgF_3$ + KF.

Data Table 20   K₂MgF₄

Tetragonal, space group $I4/mmm$, No. 139; $a = 3.995$, $c = 13.706$ Å; $Z = 2$, $V = 218.75$Å^3

Atomic Positions
Mg in 2(a):   0,0,0; $\frac{1}{2},\frac{1}{2},\frac{1}{2}$
K in 4(e): $\pm(0,0,z; \frac{1}{2},\frac{1}{2},z + \frac{1}{2})$; $z = 0.35$
F(1) in 4(c): $0,\frac{1}{2},0; \frac{1}{2},0,0; \frac{1}{2},0,\frac{1}{2}; 0,\frac{1}{2},\frac{1}{2}$
F(2) in 4(e): $z = 0.15$
Atomic Distances
Mg–F = 1.98 Å (4×), 2.06 Å (2×)
K–F = 2.82 Å (4×), 2.87 Å (2×), 2.74 Å

$a_{calc} = d\sqrt{2}$ (d = anion–anion spacing); $c_{calc} = 2d(\sqrt{2} + 1)$
$(c/a)_{calc} = 3.41$, $(c/a)_{obs} = 3.43$
$z(K)_{calc} = (\sqrt{2} + 2)/[4(\sqrt{2} + 1)] = 0.354$; $z(F)_{calc} = \sqrt{2}/[4(\sqrt{2} + 1)] = 0.146$

An Addendum on the New High-Temperature Superconductors

Very recently (1986–1987), these structure types and those of some related, anion-deficient perovskites, $ABX_{3-\delta}$, with the A sites occupied by rare earth and alkaline earth cations, the B sites by Cu, and the anion sites by O, have become the intense preoccupation of a large proportion of the world's solid state scientists (chemists, physicists, and ceramicists). An unprecedented burst of feverish research activity was stimulated by Bednorz and Müller's discovery (36) that $(La,Ba)_2CuO_4$ was probably superconducting at temperatures well above the previous record. The critical temperature (below which superconductivity, i.e., zero electrical resistivity, sets in) was $T_c \approx 30$ K, whereas the previous record (the culmination of 60 years of superconductor research) was $T_c = 23.2$ K for Nb_3Ge (with the A15 or Cr_3Si-type structure, see Chapter XIII). It was established that the compound $La_{2-x}Ba_xCuO_4$, $x = 0.1$, was isostructural with K_2NiF_4.

Initial skepticism of this result [published (36) in September 1986] was rapidly swept aside by the frenetic activity of very many people, who dropped everything to search for materials with even higher T_cs. Success was rapid (and the photocopy machine, telephone, and daily newspapers became the media for short-circuiting the delay inevitably associated with the normal communication of scientific results through scientific journals). Very quickly, compounds of similar constitution but with $Sr_3Ti_2O_7$- and $Sr_4Ti_3O_{10}$-type structures were reported to have higher T_cs. By March 1987, Chu and his group had prepared an yttrium barium copper oxide with $T_c \approx 90$ K (37). [The significance of this discovery was that superconductivity was now possible at liquid nitrogen temperature (boiling point 77 K) rather than liquid helium temperature (boiling point 4.2 K). Liquid nitrogen is very much cheaper than liquid helium.]

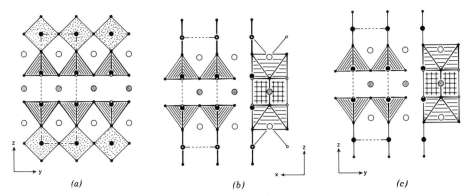

Figure 33. The orthorhombic structure of $Ba_2YCu_3O_7$ projected on (*a*) (100) and (*b*) (010). (*c*) The related tetragonal structure of $Ba_2YCu_3O_6$ projected on (100) = (010). Small, medium, and large circles are O, Cu, and Ba/Y, respectively, Y being dotted to distinguish them from Ba. In (*b*) and (*c*) the coordination "polyhedra" of Cu are drawn on the left, and those of Y and Ba on the right.

The compound responsible was found to be $Ba_2YCu_3O_7$, with $T_c = 93$ K (but Y could be replaced by almost any trivalent lanthanide with little effect on T_c). Its structure was an anion-deficient perovskite, $(Ba_2Y)Cu_3O_{9-2}$: a $1 \times 1 \times 3$ times superstructure of ABX_3 with a slight orthorhombic distortion, with Ba and Y (ordered) on A sites and Cu on B sites (Figure 33). The oxygen atoms are ordered on seven-ninths of the anion sites of perovskite, so that two-thirds of the Cu atoms are in square-pyramidal (half-octahedral) coordination and the other third in square-planar coordination. Yttrium is in an almost perfect cube, YO_8, and barium in a bicapped square antiprism, BaO_{8+2}.

Above about 350°C, oxygen is readily and reversibly lost if the ambient oxygen pressure is reduced below 1 atm, and $Ba_2YCu_3O_7$ is easily reduced, eventually to $Ba_2YCu_3O_6$. The structure of the latter is also shown in Figure 33. It differs from that of the oxidized material: Cu is now in linear 2-coordination (CuO_2, as in Cu_2O, see above), and the Ba-containing square antiprism has lost its two caps. Furthermore, $Ba_2YCu_3O_6$ is a semiconductor, not a superconductor.

Assuming the normal valences for Ba (= 2), Y (= 3), and O (= 2), the formal average valence for Cu is $7/3 = 2.33$ in $Ba_2YCu_3O_7$ and $5/3 = 1.67$ in $Ba_2YCu_3O_6$, suggesting Cu(II) in 5-coordination, Cu(III) in 4-coordination, and Cu(I) in 2-coordination. But, it should be added, this is formal: Cu^{3+} has not been experimentally confirmed (and, above T_c, $Ba_2YCu_3O_7$ is a metal). The normal valence states of Cu are, of course, I and II, while III, though not unknown, is unusual. Structural data are given in Data Tables 21 and 22.

In the last few months (to October 1987) there have been tantalizing indications of much higher T_cs (from ~105 K up to 338 K = 65°C), but no

Data Table 21 $Ba_2YCu_3O_{6.96}$ at 296 K (38)

Orthorhombic, space group $Pmmm$, No. 47; $a = 3.8240(2)$, $b = 3.8879(2)$, $c = 11.6901(8) = 3 \times 3.8967$ Å; $Z = 1$, $V = 173.80$ Å^3

*Atomic Positions**

Ba in 2(t):	$\pm(\frac{1}{2},\frac{1}{2},0.1854)$
Y in 1(h):	$\frac{1}{2},\frac{1}{2},\frac{1}{2}$
Cu(1) in 1(a):	$0,0,0$
Cu(2) in 2(q):	$\pm(0,0,0.3557)$
O(1) in 1(e):	$0,\frac{1}{2},0$; occupancy = 0.96
O(2) in 2(s):	$\pm(\frac{1}{2},0,0.3775)$
O(3) in 2(r):	$\pm(0,\frac{1}{2},0.3786)$; occupancy = 1.94
O(4) in 2(q):	$\pm(0,0,0.1582)$

Bond Lengths
Ba–O(1) = 2.89_0 Å (2×); Ba–O(2) = 2.98_0 Å (2×); Ba–O(3) = 2.94_9 Å (2×);
   Ba–O(4) = 2.74_5 Å (4×); mean Ba–O = 2.86_2 Å
Y–O(2) = 2.40_7 Å (4×); Y–O(3) = 2.38_9 Å (4×); mean Y–O = 2.39_8 Å
Cu(1)–O(1) = 1.94_4 Å (2×); Cu(1)–O(4) = 1.84_9 Å (2×); mean Cu(1)–O = 1.89_6 Å
Cu(2)–O(2) = 1.93_1 Å (2×); Cu(2)–O(3) = 1.96_1 Å (2×); Cu(2)–O(4) = 2.30_9 Å;
   mean Cu(2)–O = 2.01_9 Å

* Occupancies deduced in the structure determination (by powder neutron diffraction).

Data Table 22 $Ba_2YCu_3O_{6.00}$ at 296 K (38)

Tetragonal, space group $P4/mmm$, No. 123; $a = b = 3.8577(2)$, $c = 11.8274(8) = 3 \times 3.9427$ Å; $Z = 1$, $V = 176.01$ Å^3

Atomic Positions

Ba in 2(h):	$\pm(\frac{1}{2},\frac{1}{2},0.1954)$
Y in 1(d):	$\frac{1}{2},\frac{1}{2},\frac{1}{2}$
Cu(1) in 1(a):	$0,0,0$
Cu(2) in 2(g):	$\pm(0,0,0.3608)$
[O(1) in 2(f):	$0,\frac{1}{2},0$; $\frac{1}{2},0,0$; occupancy = 0.08(3) $\approx$ 0]*
O(2) in 4(i):	$\pm(0,\frac{1}{2},0.3795$; $\frac{1}{2},0,0.3795)$; occupancy = 4.04(3)
O(4) in 2(g):	$\pm(0,0,0.1528)$; occupancy = 2.01(2)

Bond Lengths
Ba–O(2) = 2.91_0 Å (4×); Ba–O(4) = 2.77_4 Å (4×); mean Ba–O = 2.84_2 Å
Y–O(2) = 2.39_8 Å (8×) = mean Y–O
Cu(1)–O(4) = 1.80_7 Å (2×) = mean Cu(1)–O
Cu(2)–O(2) = 1.94_2 Å (4×); Cu(2)–O(4) = 2.46_1 Å; mean Cu(2)–O = 2.04_6 Å

* Occupancies deduced as in Data Table 21 (also by powder neutron diffraction). Note that the O(1) site is empty, as a result of which the square-planar coordination of Cu(1) in $Ba_2YCu_3O_7$ has now become linear 2-coordination. O(2) here is equivalent to O(2) + O(3) in $Ba_2YCu_3O_7$, and the two O(4)'s are equivalent.

pure, stable material has been prepared with $T_c \gtrsim 93$ K. Many hundreds (thousands?) of papers and several new journals devoted to this field have already been published; and the system Y + Ba + Cu + O (and related systems) must be the most studied ever—eclipsing even studies of the uranium oxides in the 1940s. All this has culminated in the award of the Nobel prize for physics to Bednorz and Müller (in October 1987). Hence, it might be expected that the level of scientific activity in the field will now fall off. Nevertheless, the likely technical importance of these materials will ensure a high level of research activity for some years to come, especially in industry. The possible economic returns are enormous.

[Although we have not considered them, several other sorts of anion-deficient perovskites are known. Perhaps the best known (until now) are the (similar) structures of $Ca_2Fe_2O_5$ and brownmillerite, Ca_2FeAlO_5, in which half the (smaller) B cations are octahedrally coordinated (as in perovskite) and the other half are tetrahedrally coordinated (by omission of two of the six octahedral oxygens).]

Note Added in Proof (April 1988)

Further developments in the "New Superconductor" story have occurred in early 1988. Earlier indications of higher T_cs in lanthanide–alkaline earth–copper oxides (referred to above) have not been substantiated, but new materials with T_cs up to ~125 K have been synthesized. These contain bismuth or thallium instead of rare earth cations and, while their structures are also perovskite related, they are not the same as those in the earlier group. Some reported stoichiometries are ~$Bi_2(Sr,Ca)_3Cu_2O_8$, ~$Bi_2(Sr,Ca)_3Cu_3O_9$, and ~$Tl_2Ba_2CaCu_2O_8$, ~$Tl_2Ba_2Ca_2Cu_3O_{10}$.

At first it was thought that the Bi compounds had structures like those of the Aurivillius phases,* $AO_3 \cdot nABO_3$ with $n = 1$–5, but oxygen deficient. But it is now clear that this is not so: they are a new perovskite–layer intergrowth type, with Bi at the boundary of the perovskite layers in B rather than A positions. It is also apparent that the detailed solution of their structures will not be a simple matter—certainly not as straightforward as was the determination of the $Ba_2YCu_3O_7$ structure:

1. Their unit cells are even bigger: for ~$Bi_2(Sr,Ca)_3Cu_2O_8$, 5.39 × 5.39 × 30.725 Å^3 ($V \approx 893$ Å^3, compared with ~3.8 × 3.9 × 11.7 ≈ 174 Å^3 for $Ba_2YCu_3O_7$).

2. Even this is for the average structure, the true structure being the average structure modulated with a periodicity of ~5 × a. But the modulation period is, in fact, ~4.76 × a; that is, it is incommensurate with the lattice of the average structure.

3. The compound ~$Tl_2Ba_2CaCuO_8$ has a unit cell 3.855 × 3.855 × 29.318 Å^3 ($V \approx 435.7$ Å^3), and was said to be unmodulated. But strong diffuse scattering in electron-diffraction patterns suggests that it too may be modulated, although less regularly.

---

* Not described in this book: they are similar to the Ruddlesden–Popper structures, but with an additional oxygen layer between the perovskite slabs, so that AO → AO$_3$.

The accurate and detailed solution of such structures—with a large number of heavy cations in the unit cell, incomplete occupancy of the "perovskite" anion sites, incommensurate modulation (whose direction has also been seen to vary), and substitutional "disorder" on cation sites—is likely to be a formidable task. At the time of writing, it has not been achieved.

Layer Structures (Continued)

$Na_2Ti_3O_7$ and $Na_2Ti_6O_{13}$ show a relationship analogous to that between K_2MgF_4 and $KMgF_3$. [Chemically, $Na_2Ti_6O_{13}$ is easily made by heating $Na_2Ti_3O_7$ in air at 950°C: $2Na_2Ti_3O_7 \rightarrow Na_2Ti_6O_{13} + Na_2O$ (39).] The crystal structure of $Na_2Ti_3O_7$ shows sheets or layers of $Ti_3O_7^{2-}$ parallel to the ab plane (Figure 34). These sheets consist of ribbons of three edge-sharing octahedra; such ribbons share edges along b to form slabs, which join up by octahedral corner-sharing to form zigzag sheets. Each sheet is slightly distorted ccp; in fact, the whole structure becomes ccp if every second sheet and half the sodium atoms are translated by the vector $\frac{1}{2}\mathbf{b}$. If we do this operation, indicated by the arrow (and taking out sodium oxide), the sheets now share corners, and the structure of $Na_2Ti_6O_{13}$ is formed. This structure is depicted in Figure 35; the similarities to $Na_2Ti_3O_7$ are obvious. The sodium atoms are now in tunnels, and the whole structure is very close to cubic-close-packed if sodium atoms are assumed formally to occupy anion positions. However, in the center of each tunnel there is one empty position when compared with ccp. In $Na_2Ti_3O_7$ the sodiums and oxygens pack to form columns of (approximately) a primitive cubic structure, of the same kind as in K_2MgF_4. One cube is marked in Figure 34.

The structure of two bronzes $Ag_xV_2O_5$ (Figure 36) and $Na_xTi_4O_8$ (Figure 37) are also interesting to compare. Both can be idealized to ccp, although the Ag^+ ions in $Ag_xV_2O_5$ are quite off the center of the cuboctahedron.

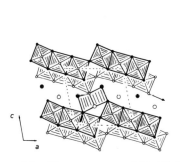

Figure 34. The structure of $Na_2Ti_3O_7$. Larger circles are sodium. Between the layers there is a primitive cubic (pc) structure of anions and cations combined.

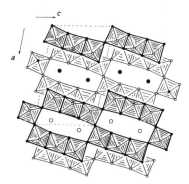

Figure 35. The structure of $Na_2Ti_6O_{13}$.

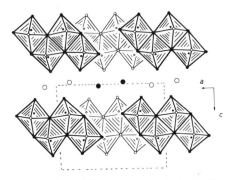

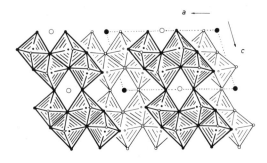

Figure 36. The structure of $Ag_xV_2O_5$, with $x = 0.68$. Larger circles are silver atoms.

Figure 37. The structure of $Na_xTi_4O_8$, $x = 0.8$.

Again, the sheets of the structure of $Ag_xV_2O_5$ can be imagined to be joined by corner-sharing, and a structure of the $Na_xTi_4O_8$ type is then obtained. It is interesting to note that the distortions in the layers of $Ag_xV_2O_5$, due to ionic sizes and cation repulsive forces, are conserved in the three-dimensional structure of $Na_xTi_4O_8$. This is obvious from the two figures, which represent the real structures. [Notice also that the octahedral framework in $Na_xTi_4O_8$ is a distortion of that in the 2×2 column structure of $AlNbO_4 = (Al,Nb)_4O_8$.]

Some crystal data for $Na_2Ti_3O_7$, $Na_2Ti_6O_{13}$, $Na_xTi_4O_8$, and $Ag_xV_2O_5$ are given in Data Tables 23–26.

Data Table 23 $Na_2Ti_3O_7$ (40)

Monoclinic, space group $P2_1/m$, No. 11; $a = 8.571$, $b = 3.804$, $c = 9.135$ Å, $\beta = 101.57°$; $Z = 2$, $V = 291.79$ Å^3

Data Table 24 $Na_2Ti_6O_{13}$ (39) [and jeppeite, $(K_{0.58}Ba_{0.36})_2(Ti_{0.93}Fe^{3+}_{0.06})_6O_{13}$ (41)]

Monoclinic, space group $C2/m$, No. 12; $a = 15.131$ [14.453(2)], $b = 3.745$ [3.8368(7)], $c = 9.159$ [9.123(2)] Å, $\beta = 99.30°$ [99.25(1)°]; $Z = 2$, $V = 512.18$ [533.87] Å^3

Data Table 25 $Na_xTi_4O_8$ (42)

Monoclinic, space group $C2/m$, No. 12; $a = 12.146$, $b = 3.862$, $c = 6.451$ Å, $\beta = 106.85°$; $Z = 2$, $V = 289.61$ Å^3

Occupancy parameter x deduced from structure determination to be 0.8.*

Atomic Distances
Ti(1)–O: 1.71 Å, 1.83 Å, 1.92–2.17 Å (4×)
Ti(2)–O: 1.90 Å, 2.00–2.04 Å (5×)

* Isostructural compounds: $AlNbO_4$ (with Na sites empty) and some minerals of complex compositions

Data Table 26   Ag$_x$V$_2$O$_5$ (43)

Monoclinic, space group $C2/m$, No. 12; $a = 11.742$, $b = 3.667$, $c = 8.738$ Å, $\beta = 90.48°$; $Z = 4$, $V = 376.23$ Å^3

Occupancy parameter x was refined with least squares to 0.68.

Atomic Distances

V(1)–O: five shorter—1.49 Å(terminal), 1.90 Å (2×), 1.85 Å, 1.95 Å—and one
            longer—2.43 Å

V(2)–O: 1.54 Å (terminal), 1.78 Å, 1.89 Å (2×), 2.09 Å, 2.35 Å

Some Compounds Projected along an Axis Corresponding to Two Octahedral Diagonals

Such structures demand bounded projections in order to avoid confusion in trying to depict four layers of anions (and usually four layers of cations also).

Spinel, Al$_2$MgO$_4$

The oxygens in spinel are in almost perfect ccp (it would be perfect for $x = 0.375$). Magnesium atoms are in tetrahedra and the aluminum atoms are in octahedra. The structure is described by means of two bounded projections in Figures 38*a* and *b*‡ and in Figure 39 (see color plates). Strings of edge-sharing octahedra, parallel to each other on the same level and perpendicular on adjacent levels, are joined (between adjacent levels) by edge-sharing. Magnesium atoms lie in tetrahedra between these strings, tetrahedra and octahedra being joined only by corner-sharing.

Crystal data are given in Data Table 27.

Data Table 27   Spinel (Al$_2$MgO$_4$)

Cubic, space group $Fd3m$ No. 227; $a = 8.0800$ Å; $Z = 8$, $V = 527.5$ Å^3

Atomic Positions

Mg in 8(a): $0,0,0;\ \frac{1}{4},\frac{1}{4},\frac{1}{4}$; fc* (tetrahedra)

Al in 16(d): $\frac{5}{8},\frac{5}{8},\frac{5}{8};\ \frac{5}{8},\frac{7}{8},\frac{7}{8};\ \frac{7}{8},\frac{5}{8},\frac{7}{8};\ \frac{7}{8},\frac{7}{8},\frac{5}{8}$; fc (octahedra);

O in 32(e): $\pm(x,x,x;\ x,\frac{1}{4}-x,\frac{1}{4}-x;\ \frac{1}{4}-x,x,\frac{1}{4}-x;\ \frac{1}{4}-x,\frac{1}{4}-x,x)$; $x = 0.387$,
            $x_{\text{calc}} = 3/8 = 0.375$†

* fc indicates face-centered; to the given coordinates also add $\frac{1}{2},\frac{1}{2},0;\ \frac{1}{2},0,\frac{1}{2};\ 0,\frac{1}{2},\frac{1}{2}$.

† For perfect ccp anions.

‡ Using a different projection it *is* possible to show the complete spinel structure in a single figure. This is done in Chapter VI.

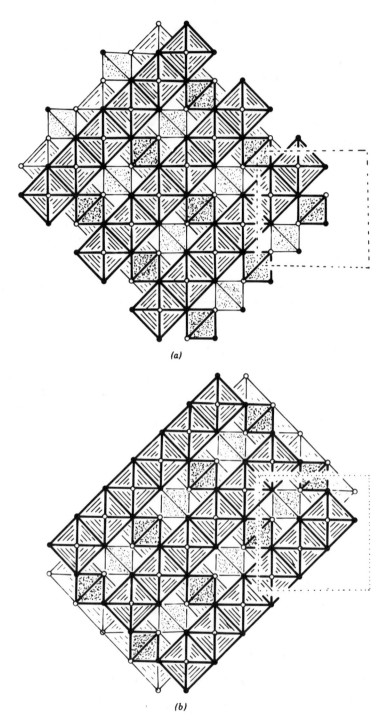

Figure 38. The slightly idealized structure of spinel, $MgAl_2O_4$, projected on to (001): (*a*) Bounded projection $-\frac{1}{8} \leqq z \leqq \frac{3}{8}$, Mg in $z = 0$ or $\frac{1}{4}$ (tetrahedra); Al in $z = \pm\frac{1}{8}$ (octahedra); (*b*) bounded projection, $\frac{3}{8} \leqq z \leqq \frac{7}{8}$. Mg in $z = \frac{1}{2}$ or $\frac{3}{4}$; Al in $z = \frac{3}{8}$ or $\frac{5}{8}$.

In Fe_2MgO_4, on the other hand, half the Fe atoms are in $8(a)$ (tetrahedral sites) and the rest of the Fe plus the Mg atoms are in $16(d)$ (octahedral sites). Such a spinel is said to be *inverse* (whereas Al_2MgO_4 is *normal*, or very nearly so at room temperature). In many spinels the distribution of majority metal atoms between these two sites (and thus the inversion parameter) can often be determined using some diffraction technique. $Fe_3O_4 = Fe_2^{3+}Fe^{2+}O_4$ is also an inverse spinel, with eight Fe^{3+} in $8(a)$ and eight Fe^{3+} plus eight Fe^{2+} in $16(d)$. In this case, below 120 K there is an ordering of the 2- and 3-valent iron in the octahedra, so that the structure becomes orthorhombic. They are random at higher temperature, and so the structure is then really a normal spinel, with Fe^{3+} in $8(a)$ and "$Fe_{1/2}^{3+} Fe_{1/2}^{2+}$" in $16(d)$.

PrI_2-V is just the octahedral (Al_2O_4) part of spinel: the tetrahedra are empty. [But there are small atomic shifts due to Pr–Pr bonding into tetrahedral, Pr_4, clusters (44).] Atacamite, $Cu_2(OH)_3Cl$, in its idealized form (not distinguishing O and Cl), is similar, also with the tetrahedra empty* (45). An important antiform exists, namely, Ti_2C (47). [According to Goretzki (47), Ti_2CH also exists; it is anti-NaCl.]

A very great number of A_2BX_4 compounds (mainly oxides or sulfides) have the spinel structure. Some examples are given by Wyckoff (34), including those with tetragonal or orthorhombic distortions. Many examples with cubic symmetry (229: 149 oxides and 80 sulfides) are given by Hill et al. (48). Inverse spinels in which the octahedral cations are ordered are usually tetragonal: $^{IV}A^{VI}A^{VI}B^{IV}X_4$ is (crystallographically) $^{IV}A^{VI}C^{VI}B^{IV}X_4$ = the $^{IV}Zn^{VI}Li^{VI}Nb^{IV}O_4$ structure type.†

Two complex structures very much related to the spinel structure are shown in Figures 40 and 41 (see color plates). The first shows the ferrite $BaFe_{12}O_{19}$, which is an intergrowth of $BaFeO_3$ (of $BaNiO_3$ type, see Chapter III) and spinel, while the second shows the structure of the so-called β-alumina, a well-known "superionic" conducting material. In both these structures the spinel blocks are in twin orientations, and in the latter, Na^+ atoms partly occupy close-packed anion positions in the twin plane (49).

UCl₅

In this structure the chlorine atoms are in rather regular ccp, with uranium ions in octahedral positions so that molecules of U_2Cl_{10} are formed by means of two octahedra sharing an edge, as shown in Figures 42a and b.

The uranium atoms are slightly shifted away from each other by repulsive forces. The bridging chlorine atoms consequently approach each other so that the distance between them is about 6–7% shorter than the other Cl–Cl

---

* It is now known that the tetrahedra (MgO_4 in spinel) are occupied by groups of three H atoms in atacamite [i.e., $(H_3)O_3Cl$]; $(Mg)Al_2(O_4)$ becomes $(H_3)Cu_2(O_3Cl)$ (46).

† The prefixed Roman numeral superscripts indicate coordination numbers; for example, ^{IV}Zn = 4-coordinate Zn = ZnO_4.

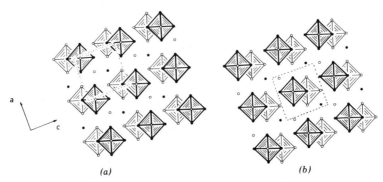

Figure 42. The structure of UCl_5. (*a*) bounded projection, $-\frac{1}{8} \le y \le \frac{1}{8}$; (*b*) bounded projection, $\frac{3}{8} \le y \le \frac{5}{8}$.

distances in the molecule. As a result of these two distortions, the variation of the bond angles within the octahedra can be explained.

The crystal data for UCl_5 are given in Data Table 28.

Data Table 28 UCl_5 (50)

Monoclinic, space group $P2_1/n$, No. 14; $a = 7.99$, $b = 10.69$, $c = 8.48$ Å, $\beta = 91.5°$; $Z = 4$, $V = 724.05$ Å^3

Atomic Positions
All atoms in fourfold general positions in $P2_1/n$ (see ref. 50).

$(NH_4)_3FeF_6$ and K_2PtCl_6

These are two very important structure types. In both structures, alkali or NH_4^+ ions together with the anions form nearly regular ccp arrangements.

The general formula for these compounds is $A_2BB'X_6$. As in the perovskite formula ABX_3, A is the larger cation occupying an anion position in ccp, but B and B' are two different cations, occurring (ordered) in the octahedral positions of perovskite (space group $Fm3m$):

> A in 8(*c*): NH_4 in $(NH_4)_3FeF_6$ and K in K_2PtCl_6
> B in 4(*a*):   Fe in $(NH_4)_3FeF_6$, Pt in K_2PtCl_6
> B' in 4(*c*): NH_4 in $(NH_4)_3FeF_6$, empty in K_2PtCl_6

In K_2PtCl_6 (Figure 43), the octahedra are isolated and not joined by corner-sharing.* In Figure 44 [$(NH_4)_3FeF_6$], B_1 and B_2 are, respectively, Fe and NH_4^+ ordered in octahedra (= B and B' above), doubling the perovskite cube axes. Crystal data for these two types are in Data Tables 29 and 30.

* Alternatively, one may regard it as an array of corner-connected octahedra (as in perovskite) but with alternate octahedra being empty.

Data Table 29 $(NH_4)_3FeF_6$

Cubic, space group $Fm3m$, No. 225; $a = 9.10$ Å; $Z = 4$, $V = 753.57$ Å^3

Atomic Positions
Fe in $4(a)$:   0,0,0; fc
F in $24(e)$: $\pm(x,0,0;\ 0,x,0;\ 0,0,x)$; $x = 0.20$–0.25; fc
$NH_4(1)$ in $4(b)$: $\frac{1}{2},\frac{1}{2},\frac{1}{2}$; fc
$NH_4(2)$ in $8(c)$: $\pm(\frac{1}{4},\frac{1}{4},\frac{1}{4})$; fc

A great number of compounds have this structure. Some examples are Cs_3CoF_6, Cs_3FeF_6, K_3CrF_6, K_3CuF_6, Li_3FeF_6, and α-Na_3AlF_6. Several oxides are reported to have this structure in principle, but with minor distortions. Examples are Ba_3UO_3, Ba_3WO_6, Ca_3WO_6, $CaSr_2WO_6$, and Ba_2MgWO_6. See Wyckoff (34).

Data Table 30 K_2PtCl_6

Cubic, space group $Fm3m$, No. 225; $a = 9.755$ Å; $Z = 4$, $V = 928.28$ Å^3

Atomic Positions
Pt in $4(a)$:   0,0,0; fc
Cl in $24(e)$: $\pm(x,0,0;\ 0,x,0;\ 0,0,x)$; $x = 0.240$; fc
K in $8(c)$: $\frac{1}{4},\frac{1}{4},\frac{1}{4}$; fc

A very great number of halides, especially fluorides and chlorides, have this structure. For a long list, see Wyckoff (34).

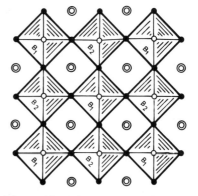

Figure 43. The structure of K_2PtCl_6. Light Pt-centered octahedra on 0, heavy Pt-centered octahedra on $\frac{1}{2}$. Directly above and below these filled octahedra there are empty ones; if they are filled, the perovskite structure is obtained.

Figure 44. The structure of $(NH_4)_3FeF_6$: $A_2BB'F_6 = (NH_4)_2NH_4FeF_6$. A = NH_4, B = Fe, B' = NH_4. Double circles are two overlapping NH_4 (A).

Very closely related to these two structures is that of elpasoite, K_2NaAlF_6, also $A_2BB'X_6$. Slight anion shifts lower the symmetry from $Fm3m$ to $Pa3$. K atoms are in the A sites, with Na and Al in B and B', respectively. There are many isostructural compounds and very many more [including nitrites such as $K_3Co(NO_2)_6$] that are almost identical (34).

Other Perovskite Derivatives

There are three important classes of structures simply derived from the cubic perovskite structure type ABX_3. Geometrically, their origin may be described as follows.

The cubic (space group $Pm3m$) perovskite ABX_3 structure may be said to be "overdetermined": The A atoms are in position 1(b), $\frac{1}{2},\frac{1}{2},\frac{1}{2}$; the B atoms in 1($a$), 0,0,0; and the X atoms in 3(c), $\frac{1}{2}$,0,0; 0,$\frac{1}{2}$,0; 0,0,$\frac{1}{2}$—all special positions. Hence the only adjustable parameter is the unit cell edge a. But two bond lengths (which, to a first approximation, may be taken as constants) have to be accommodated, $l(A–X) = l_A$ and $l(B–X) = l_B$, say.* In this structure the former is $a/\sqrt{2}$ and the latter is $a/2$, so that their ratio is $l_A/l_B = \sqrt{2}$. It is unlikely that this condition will be exactly satisfied for any combination of A, B, and X. Hence, in the older literature there is a great deal of discussion of Goldschmidt's "tolerance factor," $t = l_A/(\sqrt{2}\, l_B)$. One might expect, as is observed, the *cubic* perovskite structure for t values within a few percent of unity. [Of course, X–X distances must not be too small either; in conventional terms, B must be large enough for 6-coordination by X, or A for 12-coordination, and $l(A–X) = d(X\cdots X)$, or A and X have the same radii.] The three important types of perovskite structure deviations then arise when (1) $t > 1$ and B is slightly too small for BX_6 coordination, (2) $t < 1$ but B is large enough for BX_6 coordination, or (3) $t > 1$ and B is large enough for BX_6 coordination. (By implication, we are now treating the structure as a BX_3 framework of corner-connected BX_6 octahedra of the ReO_3 type into which A atoms are inserted. There are also other ways of regarding the structure: B atoms inserted into an AX_3 ccp array or even X atoms inserted into a β-brass-type AB structure.)

In the first class, the B cations are slightly too small for their X_6 octahedra and move off-center (as stated earlier) in a cooperative way. The structure is then polar, but the polarity is easily switched by an electric field. Such crystals are ferroelectric and are important in modern electronic devices. An example is $BaTiO_3$.

In the second class, the A cations are slightly too small for their X_{12} cuboctahedra, and the ReO_3-like framework collapses (reducing some A–O bond lengths) by a correlated tilting of the BX_6 octahedra in their corner-connected array (51–53). The symmetry of the structure is reduced (as it is in

---

* Of course, these have to be for the appropriate coordination numbers 12 for A, 6 for B, 6 for X.

the previous case), and in the simplest examples, with a unit cell no larger than $2 \times 2 \times 2$ cubic perovskite cells, there are four resulting structural families.

1. Space group $Im3$ (cubic) as in $NaMn_7O_{12}$
2. Space group $R\bar{3}c$ (rhombohedral) as in $LiNbO_3$
3. Space group $I4/mmm$ (tetragonal)
4. Space group $Pnma$ (orthorhombic) as in the lanthanide orthoferrites, $LnFeO_3$

Only one of these could be dealt with adequately here; the others need different projections that have not yet been introduced. We therefore postpone their discussion until the end of Chapter XI.

Structures of Compounds Described by Parallel (Reflection) and Cyclic Translations

This is the third class of perovskite-derived structures (3, above), in which the ReO_3 framework is transformed (without any composition change) to create A interstices larger than the cuboctahedron. Some (60%) of the cuboctahedra (which are "tetracapped square prisms") are transformed to pentacapped pentagonal prisms by a rotation operation on ReO_3 columns 2×2 octahedra in cross section. The remainder (40%) become (smaller) tricapped trigonal prisms (Figure 45). If, in the resulting "tetragonal tungsten bronze" (TTB) framework, only the former are occupied, the stoichiometry is $A_6(BO_3)_{10}$ or $A_3B_5X_{15}$. The crystal data in Data Table 31 are for $Ba_3(TiNb_4)O_{15}$.

Data Table 31 $Ba_3TiNb_4O_{15}$ (A "Tetragonal Tungsten Bronze") (54)

Tetragonal, space group $P4bm$, No. 100; $a = 12.54$, $c = 4.01$ Å; $Z = 2$, $V = 630.6$ Å^3

Atomic Positions

Ba(1) in 2(a): $\quad 0,0,z; \frac{1}{2},\frac{1}{2},z; z = 0.9937$

Ba(2) in 4(c): $\quad x,\frac{1}{2} + x,z; \bar{x},\frac{1}{2} - x,z; \frac{1}{2} + x,\bar{x},z; \frac{1}{2} - x,x,z; x = 0.1721,$
$\qquad z = 0.9839$

Ti + Nb(1) in 2(b): $\quad 0,\frac{1}{2},z; \frac{1}{2},0,z; z = 0.4817$

Ti + Nb(2) in 8(d): $\quad x,y,z; \bar{x},\bar{y},z; \frac{1}{2} + x,\frac{1}{2} - y,z; \frac{1}{2} - x,\frac{1}{2} + y,z; \bar{y},x,z; y,\bar{x},z;$
$\qquad \frac{1}{2} + y,\frac{1}{2} + x,z; \frac{1}{2} - y,\frac{1}{2} - x,z; x = 0.0748, y = 0.2159,$
$\qquad z = 0.4562$

O(1) in 2(b): $\quad z = 0.021$

O(2) in 4(c): $\quad x = 0.279, z = 0.500$

O(3) in 8(d): $\quad x = 0.064, y = 0.218, z = 0.000$

O(4) in 8(d): $\quad x = 0.345, y = 0.007, z = 0.500$

O(5) in 8(d): $\quad x = 0.146, y = 0.066, z = 0.500$

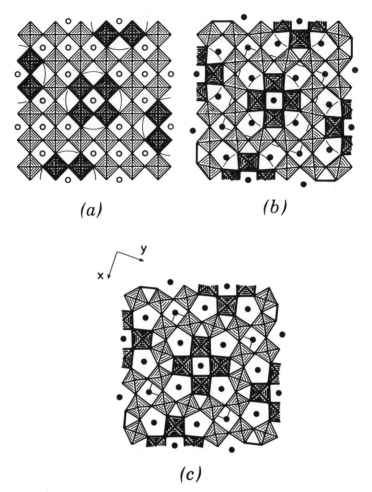

(a)   *(b)*

(c)

Figure 45. Production of the TTB structure from an ReO_3-like framework. (*a*) ReO_3 with groups to be rotated heavily outlined, and sites to be occupied by alkali or alkaline earth metal atoms shown as open circles. (*b*) After the rotation and filling of A sites. (*c*) The real tetragonal potassium tungsten bronze structure. It is apparent that (*b*) and (*c*) are topologically identical.

It was via this mechanism that the rotary stacking fault or cylindrical antiphase boundary defects were defined and described (55–57) and were also found experimentally (58,59).

Notice that the tetragonal tungsten bronze structure consists of two sets of 2×2 octahedral columns in twin orientations (plus an additional octahedral column with intermediate orientation), as depicted in Figure 46. This figure also shows that macroscopic twinning of the ReO_3 type is readily achieved by the rotation operation (e.g., left-hand to right-hand parts of the figure).

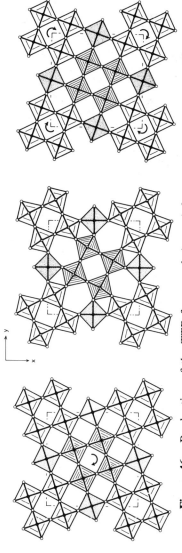

Figure 46. Production of the TTB framework (center) from ReO₃ type in two different (twin-related) orientations (shown on the extreme left and right). In each case 45° rotations (indicated by the curved arrows) of 2 × 2 octahedral columns accomplish the transformation; those of lightly shaded octahedra in one case, and heavily shaded ones in the other. The dotted octahedra have an orientation between those of the twins. The outlined unit cell is that for TTB.

(Note also that the sequence left → center → right, or vice versa, represents a macroscopic twinning mechanism for ReO₃ type.)

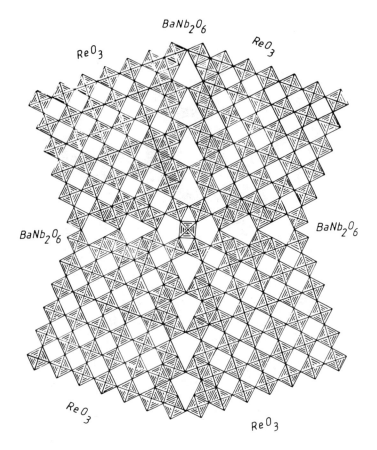

Figure 47. Cyclic reflection (fourling) operation generates a part of the tetragonal bronze structure in the center of the drawing. In the twin planes, the structure of $BaNb_2O_6$ is formed.

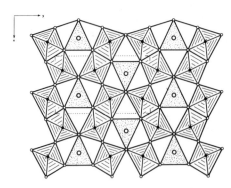

Figure 48. The real structure of $BaNb_2O_6$ projected on (001) and showing NbO_6 octahedra and BaO_6 trigonal prisms. O atoms are at height $z = 0$ (filled circles) and $z = \frac{1}{2}$ (small open circles), Nb atoms (in the octahedra) are at $z = \frac{1}{2}$, and Ba atoms (large open circles) at $z = 0$.

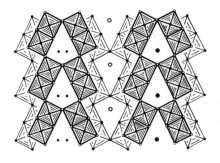

Figure 49. The idealized structure of $BaTi_4O_9$ or KTi_3NbO_9. The pairs of dots on the left are A sites for cubic perovskite: the occupied A sites in the twinned structure (centre and right) are midway between them.

Analogously, the same structure element may be derived by a cyclic reflection operation giving the fourling construction shown in Figure 47. In the twin plane a $BaNb_2O_6$ structure element is formed (cf. Figure 48), and in the fourling center a large unit of the tetragonal bronze structure is shown.

The tetragonal bronze (TTB) structure and the $BaNb_2O_6$ and $BaTi_4O_9$ structures are examples that can be described as being derived by cyclic and reflection operations on an ReO_3 type of structure. In the case of $BaTi_4O_9$ (Figure 49), it is easy to see ccp blocks, containing cations in octahedral coordination, that can be traced back to the sodium titanate layer structures; but here they are repeated by a reflection operation. In Data Table 32 we give the crystal data for the isostructural $KTiNb_3O_9$.

Data Table 32 KTi_3NbO_9 (Structure Type $BaTi_4O_9$) (60)

Orthorhombic, space group *Pmmn*, No. 59; $a = 14.865$, $b = 3.785$, $c = 6.392$ Å; $Z = 2$, $V = 359.6$ Å^3

Atomic Positions

K in 2(b):	$\pm(\frac{1}{4},\frac{3}{4},z)$; $z = 0.778$;
Ti,Nb(1) in 4(f):	$\pm(x,\frac{1}{4},z; \frac{1}{2} - x,\frac{1}{4},z)$; $x = 0.033$, $z = 0.685$
Ti,Nb(2) in 4(f):	$x = 0.126$, $z = 0.241$
O(1) in (4f):	$x = 0.53$, $z = 0.34$
O(2) in 4(f):	$x = 0.62$, $z = 0.77$
O(3) in 4(f):	$x = 0.10$, $z = 0.91$
O(4) in 4(f):	$x = 0.17$, $z = 0.55$
O(5) in 2(a):	$\pm(\frac{1}{4},\frac{1}{4},z)$; $z = 0.17$

References

1. S. Andersson, in *The Chemistry of Extended Defects in Non-metallic Solids,* L. Eyring and M. O'Keeffe, Eds. North-Holland, Amsterdam, 1970.

2. S. Andersson and A. D. Wadsley, *Acta Chem. Scand.* **15**, 663 (1961).

3. J. Galy and B. Darriet, *Rev. Chim. Miner.* **11**, 513 (1974).

4. S. Andersson, *Acta Chem. Scand.* **18**, 2339 (1964).

5. A. D. Wadsley and S. Andersson, *Perspect. Struct. Chem.* **1970,** 1.

6. D. Watanabe, O. Terasaki, A. Jostsons, and J. R. Castles, in *The Chemistry of Extended Defects in Non-metallic Solids,* L. Eyring and M. O'Keeffe, Eds. North-Holland, Amsterdam, 1970, p. 238.

7. S. Andersson, *J. Solid State Chem.* **1,** 306 (1970).

8. M. O'Keeffe and J.-O. Bovin, *Am. Mineral.* **63,** 180 (1978).

9. L. Pauling, *Proc. Natl. Acad. Sci. USA* **15,** 709 (1929).

10. M. O'Keeffe and B. G. Hyde, *Acta Cryst. B* **32,** 2923 (1976).

11. J. Brynestad and H. L. Yakel, *Inorg. Chem.* **17,** 1376 (1978); H. L. Yakel and J. Brynestad, *ibid., 3294.*

12. P. T. Cromer and K. Herrington, *J. Am. Chem. Soc.* **77,** 4708 (1955).

13. L. Kihlborg, *Arkiv Kemi* **21,** 357 (1963).

14. H. G. Bachmann, F. R. Ahmed, and W. H. Barnes, *Z. Kristallogr.* **115,** 110 (1961).

15. R. Hoppe and W. Dähne, *Naturwiss.* **49,** 254 (1962).

16. F. Meller and I. Fankuchen, *Acta Cryst.* **8,** 343 (1955).

17. D. E. Sands, A. Zalkin, and R. E. Elson, *Acta Cryst.* **12,** 21 (1959).

18. W. H. Zachariasen, *Acta Cryst.* **2,** 296 (1949).

19. L. Kihlborg, *Acta Chem. Scand.* **20,** 722 (1966).

20. W. Büssem, H. Fischer, and E. Gruner, *Naturwiss.* **23,** 740 (1935).

21. E. Zintl and K. Loosen, *Z. Physik. Chem.* **174A,** 301 (1935).

22. J. G. Allpress, J. V. Sanders, and A. D. Wadsley, *Phys. Stat. Sol.* **25,** 541 (1968).

23. L. A. Bursill, B. G. Hyde, and D. K. Philp, *Phil. Mag.* **23,** 150 (1970); L. A. Bursill and B. G. Hyde, *Prog. Solid State Chem.* **7,** 177 (1972).

24. J. B. Parise, E. M. McCarron, and A. W. Sleight, *Mater. Res. Bull.* **22,** 803 (1987).

25. E. M. McCarron, *J. Chem. Soc., Chem. Commun.* **1986,** 336.

26. R. Gruehn, *J. Less-Common Metals* **11,** 119 (1966).

27. J. Darriet and J. Galy, *J. Solid State Chem.* **4,** 357 (1972).

28. G. Heuring and R. Gruehn, *Z. Anorg. Chem.* **491,** 101 (1982).

29. W. Mertin, S. Andersson, and R. Gruehn, *J. Solid State Chem.* **1,** 419 (1970).

30. A. D. Wadsley, *Acta Cryst.* **14,** 660 (1961).

31. S. Andersson, *Z. Anorg. Chem.* **351,** 106 (1967).

32. R. S. Roth, A. D. Wadsley, and S. Andersson, *Acta Cryst.* **18,** 643 (1965).

33. B. M. Gatehouse and A. D. Wadsley, *Acta Cryst.* **17,** 1545 (1964).

34. R. W. G. Wyckoff, *Crystal Structures,* Vol. 3. Wiley, New York, 1965.

35. S. N. Ruddlesden and P. Popper, *Acta Cryst.* **11,** 54 (1958).

36. T. G. Bednorz and K. A. Müller, *Z. Phys. B* **64,** 189 (1986).

37. M. K. Wu, J. R. Ashburn, C. J. Torng, P. H. Hor, R. C. Meng, L. Gao, Z. J. Huang, Y. Q. Wang, and C. W. Chu, *Phys. Rev. Lett.* **58,** 908 (1987).

38. D. C. Johnston, A. J. Jacobson, J. M. Newsam, J. T. Lewandowski, D. P. Goshorn, D. Xie, and W. B. Yelon, in *Chemistry of High T_c Superconducting*

Oxides (ACS Symp. Ser., 351) D. L. Nelson, M. S. Whittingham, and T. F. George, Eds. American Chemical Society, Washington, DC, 1987, p. 136.

39. S. Andersson and A. D. Wadsley, *Acta Cryst.* **15,** 194 (1962).

40. S. Andersson and A. D. Wadsley, *Acta Cryst.* **14,** 1245 (1961).

41. A. N. Bagshaw, B. H. Doran, A. H. White, and A. C. Willis, *Aust. J. Chem.* **30,** 1195 (1977).

42. S. Andersson and A. D. Wadsley, *Acta Cryst.* **15,** 201 (1962).

43. S. Andersson, *Acta Chem. Scand.* **19,** 1361 (1965).

44. E. Warkentin and H. Bärnighausen, *Z. Anorg. Chem.* **459,** 187 (1979).

45. A. F. Wells, *Structural Inorganic Chemistry,* 4th ed. Clarendon Press, Oxford, 1975, p. 143.

46. J. B. Parise and B. G. Hyde, *Acta Cryst. C* **42,** 1277 (1986).

47. H. Goretzki, *Phys. Stat. Sol.* **20,** K141 (1967).

48. R. J. Hill, J. R. Craig, and G. V. Gibbs, *Phys. Chem. Miner.* **4,** 317 (1979).

49. J.-O. Bovin, *Acta Cryst. A* **35,** 572 (1979).

50. G. S. Smith, Q. Johnson, and R. E. Elson, *Acta Cryst.* **22,** 300 (1967).

51. H. D. Megaw, *Acta Cryst. A* **24,** 583 (1968).

52. A. M. Glazer, *Acta Cryst. B* **28,** 3384 (1972).

53. M. O'Keeffe and B. G. Hyde, *Acta Cryst. B* **33,** 3802 (1977).

54. V. K. Trunov, I. M. Averina, and Yu. A. Velikodnii, *Russ. J. Inorg. Chem.* **25,** 632 (1980).

55. B. G. Hyde, in *Reactivity of Solids,* J. S. Anderson, M. W. Roberts, and F. S. Stone, Eds. Chapman & Hall, London, 1972, p. 23.

56. L. A. Bursill and B. G. Hyde, *Nature Phys. Sci.* **240,** 122 (1972).

57. B. G. Hyde and M. O'Keeffe, *Acta Cryst. A* **29,** 243 (1973).

58. S. Iijima and J. G. Allpress, *Acta Cryst. A* **30,** 22, 29 (1974).

59. K. Hiraga and B. G. Hyde, 1981 (unpublished).

60. A. D. Wadsley, *Acta Cryst.* **17,** 623 (1964).

CHAPTER III

HCP I
Hexagonal-Close-Packed Arrays (Usually of Anions) Projected along the Three-fold Axis (= $c \approx 2.5$ Å for Many Oxides) Normal to the Close-Packed Layers

The following structure types are described in this chapter: orthorhombic HgO, BCl_3, PI_3, wurtzite (ZnS), zincite (ZnO), lonsdaleite (C), cubanite (Fe_2CuS_3), NiAs, Si_2N_2O, $Be_4(Si_2O_7)(OH)_2$, B_2O_3(II), Na_2SiO_3, Li_2SiO_3, $LiGaO_2$, Li_3PO_4, chrysoberyl (Al_2BeO_4) and other olivines, $Cd(OH)_2$, rutile (TiO_2) and cassiterite (SnO_2), stishovite (high-pressure SiO_2), $CaCl_2$ and Co_2C, marcasite (FeS_2), trirutile, α-PbO_2, columbite, wolframite, baddeleyite (ZrO_2), fluorite (CaF_2), diaspore (α-AlOOH), Li_2ZrF_6, ε-Fe_2N, $BaNiO_3$, β-$(NH_4)_2SiF_6$, β-$TiCl_3$, ε-Fe_3N, PdF_3, ReO_3, $MoOCl_3$, $NbCl_5$, UCl_6, $PbCl_2$, PbFCl (and Cu_2Sb), Fe_2P and β_1-K_2ThF_6, $Y(OH)_3$ = UCl_3, NbCoB, Co_2Si, $PbSbO_2Cl$, gagarinite [$Na(Ca,Ln)_2F_6$], $Al_8FeMg_3Si_6$, $La_7(OH)_{18}I_3$, $TlFe_3Te_3$, the humite (or chondrodite) series, "reduced rutiles" (Ti_nO_{2n-p}). All these structures are hexagonally close-packed (i.e., contain hcp ions of one type, usually anions) or exhibit a small distortion of hexagonal close-packing.

For hexagonal close-packing, this projection down the **c** axis—which we will call HCP I—is the most familiar and probably the most easily comprehended. The analogous projection of ccp is not very useful or convenient: being three anion layers in depth, the repeat along the projection axis is

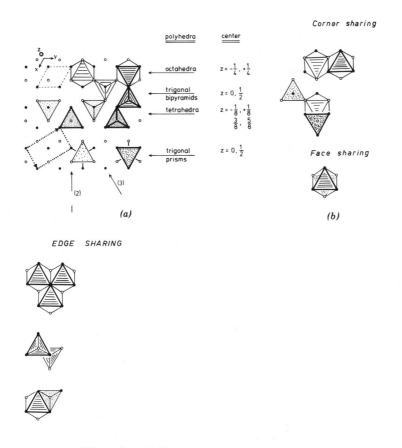

Figure 1. (*a*) Hexagonal close packing, hcp, projected down the **c** axis; for description, see text. This projection is called HCP I. (*b*), (*c*). Various combinations of polyhedra united by common vertices, faces, and edges. These can serve as a key when dissecting or building structures.

(usually) too long; especially for more complex cp stackings, such as **hc.** It is shown in Figure 1*a*. The close-packed layers (which contain the *a* axes) are parallel to the plane of the page: open circles represent atoms at $z = 0$, 1 (layer *a*); filled circles represent atoms at $z = \frac{1}{2}$ (layer *b*).* Layers *a* and *b* alternate, giving the hcp sequence $\cdots ababab \cdots$. The broken lines outline a projected hexagonal unit cell. Dotted lines outline an orthorhombic ("orthohexagonal") unit cell of twice the volume, which is often useful. Arrows numbered 2 and 3 are the projection axes for HCP II (Chapter IV) and HCP III (Chapter V).

---

* In a few cases the unit-cell origin is translated parallel to **c.** For example, in the case of ZnO (p. 55) the layers are at $z_a = 0.3825$ and $z_b = \frac{1}{2} + 0.3825$.

This figure also illustrates, in projection, various simple coordination polyhedra, with vertices defined by the close-packed atoms. From top to bottom are depicted octahedra, trigonal bipyramids, tetrahedra, and tricapped trigonal prisms (sometimes called tetrakaidecahedra),* each at two possible levels. The coordination numbers of cations at the centers of these polyhedra are respectively 6, 5, 4, and 9 (6 + 3). In octahedral and tetrahedral coordination the cations lie between the close-packed anion layers. In trigonal bipyramidal and tricapped trigonal prismatic coordination, the cation is inserted into triangular holes within a close-packed layer. It therefore forces apart the three nearest-neighbor anions and distorts these idealized figures. In these cases there are several common types of distortion that will be discussed later. In perfect hcp the octahedra and tetrahedra are perfectly regular, and for the hexagonal unit cell $c/a = \sqrt{8/3} = 1.633$. In the case of trigonal bipyramidal or tricapped trigonal prismatic coordination, the distortions will reduce the c/a ratio. [This reduces the differences in bond lengths that would otherwise occur, for example, 0.816 (= $\sqrt{2/3}$) and 0.577 (= $1/\sqrt{3}$) for axial and equatorial bonds in trigonal bipyramids of unit edge length.]

In order to visualize three-dimensional structures it is important to be able to recognize the different polyhedral projections: at different levels and in different orientations (e.g., tetrahedra pointing up and down). In the figures, ideal trigonal bipyramids and tetrahedra pointing up are distinguished only by the different types of shading: hatching and stippling, respectively. In all cases, different densities of shading distinguish identical polyhedra at different heights (denser = higher). Figures 1b and c depict various combinations of polyhedra united by common vertices, faces, and edges.

Figure 2 is the projected (total) anion net for HCP I. It is useful as a template for plotting ideal structures.

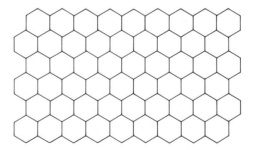

Figure 2.   Honeycomb net, which is the projected atomic net of HCP I. Useful for plotting ideal structures when tracing hcp.

---

* Although not obvious in this projection, in perfect hcp these are very elongated along the projection axis **c**; but see below.

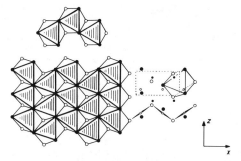

Figure 3. The real structure of HgO projected along **b** of its orthorhombic unit cell. Smaller circles are Hg. Regular octahedra are shown in the upper part.

Mercuric Oxide

The exact crystal structure of the orthorhombic form of HgO is depicted in Figure 3, and its data are given in Data Table 1. It is a rare example of 2-coordination in hcp and is usually described as consisting of planar, zigzag $\cdots$ O–Hg–O–Hg–O $\cdots$ chains, shown on the right of the figure. It can also be described as a slightly distorted hcp arrangement of oxygens, with mercury atoms sitting at the centers of two opposite edges of each octahedron, which are therefore elongated (4.11 Å compared with a mean of 3.51 Å for the other edges). For comparison, the top part of the diagram shows three octahedra in exact hcp. That the distortion is not very great is emphasized by the fact that the ratio of b to the mean octahedral edge length is $5.520/3.612 = 1.528$. This is equivalent to c/a for the hexagonal unit cell of Figure 1a, which is 1.633 in perfect hcp. The unit cell of HgO is equivalent to the orthohexagonal cell introduced above. Note that for perfect hcp this

Data Table 1   HgO (1)

Orthorhombic, space group *Pnma,* No. 62; $a = 6.6129$, $b = 5.5200$, $c = 3.5219$ Å; $Z = 4$, $V = 128.56$ Å^3

Atomic Positions
Hg in 4(c): $\pm(x,\frac{1}{4},z; \frac{1}{2} + x,\frac{1}{4},\frac{1}{2} - z)$; $x = 0.1136$, $z = 0.2456$
O in 4(c): $x = 0.3592$, $z = 0.5955$
Atomic Distances
Hg–O: 2.07 Å, 2.04 ($\pm$ 0.03) Å
Next Hg–O: 2.82 $\pm$ 0.03 Å (between the two nearest adjacent
                              chains), 2.87 Å (to the next nearest chain)
Octahedral edges O–O: 3.40 Å (2×); 3.48 Å (2×); 3.52 Å (2×); 3.58 Å (4×);
                              O–Hg–O = 4.11 Å (2×)

Bond Angles
Hg–O–Hg = 107.3° $\pm$ 1°; O–Hg–O = 179.5° $\pm$ 1°

orthohexagonal cell has axial ratios $c/a = 1.633$ and $b/a = \sqrt{3} = 1.732$, whereas here the equivalent values are $b/c = 1.567$ and $a/c = 1.878$, respectively.

The reality of the chains is shown by the shortest O–O distance being 3.40 Å, compared with a normal 2.6–2.8 Å in ionic oxides. The Hg–O separation in the chains is 2.04 and 2.07 Å, the next-nearest Hg–O distance being 2.82 Å.

Boron Trichloride and Phosphorus Triiodide

In Figure 4 the structure of BCl_3 is shown projected along **c**, idealized so that the chlorine atoms are in undistorted hcp. It is an example of 3-coordination in hcp, and the crystal data are given in Data Table 2. The dotted lines indicate bonds between boron (smaller circles) and chlorine atoms. The BCl_3 triangular molecules form one face on an octahedron of anions. This triangle is smaller than the other triangles in the basal plane (0001), this being the main distortion in the close-packed arrangement. BBr_3 and BI_3 are isostructural, and PI_3 is nearly so.

Data Table 2 BCl_3 (3)

Hexagonal, space group $P6_3$, No. 173; $a = 6.08$, $c = 6.55$ Å, $3c/a = 1.866$ (ideal 1.633); $Z = 2$, $V = 209.69$ Å^3

Atomic Positions

B in 2(b): $\frac{1}{3},\frac{2}{3},z$; $\frac{2}{3},\frac{1}{3},\frac{1}{2} + z$; $z = 0$

Cl in 6(c): x,y,z; $y - x,\bar{x},z$; $\bar{y},x - y,z$; $\bar{x},\bar{y},z + \frac{1}{2}$; $x - y,x,z + \frac{1}{2}$; $y,y - x,z + \frac{1}{2}$;
   $x = 0.045$, $y = 0.376$, $z = 0.00$

Atomic Distance

B–Cl = 1.75 Å

In PI_3 the iodine layers are almost perfectly close-packed (although $3c/a = 1.800$), but the P atoms lie just above the I_3 triangles (at $z = 0.146$), forming PI_3 pyramids. One can imagine that each P atom is octahedrally coordinated (cf. ε-Fe_3N, below) but displaced from the center of its octahedron (at $z = \frac{1}{4}$) by its stereochemically active lone pair of electrons (cf. Chapter X and also AsI_3, etc., in Chapter IV). Its true coordination is

Figure 4. The idealized structure of BCl_3. Smaller circles are B.

therefore PEI_3 (tetrahedral, E = lone pair), and it is really isostructural with iodoform (triiodomethane), CHI_3 (2). The bond lengths $l(P-I)$ are three of 2.46 and three of 3.67 Å (cf. three of 2.43 Å in PI_3 gas molecule).

Wurtzite and Zincite

Wurtzite is a ZnS mineral. The structures of BeO (low-temperature form) and ZnO (zincite), isostructural with wurtzite, have both been refined in great detail (4,5). The data given in Data Table 3 are for ZnO.

Data Table 3 ZnO (4,5)

Hexagonal, space group $P6_3mc$, No. 186; $a = 3.250$, $c = 5.207$ Å, $c/a = 1.602$; $Z = 2$, $V = 47.63$ Å^3

Atomic Positions
Zn and O in 2(b): $\frac{1}{3},\frac{2}{3},z$; $\frac{2}{3},\frac{1}{3},\frac{1}{2} + z$
Zn: $z = 0$
O: $z = 0.3825 \pm 0.0014$*
Atomic Distances
Zn–O: 1.992 ± 0.007 Å, 1.973 ± 0.002 Å (3×); mean = 1.978 Å†
Bond Angles
O–Zn–O = 108.1° and 110.8°, both ± 0.2°

* The ideal $z = 3/8 = 0.3750$. The ideal $c/a = 1.633$ is to be compared with the observed 1.602. An explanation of the distortion has been suggested elsewhere (6).
† These may be compared with the normal Zn–O distances for various geometries,

Coordination number	4	5	6
l(Zn–O) (Å)	1.978	2.013	2.078

In Figure 5a the structure is projected along the **c** axis. Tetrahedra share corners only. (The unit cell is outlined in Figure 7.) Layers of tetrahedra with coplanar bases all point in the same direction. Their apices are also the corners of the bases in the next highest layer, and so on. In this structure the stacking sequence is $\cdots$ $a\beta$ $b\alpha$ a $\cdots$,* but clearly it would be equally feasible to place the second layer so that the sequence was $\cdots$ $a\beta$ $b\gamma$ $c\alpha$ $\cdots$. (The bases of the second layer of tetrahedra then sit in the empty triangles in that layer in Figure 5a; cf. Figure 5b.) Regular repetition of these two sequences leads, in the first case, to the wurtzite or

* Roman letters normally indicate anions, Greek letters cations (although in this case the distinction is unimportant because this structure is its own antitype, i.e., it is unaltered if cations and anions are interchanged). The letter spacing represents schematically the layer spacing along **c**, the stacking direction.

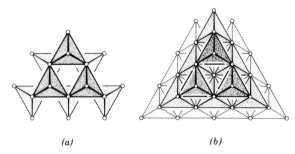

Figure 5. (*a*) The hexagonal structure of wurtzite; (*b*) (for comparison) the cubic structure of zinc blende (with hexagonal unit cell outlined).

zincite structure, Figure 5*a*, with hcp anions (and cations):

$$\cdots \quad \alpha\beta \quad b\alpha \quad a\beta \quad b\alpha \quad a \quad \cdots$$

and, in the second case, to the sphalerite or zinc blende (cubic ZnS) structure (Figure 5*b*) with ccp anions (and cations):

$$\cdots \quad a\beta \quad b\gamma \quad c\alpha \quad a\beta \quad b\gamma \quad c\alpha \quad a \quad \cdots$$

Furthermore, mixtures of these stacking sequences are equally feasible. There are an infinite number of possible regular stacking sequences with rhombohedral or hexagonal unit cells, all with the same sort of tetrahedral layers similarly oriented in the **c** direction. This is the phenomenon known as polytypism (7a,7b). Very many ordered stackings are known for ZnS and, especially, the isostructural SiC (see below). The phenomenon is not restricted to the case of tetrahedral coordination. For example, many cadmium iodide and lead iodide structures, MI_2, are ordered polytypic stacking sequences of layers of edge-shared octahedra.

Compounds isostructural with ZnO include AgI, BeO, Cd(S and Se), Cu(H, Cl, Br, and I), MgTe, Mn(S, Se, and Te), (Al, Ga, In, Nb, and Ta)N, NH_4F, SiC, Zn(O, S, Se, and Te). Notice that many consist of two elements symmetrically disposed with respect to group IV in the periodic table. Especially prevalent are the so-called II–VI and III–V compounds. If both Zn and S are replaced by C, then we have the IV–IV "compound" hexagonal diamond, or lonsdaleite [which is very rare compared with the cubic (zinc blende) analogue, the common gem diamond]. Many of these materials are semiconductors of technological interest for electronic devices. This is even more true for the II–VI and III–V analogues of cubic ZnS.

Cubanite

The structure of cubanite, Fe_2CuS_3, is an interesting variation of the wurtzite (zincite) structure; the crystal data are in Data Table 4. (CuFeS$_2$, chalco-

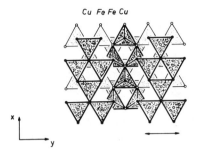

Cu Fe Fe Cu

x
y

Figure 6. The structure of cubanite, Fe_2CuS_3, projected along **c**. The arrow in the lower right indicates the width of a unit slab.

pyrite, is a simple superstructure of zinc blende.) Cubanite is shown idealized and projected along **c** ($\equiv$ **c**$_{hcp}$) in Figure 6. Microtwinned slabs of wurtzite type,* parallel to (010) and of width indicated by an arrow in the figure, are joined by common tetrahedral edges. Copper and iron atoms are ordered onto (010) planes as indicated, giving a short Fe–Fe distance of 2.81 Å across the shared tetrahedral edge. (Cf. 2.54 Å in fcc Fe metal.)

Geometrically, the cubanite structure is easily transformed into a wurtzite superstructure by shuffling all the metal atoms in alternate slabs $\pm\frac{1}{4}$ **c** so that they pass through the base of the tetrahedron (the trigonal bipyramid site) into the tetrahedron immediately above or below. (It is $+\frac{1}{4}$ **c** for the central slab in Figure 6 or $-\frac{1}{4}$ **c** for the slabs to the left and right.) The ideal positions are the same in both structures for all the anions and half the cations.

Data Table 4   Cubanite, Fe_2CuS_3 (8)

Orthorhombic, space group *Pcmn*, No. 62; $a = 6.46$, $b = 11.117$, $c = 6.233$ Å, $Z = 4$, $V = 447.63$ Å^3

Atomic Positions
Cu in 4(*c*): $\pm(x,\frac{1}{4},z; \frac{1}{2} - x,\frac{1}{4},z + \frac{1}{2})$; $x = 0.583$, $z = 0.127$
Fe in 8(*d*): $\pm(x,y,z; x + \frac{1}{2},y + \frac{1}{2},\frac{1}{2} - z; x,\frac{1}{2} - y,z; \frac{1}{2} - x,y,z + \frac{1}{2})$; $x = 0.0875$,
        $y = 0.088$, $z = 0.134$
S(1) in 4(*c*): $x = 0.913$, $z = 0.2625$
S(2) in 8(*d*): $x = 0.413$, $y = 0.0835$, $z = 0.274$
Atomic Distances
Fe–S = 2.25–2.29 Å
Cu–S = 2.27–2.34 Å

Nickel Arsenide

The NiAs structure may be described as consisting of columns of face-sharing NiAs$_6$ octahedra joined at octahedral edges, or columns of face-

---

* Twinned by rotating alternate (010) slabs by 180° about *b*; that is, their **c** ($\equiv$ **c**$_{hcp}$) axes are inverted.

sharing $AsNi_6$ trigonal prisms joined by sharing edges (cf. Chapter IX, Figure 2).

The arrangement of the arsenic atoms is close to hcp, and there is a nickel atom in each octahedron. Each As is surrounded by six equidistant Ni situated at the corners of a triangular prism. Each Ni has eight close neighbors; six are As, while the other two are the Ni atoms immediately above and below. (Since the As cp layer stacking is $\cdots ababab \cdots$ and the Ni are in the octahedra, the Ni cp layer stacking is $\cdots \gamma\gamma\gamma\gamma \cdots$, that is, primitive hexagonal. This facilitates Ni–Ni bonding, which probably accounts for the reduced c/a.) Crystal data for NiAs and VS are given in Data Table 5. The relation between the wurtzite and nickel arsenide structures is given in Figure 7. The cations are shifted from the centers of the tetrahedra in wurtzite to the centers of the octahedra in NiAs. (This is indicated by arrows in Figure 7.)

$$\cdots \; a\beta \quad b\alpha \quad a\beta \quad b\alpha \quad a\beta \; \cdots$$

$$\downarrow \; \text{Cation ``shuffles''}$$

$$\cdots \; a \; \gamma \; b \; \gamma \; a \; \gamma \; b \; \gamma \; a \; \gamma \; \cdots$$

(The variation in the spacing of the letters—Roman for cp anion, Greek for cp cation layers—again indicates the variation in the layer spacing. The cation in an anion tetrahedron is closer to the base of the tetrahedron than to its apex; but the cation in an anion octahedron is equidistant from the top and bottom faces of the octahedron.)

The transformation may also be achieved by transposing alternate anion–cation layer pairs, for example, $a\beta \rightarrow c\alpha$:

$$\cdots \; a\beta \quad b\alpha \quad a\beta \quad b\alpha \quad a\beta \; \cdots$$

$$\downarrow \; \text{slip}$$

$$\cdots \; c\alpha \quad b\alpha \quad c\alpha \quad b\alpha \quad c\alpha \; \cdots$$

$$\downarrow \; \text{cation shifts only}$$

$$\cdots \; c \; \alpha \; b \; \alpha \; c \; \alpha \; b \; \alpha \; c \; \alpha \; \cdots$$

The last step is a relaxation: the cation array shifts $c/8$ to bring the cations from the tetrahedral level (closer to one anion layer) to the octahedral level (midway between anion layers). The latter, of course, is simply another description of the first method (indicated by arrows in Figure 7). Both transformations are *slip* processes (the first involving only the cations), and slip is a plausible, cooperative crystal-mechanical process or mechanism. In all cases the anion layer stacking remains hcp, but the cation stacking previously hcp (in wurtzite) changes to primitive hexagonal (in NiAs).

At least 80 different compounds have the NiAs structure. Many have related structures, with more complicated compositions, for example, the

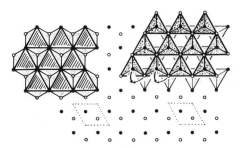

Figure 7. The NiAs structure (left) compared with the wurtzite structure (right). Dotted lines in the NiAs structure indicate octahedra sharing faces (the one underneath is invisible in this projection). Arrows indicate how cations have to move in the wurtzite structure in order to transform it into the NiAs structure. (Only the anions are shown; cations occupy tetrahedra or octahedra as indicated by stippling and hatching.)

chromium sulfides (9,10). Nominally NiAs-type compounds are often cation-deficient.*

Data Table 5   NiAs (VS†)

Hexagonal, space group $P6_3mc$, No. 186; $a = 3.602$, $c = 5.009$ Å; $c/a = 1.391$; $Z = 2$, $V = 56.28$ Å^3

Atomic Positions
Ni(V) in 2(a): $0,0,0$; $0,0,\frac{1}{2}$
As(S) in 2(b): $\frac{1}{3},\frac{2}{3},z$; $\frac{2}{3},\frac{1}{3},z + \frac{1}{2}$

Ideal $z = \frac{1}{4}$

† Unit cell for VS: $a = 3.360$, $c = 5.813$ Å; $c/a = 1.730$, $Z = 2$, $V = 56.83$ Å^3.

The Three Compounds Silicon Oxynitride (Si_2N_2O) (12), Bertrandite ($Be_4(Si_2O_7)(OH)_2$)(13), and High-Pressure Boron Sesquioxide ($B_2O_3(II)$) (14), Which Are All Isostructural; and Some Other Wurtzite-Related Structures

The structure of $B_2O_3(II)$ is projected along the **c** axis in Figure 8a; crystal data are given in Data Table 6. [II stands for the high-pressure form of B_2O_3,

* Recently it has become obvious that, in several instances at least, the NiAs-type structure is unstable with respect to small correlated displacements of the atoms (condensed soft modes), leading to several sorts of (incommensurate) modulated structures. (See Wyckoff's comments in ref. 11.) Nevertheless, it is still a useful structure type and often the stable form at high temperature. Indeed, the data in Data Table 5 are now known to correspond to an averaged, slightly idealized structure. At room temperature, in the real structure, there are very small displacements of both Ni and As from the ideal average positions. References are given in Chapter IX. Larger, more obvious displacements have long been known, for example in MnP and FeS (troilite).

prepared at 1100°C and 65 kbar (1373 K and 6.5 GPa).] It is built up of BO_4 tetrahedra sharing corners. The anion arrangement is a distorted hcp. If this is made perfect, as in Figure 8b, the relationship with the wurtzite structure (Figure 5a) is obvious: an ordered array of one-third of the cation positions in every wurtzite layer is empty.

Arrows in the central part of Figure 8b show the necessary anion movements (projected, of course) needed to convert the ideal to the real structure. (The anion positions in the real structure are at the heads of the arrows.) These we will call *distortion vectors*.

In this case a single *form* (symmetry-related set) of distortion vectors explains the main features of the real structure. The "distortion" is easily understood as a consequence of repulsion forces. If a cation is removed from the center of a tetrahedron, the cation–anion attraction is lost, and the nearest anions relax in a direction away from the empty cation site. This vector form, "umbrella distortion," will be considered in more detail later.

An additional distortion is not given by this simple form. It arises from the fact that a B–O bond length is longer when the oxygen [O(2)] is shared by three B atoms than when it is shared by two [O(1)]. This is because the B–O(2) bond strength is $\frac{2}{3}$, which is less than the B–O(1) bond strength of $\frac{2}{2} = 1$.

Data Table 6 B_2O_3(II) (14)

Orthorhombic, space group $Ccm2_1$, No. 36; $a = 4.613$, $b = 7.803$, $c = 4.129$ Å; $Z = 4$, $V = 148.64$ Å^3; $b/a = 0.977 \times \sqrt{3}$ (ideal $= \sqrt{3}$), $3c/b = 0.972 \times \sqrt{8/3}$ (ideal $= \sqrt{8/3}$), $\sqrt{3}c/a = 0.949 \times \sqrt{8/3}$ (ideal $= \sqrt{8/3}$)

Atomic Positions: $(0,0,0)$ and $(\frac{1}{2},\frac{1}{2},0)$ +

B in 8(b): $x,y,z; x,\bar{y},z; \bar{x},\bar{y},\frac{1}{2} + z; \bar{x},y,\frac{1}{2} + z$; $x = 0.1606$, $y = 0.16464$, $z = 0.4335$

O(1) in 4(a): $x,0,z; x,0,\frac{1}{2} + z$; $x = 0.2475$, $z = 0.5$

O(2) in 8(b): $x = 0.3698$, $y = 0.2911$, $z = 0.5802$

Atomic Distances

B–O(1) = 1.373 Å

B–O(2) = 1.506 Å, 1.507 Å, and 1.512 Å

The structure of sodium metasilicate, Na_2SiO_3, is projected along its **c** axis in Figure 9a. It is an example of the occurrence of trigonal bipyramids in hcp. It may be derived from a wurtzite-type (A_2BX_3) superstructure by shifting the Na atoms from the center of a tetrahedron (almost) into its base. (The two Na–O apical distances are 2.370 and 2.549 Å, a difference of 7.2%.) The sodium atoms thus have a triangular bipyramidal coordination and a coordination number (CN) of 5, while the silicon atoms have the normal tetrahedral arrangement of their surrounding oxygens, and a CN of 4. The triangular bipyramids share edges and corners with each other and with the tetrahedra: corners within a (001) layer, and edges between layers. Data Table 7 gives the crystallographic data.

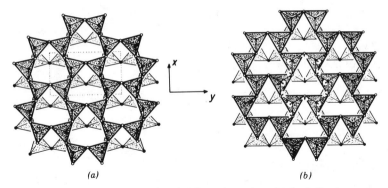

(a)   (b)

Figure 8. (*a*) The real structure of B$_2$O$_3$(II) projected on (001). Only the oxygens are shown; boron atoms are in the stippled O$_4$ tetrahedra. (*b*) The idealized structure of B$_2$O$_3$(II), in which it is easy to see the relation to wurtzite (Figure 7). Vectors show the distortions for transforming to the real structure.

As mentioned earlier, the ideal bipyramid must be distorted to accommodate the Na atom in the plane of the O$_3$ triangle. Figure 9*b* shows the structure idealized so that the anions are in exact hcp. In the lower central part of this diagram, the arrows indicate the magnitude and direction of the distortion (distortion vector form) to produce the real structure from the idealized structure. (Compare Figure 9*b* with Figure 5*b*, wurtzite.) In order to enlarge the basal plane of the triangular bipyramid, the anions move toward the center of the SiO$_4$ tetrahedra, so that these become smaller. This is the exact opposite of the distortion in B$_2$O$_3$(II), where the tetrahedron equivalent to SiO$_4$ is empty, and therefore expanded. Hence, here we have an "umbrella distortion" of opposite sign. [Note that the two structures Na$_2$SiO$_3$ and B$_2$O$_3$(II) have the same space group and similar atomic parameters.]

There are several isostructural compounds such as Li$_2$SiO$_3$ (Figure 10) in which the bond length differences (between, for example, Li–O $\approx$ 2.0 Å and

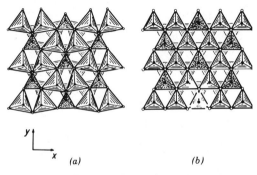

(a)   (b)

Figure 9. (*a*) The real structure of Na$_2$SiO$_3$; (*b*) the idealized structure.

Si–O $\approx$ 1.6 Å) are not as great as in sodium metasilicate (Na–O $\approx$ 2.4 Å, Si–O $\approx$ 1.6 Å). The larger cation (Li) is then 4-coordinate rather than 5-coordinate. The two apical Li–O distances are 1.954 and 2.758 Å, a difference of 34% (compared with only 7% for Na_2SiO_3), so that the longer distance does not correspond to a bond. But the close similarity between the two structures is clear if Figure 10 is compared with Figure 9a.

Data Table 7 Na_2SiO_3 (15) and Li_2SiO_3 (16)*

Orthorhombic, space group $Cmc2_1$, No. 36; a = 10.48 (9.392), b = 6.07 (5.397), c = 4.82 (4.660) Å; Z = 4, V = 306.6 (236.21) Å^3; a/b = 0.997 (1.005) $\times$ $\sqrt{3}$ (ideal = $\sqrt{3}$), $3c/a$ = 0.845 (0.912) $\times$ $\sqrt{8/3}$ (ideal = $\sqrt{8/3}$), $\sqrt{3}c/b$ = 0.842 (0.916) $\times$ $\sqrt{8/3}$ (ideal = $\sqrt{8/3}$)

Atomic Positions: (0,0,0) and $(\frac{1}{2},\frac{1}{2},0)$ +

Na in 8(b): x,y,z; $\bar{x},y,z$; $\bar{x},\bar{y},\frac{1}{2} + z$; $x,\bar{y},\frac{1}{2} + z$; x = 0.1656 (0.174), y = 0.3388 (0.345), z = 0 (-0.002)

Si in 4(a): $0,y,z$; $0,\bar{y},\frac{1}{2} + z$; y = 0.1574 (0.1703), z = 0.5368 (0.491)

O(1) in 8(b): x = 0.1295 (0.1446), y = 0.2873 (0.3077), z = 0.4811 (0.4108)

O(2) in 4(a): y = 0.0844 (0.114), z = 0.8722 (0.846)

Atomic Distances†

Average Si–O(2) (bridging) = 1.672 (1.680) Å
Si–O(1) (nonbridging) = 1.592 (1.592) Å
Na(Li)–O (average) = 2.38 (2.00) Å

* Values for Li_2SiO_3 are given in parentheses.
† Bridging oxygens are those linking two Si atoms: Si–O(2)–Si.

If Na_2SiO_3 is A_2BX_3, then B_2O_3(II) is $A_2\square X_3$ (where $\square$ represents an empty cation site). Also related is the complementary type $\square_2BX_3$—the structure of the so-called asbestos-like form of β-SO_3. The structure of CrO_3 is somewhat similar, and its anion array is deformed hcp (17).

There are also other "wurtzite" superstructures: ABX_2, such as $LiGaO_2$ (Data Table 8, Figure 11), and A_3BX_4, such as Li_3PO_4 (Data Table 9, Figure 12). These structures have been discussed in some detail elsewhere (6).

Data Table 8 $LiGaO_2$ (18)

Orthorhombic, space group $Pna2_1$, No. 33; a = 5.402, b = 6.372, c = 5.007 Å; Z = 4, V = 172.35 Å^3

Atomic Positions

All atoms in 4(a): x,y,z; $\bar{x},\bar{y},\frac{1}{2} + z$; $\frac{1}{2} - x,\frac{1}{2} + y,\frac{1}{2}+ z$; $\frac{1}{2} + x,\frac{1}{2} - y,z$; with

	x	y	z
Li	0.4207	0.1267	0.4936
Ga	0.0821	0.1263	0
O(1)	0.4066	0.1388	0.8927
O(2)	0.0697	0.1121	0.3708

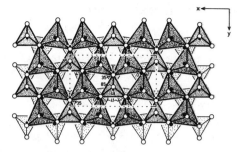

Figure 10. The real structure of Li_2SiO_3 projected on (001). Filled circles are cations, Li (larger) and Si (smaller); open circles are O atoms. Atomic heights are in units of $c/100$.

Data Table 9 Li₃PO₄ (19)

Orthorhombic, space group $Pmn2_1$, No. 31; $a = 6.115$, $b = 5.239$, $c = 4.855$ Å; $Z = 2$, $V = 155.54$ Å^3

Atomic Positions

Li(1) in 4(b): $x,y,z; \bar{x},y,z; \frac{1}{2} - x,\bar{y},\frac{1}{2} + z; \frac{1}{2} + x,\bar{y},\frac{1}{2} + z;$ $x = 0.248$, $y = 0.328$, $z = 0.986$

Li(2) in 2(a): $0,y,z; \frac{1}{2},\bar{y},\frac{1}{2} + z;$ $y = 0.157$, $z = 0.489$

P in 2(a): $y = 0.8243$, $z = 0$

O(1) in 4(b): $x = 0.2078$, $y = 0.6868$, $z = 0.896$

O(2) in 2(a): $y = 0.105$, $z = 0.900$

O(3) in 2(a): $y = 0.819$, $z = 0.317$

The Olivine Group, A₂BO₄

Silicate members of this group include Mg_2SiO_4 (forsterite), Fe_2SiO_4 (fayalite), Mn_2SiO_4 (tephroite), and $CaMgSiO_4$ (monticellite), and many other minerals and synthetic compounds, often with more than one type of cation

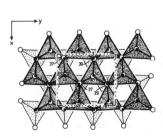

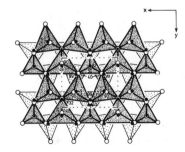

Figure 11. $LiGaO_2$ projected on (001). Filled circles are Li (larger) and Ga (smaller).

Figure 12. Li_3PO_4 projected on (001). Filled circles are Li (larger and P (smaller).

in each type of site. [Olivine itself is $(Mg,Fe)_2SiO_4$.] Other isostructural compounds, which are also minerals, include $LiMnPO_4$, $LiFePO_4$, $NaMnPO_4$, and Al_2BeO_4. The structure of the last, the mineral chrysoberyl, has been refined, and the data given in Data Table 10 come from this refinement. Many more oxides, sulfides, and fluorides with this structure type have also been prepared. (And there is a series of related structures, the *humite series*, which we consider later.)

In Figure 13*a* the idealized olivine structure of Al_2BeO_4 is projected along its **c** axis. Al are in the octahedra, Be in the tetrahedra. The real structure (Figure 13*b*) is fairly close to the ideal: oxygens are almost in hcp, and the cations are close to the ideal positions. Octahedra share edges (and some corners) with other octahedra. That tetrahedra share edges with octahedra seems to be an important feature of the structure; it probably explains the instability of the structure at high pressure.

For reasons already adduced, O–O distances are shorter in shared polyhedral edges than in unshared edges. In this structure the former average 2.53 Å and the latter 2.80 Å, the average being 2.73 Å.

Data Table 10 Al_2BeO_4, Chrysoberyl (20)

Orthorhombic, space group *Pnma*, No. 62; $a = 9.4041$, $b = 5.4756$, $c = 4.4267$ Å, $2\sqrt{3}\,c/a = 1.631$ [$\equiv (c/a)_{hcp}$], $2c/b = 1.617$ [$\equiv (c/a)_{hcp}$], $a/b = 0.992\sqrt{3}$ ($\sqrt{3}$ for hcp); $Z = 4$, $V = 227.94$ Å^3

Atomic Positions

Al(1) in 4(a): $0,0,0;\ 0,\frac{1}{2},0;\ \frac{1}{2},0,\frac{1}{2};\ \frac{1}{2},\frac{1}{2},\frac{1}{2}$

Al(2) in 4(c): $\pm(x,\frac{1}{4},z;\ x + \frac{1}{2},\frac{1}{4},\frac{1}{2} - z);\ x = 0.2732,\ y = \frac{1}{4},\ z = -0.0060$

Be in 4(c): $x = 0.0929,\ y = \frac{1}{4},\ z = 0.4335$

O(1) in 4(c): $x = 0.0905,\ y = \frac{1}{4},\ z = 0.7902$

O(2) in 4(c): $x = 0.4334,\ y = \frac{1}{4},\ z = 0.2410$

O(3) in 8(d): $\pm(x,y,z;\ \frac{1}{2} + x,\frac{1}{2} - y,\frac{1}{2} - z;\ x,\frac{1}{2} - y,z;\ \frac{1}{2} + x,y,\frac{1}{2} - z);$
$x = 0.1632,\ y = 0.0172,\ z = 0.2585$

In the earth's mantle, the increase in seismic velocity with depth is believed to be due to the presence of denser modifications of common minerals. The transition of the olivine structure into the spinel structure (with ccp anions; Chapter II) under high pressure, with a density increase of about 10% (and elimination of edge-sharing by tetrahedra), has stimulated syntheses in this field. Fe_2SiO_4, Ni_2SiO_4, Co_2SiO_4, and Mg_2GeO_4* are examples of substances that transform in this way. [Compare also with the transformation to the Sr_2PbO_4 structure type (Chapter IX) at even higher pressures reported by Wadsley et al. (21). We may also note that some olivines transform to spinel via an intermediate (β) phase or, perhaps, phases (22).]

---

* Mg_2GeO_4 is the only compound known to be able to exist in both forms at 1 atm pressure.

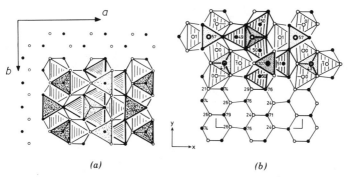

(a)   (b)

Figure 13. (a) The idealized structure of olivine. (b) The real structure of chryso-beryl, Al_2BeO_4, also projected on (001). Large circles are Be (in tetrahedra), medium circles are Al (in octahedra), and small circles are O. In the upper half the cation heights are shown (open at $z \simeq 0$, filled at $z \simeq \frac{1}{2}$), and in the lower half the anion heights (open at $z \simeq \frac{1}{4}$, filled at $z \simeq \frac{3}{4}$); in both cases the units are $c/100$. The departures of these heights from the exact values, and the distortion of the hexagonal net of anions, give an idea of the small distortions from the ideal structure. [Note that the axes in (a) form a left-handed set, those in (b) a (standard) right-handed set.]

Olivine is believed to be the main constituent of the upper mantle, and spinel, the main constituent of the lower mantle. The theory of plate tectonics explains (among other things) the origin of mountains via continental drift—plates moving and being driven into or under one another cause the elevation of mountain ranges. At midocean ridges, material rises from the deep mantle, the pressure falls, and spinel transforms to olivine with a volume increase (hence the ridges). Gravity causes ocean floor spreading, driving the plate movement. Plates driven down under the continents are caused to go through the reverse cycle: olivine to spinel. Thus there is a slow circulation of the viscous work fluid. This transition may be responsible for mountains and for earthquakes!

Cadmium Hydroxide

Once known as the cadmium iodide type, this type of structure was renamed because CdI_2 displays polytypism, whereas $Cd(OH)_2$ appears not to. This structure is also known as the brucite, $Mg(OH)_2$, type. An accurate structure has been determined for portlandite = calcium hydroxide, $Ca(OH)_2$ (see Data Table 11).

Perfect hcp anions would require $c/a = 1.633$ and $z(O) = \frac{1}{4}$. These are almost the values for CdI_2 and several other halides, but the hydroxides deviate considerably.

Figure 14 shows the idealized structure (which, in this projection, appears

Data Table 11   Ca(OH)₂ (23)

Hexagonal, space group $P\bar{3}m1$, No. 164; $a = 3.5918$, $c = 4.9063$ Å, $c/a = 1.366$; $Z = 1$, $V = 54.82$ Å³

Atomic Positions
Ca in 1(*a*):   0,0,0
O in 2(*d*): $\frac{1}{3},\frac{2}{3},z$; $\frac{2}{3},\frac{1}{3},\bar{z}$; $z = 0.2341$
H in 2(*d*): $\frac{1}{3},\frac{2}{3},z$; $\frac{2}{3},\frac{1}{3},\bar{z}$; $z = 0.4248$
*Atomic Distances**
O–H = 0.984 Å, 2.665 Å (Hence there is no H-bonding.)
Ca–O = 2.371 Å
O–O = 3.592 Å (in basal plane), 3.333 Å (out of the basal plane: the other edges
        of the CaO₆ octahedron), 3.095 Å (out of the basal plane: the H-containing
        O₄ tetrahedron)

* These are the results of neutron diffraction measurements.

to be the same as the ideal one). It consists of (0001) layers of octahedra sharing edges. Between these layers the octahedral sites are empty (but all the tetrahedra are occupied by H). In the nickel arsenide structure they are filled. Thus (if we ignore the H atoms), the NiAs and Cd(OH)₂ structure types are related in the same way as BaNiO₃ and (NH₄)₂SiF₆, and β-TiCl₃ and UCl₆ (which are described later). The relationship is important; many transition metal chalcogenides appear to show a homogeneity range between MX₂ of the Cd(OH)₂ type and MX of the NiAs type, apparently due to partial filling of these intermediate cation sites. In at least some cases, this partial occupancy is in a strictly ordered fashion, and a sequence of stoichiometric, ordered phases occurs [e.g., the chromium sulfides (9)].

Polytypism is very common among substances exhibiting this structure type, such as CdI₂, PbI₂, and TaS₂. The geometry of this will be discussed later. Its prevalence is easily understood in terms of the weak bonding between adjacent X–M–X layers, which also accounts for the very easy cleavage of these materials on (0001).

There are at least 70 isostructural substances; examples are Co(OH)₂, Fe(OH)₂, Mg(OH)₂, Mn(OH)₂, Ni(OH)₂, CaI₂, FeBr₂, MgBr₂, MgI₂, TiCl₂, VCl₂, TiI₂, SnS₂, α-TaS₂, and ZrS₂. Some antitypes [with cations and anions interchanged by comparison with Cd(OH)₂] are Ti₂O, Cs₂O, W₂C, and Ag₂F.

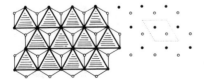

Figure 14. The idealized structure of Cd(OH)₂.

Rutile and Some Related Structures

The mineral rutile is TiO_2, and the isostructural cassiterite is SnO_2. [Two other mineral forms of TiO_2, anatase and brookite, are considered elsewhere. In these the oxygens are in ccp (Chapter II) and in mixed hc packing, respectively]. Structural data for rutile-type TiO_2 are given in Data Table 12.

Data Table 12   Rutile, TiO_2 (24)

Tetragonal, space group $P4_2/mnm$, No. 136; $a = 4.5937$, $c = 2.9587$ Å, $a/c = 1.55$; $Z = 2$, $V = 62.433$ Å^3

Atomic Positions
Ti in 2(a): $0,0,0; \frac{1}{2},\frac{1}{2},\frac{1}{2}$
O in 4(f): $\pm(x,x,0; \frac{1}{2} + x, \frac{1}{2} - x, \frac{1}{2})$; $x = 0.3048$
Atomic Distances
Ti–O = 1.9485 Å (4×), 1.9800 Å (2×); mean = 1.959 Å
Ti–Ti = 2.96 Å (parallel to **c**), 3.57 Å (parallel to ⟨111⟩)

The rutile structure is projected along the **a** axis in Figure 15. Octahedra share edges in the **c** direction and corners in all other directions. [Notice how the strings of octahedra are tilted about **c**, as indicated by the form of the (100) anion nets.] Many other compounds have this structure: CoF_2, FeF_2, MgF_2, MnF_2, NiF_2, PdF_2, ZnF_2, CrO_2, GeO_2, IrO_2, β-MnO_2, β-PbO_2, SnO_2, and so on, and many ternary compounds. δ-Co_2N and Ti_2N are antitype structures.

At very high pressure (>100 kbar), the stable, tetrahedral SiO_2 (quartz) transforms into the rutile structure. This form of SiO_2, in which silicon has a CN of 6, is called *stishovite*. (It has since been discovered in meteorites and meteor crater sites, where the pressure was presumably provided by the impact of the meteorite on the earth's surface.)

The cadmium hydroxide structure is one way of filling half the octahedral interstices in an hcp anion array with cations: *all* the interstices are occupied in *alternate* layers. A second arrangement appears in rutile: alternate rows of interstices are occupied in *every* layer. This gives the characteristic "strings" of edge-sharing octahedra parallel to **c** and united by corners.

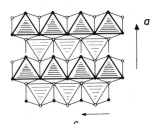

c

Figure 15.   The real structure of rutile (TiO_2).

CaCl$_2$ and CaBr$_2$ have structures that are only a slight distortion of the rutile type (see Chapter V): they are orthorhombic with $a \approx b$. (For CaCl$_2$: $a = 6.24$, $b = 6.43$, $c = 4.20$ Å, $a:b:c = 1.43:1.53:1$, compared with $1.63:1.73:1$ for perfect hcp and $1.55:1.55:1$ for rutile.) They may perhaps be more conveniently regarded as being closer to the "ideal rutile" type than rutile itself. "Ideal rutile," with perfect hcp anions, is also orthorhombic, space group *Pnnm*, No. 58, with anions in $4(g)$: $\pm(x,y,0; x + \frac{1}{2}, \frac{1}{2} - y, \frac{1}{2})$, $x = 0.25$, $y = 0.33$. For CaCl$_2$, $x = 0.275$, $y = 0.325$. (Compare real rutile, $x = y = 0.3048$, and $a = b$.) Co$_2$C is an antitype (also *Pnnm*), with cations at $x = 0.258$, $y = 0.347$. Some other M$_2$X carbides and nitrides are similar.

The close-packed anion planes, which are flat in the ideal rutile, are puckered in the real structure, with ridges parallel to **c**. In our projection (down [100]), for perfect hcp anions, the puckered (100) planes would be flattened but the puckering in (010) increased. This shortens a and lengthens b. (The effect is seen very clearly in the projection used in Chapter V.) The real structure is more symmetrical than the ideal. [The anion packing is, in fact, about halfway between hcp and ccp, but the topology of the octahedral array conforms with that in ideal hcp but *not* with that in ideal ccp (25); see Figure 4 of Chapter V.]

Marcasite, FeS$_2$, is also orthorhombic and is close to the ideal rutile type in the sense that its (001) anion planes [equivalent to (100) in rutile] are almost flat. However, this is caused by the bonding between S atoms on adjacent FeS$_6$ octahedra, forming S$_2$ groups. The S–S distance in each pair is 2.21 Å compared with distances of 2.97, 3.22, and 3.38 Å along octahedral edges.* A number of chalcogenides and pnictides of group VIII transition metals also crystallize with this structure, as does one sodium superoxide polymorph, NaO$_2$(III). The structure of the mineral arsenopyrite, FeAsS, is very similar.

Certain transition metal oxides, such as VO$_2$, MoO$_2$, WO$_2$, ReO$_2$, and TcO$_2$, have structures in which the rutile type is distorted in a monoclinic way (orthorhombic in NbO$_2$) due to metal–metal bonds between successive pairs of cations in the strings of edge-shared octahedra. In MoO$_2$ (26), the Mo–Mo distances are alternately shorter (2.51 Å) and longer (3.11 Å) than in molybdenum metal itself (2.94 Å).

Many reduced transition metal oxides and halides have metal–metal bonding of this kind (27, 28). At higher temperatures these bonds are often broken. The bonding electrons become free, and the material changes from a semiconductor to a metal (similar to a Mott transition) with a great increase in its electrical conductivity (29).

---

* These topological distortions are seen more clearly if we examine the rows of *empty* "octahedral" interstices parallel to the rutile **c** axis in the HCP III projection of Chapter V. These are truly octahedral in "ideal rutile" but enlarged to "square" tunnels (trigonal antiprismatic interstices) in real rutile (hence the easy diffusion of cations such as Li in the **c** direction). In marcasite, these tunnels are very narrow indeed.

Several ternary oxides (e.g., $AlSbO_4$ and $FeNbO_4$) have the rutile structure, with disordered cations. But others have ordered superstructures; an example is tapiolite, Ta_2FeO_6, which has the trirutile structure with a tripled c axis.

α-Lead Dioxide

Several high-pressure forms of fluorides and oxides, as well as $ZrTiO_4$, $ZrSnO_4$, and $HfTiO_4$, have this structure; ζ-Fe_2N and some other nitrides and carbides (e.g., Co_2C and Co_2N) are antitypes. There is a superstructure of the α-PbO_2 type analogous to the trirutile superstructure of rutile; it is the columbite type of $Nb_2(Fe,Mn)O_6$, with a tripled a axis. There is also an ABO_4 superstructure (the wolframite type; examples are $GaTaO_4$ and $CoWO_4$) with ordered cations.

A high-pressure polymorph of titanium dioxide, $TiO_2(II)$, also has this structure, and we give its data in Data Table 13.

Data Table 13 $TiO_2(II)$ (30)

Space group *Pbcn*, No. 60; $a = 4.563$, $b = 5.469$, $c = 4.911$ Å; $Z = 4$, $V = 122.55$ Å^3*

Atomic Positions
Ti in 4(c): $\pm(0,y,\frac{1}{4}; \frac{1}{2},\frac{1}{2} + y,\frac{1}{4})$; $y = 0.171$
O in 8(d): $\pm(x,y,z; \frac{1}{2} - x,\frac{1}{2} - y,\frac{1}{2} + z; \frac{1}{2} + x,\frac{1}{2} - y,\bar{z}; \bar{x},y,\frac{1}{2} - z)$; $x = 0.286$,
       $y = 0.376$, $z = 0.412$

Atomic Distances
Ti–O = 2.05 Å (2×), 1.91 Å (4×); mean = 1.96 Å, the same as in rutile)
Ti–Ti = 3.10–3.56 Å†

* Cf. 124.866/2 Å^3 for the rutile form.
† The mean Ti–Ti distance is 3.43 Å. In rutile it is 3.45 Å, but the shortest is 2.96 Å.

In Figure 16 the structure of $TiO_2(II)$ (idealized to perfect hcp) is projected along a. It consists of a planar arrangement of octahedra sharing edges so that they form zigzag strings in the **c** direction. The structure is completed by joining the chains at two levels through common octahedral corners. (This is a fourth way of filling half the octahedral sites in an hcp array: half the sites in each edge-sharing row are now occupied.) A high-temperature polymorph of ReO_2 and several fluorides are isostructural. The cations are displaced from the centers of the octahedra, which, in turn, are slightly distorted.

Of all the ionic octahedral AX_2 structures, this one seems to be the most stable type at high pressure. Most substances having the rutile structure—rutile itself and several fluorides—transform into the α-PbO_2 type when

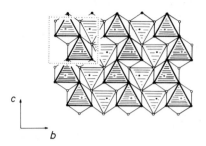

c

b

Figure 16. The real structure of TiO$_2$(II) (α-PbO$_2$ type) idealized to hcp anions but showing the cation displacements from the centers of the octahedra.

exposed to pressure. This can probably be explained in simple repulsion terms: straight chains of edge-sharing TiO$_6$ octahedra, present in the rutile structure, must become unstable when compressed because the Ti–Ti distance is reduced. The instability is reduced by transformation to the zigzag chains of edge-sharing octahedra, characteristic of the α-PbO$_2$ type, which allows the observed cation displacements (cf. Figure 16) that are impossible in straight chains of octahedra. It may be noted that, consistent with this, the Ti–Ti distance across the common octahedral edge is 2.96 Å in rutile, whereas it is 3.10 Å in TiO$_2$(II).

The transformation rutile $\rightarrow$ TiO$_2$(II) may be described simply in terms of cation jumps (31).* This will be considered in the next chapter. An alternative mechanism is by slip on {101} of rutile (32). The α-PbO$_2$-type structure may be resolved into rutile-type slabs two octahedra wide, bounded by (001) planes [in TiO$_2$(II); {101} in rutile], as is clear from Figure 16. These slabs may be said to be in antiphase, since the linear chains of filled and empty octahedra, characteristic of rutile, are interchanged across each boundary.

It is often observed that rutile structures transform to the fluorite type under pressure and that the α-PbO$_2$ form rather than the rutile type is recovered when the pressure is released. This is readily understood in terms of a third displacive (martensitic) mechanism [depicted in Figure 17 (33, 34)], the last stages (fluorite $\rightarrow$ α-PbO$_2$) being a simple, unquenchable topological distortion (cf. Chapter XI). In this context the axial ratios in Table 1 are interesting. They show (and the structure confirms) that the real α-PbO$_2$ is already part way between the ideal α-PbO$_2$ type and the fluorite type.

Diaspore

Diaspore, or α-AlO(OH), projected along the a axis, is shown (idealized to perfect hcp) in Figure 18. (The data are in Data Table 14.) It is easily understood if compared with that of rutile (Figure 16). The chains of octahedra sharing edges, which are single-width in the latter, are double width in the former; and so, like α-PbO$_2$, the diaspore structure is also rutile type with regular APBs—but now they are parallel to {100}$_{rutile}$ $\equiv$ (010)$_{diaspore}$ instead of

---

* By $\pm\frac{1}{2}\mathbf{c}_{hcp} = \frac{1}{2}\mathbf{a}_{PbO_2} = \frac{1}{2}\mathbf{a}_{rutile}$ or $\frac{1}{2}\mathbf{b}_{rutile}$.

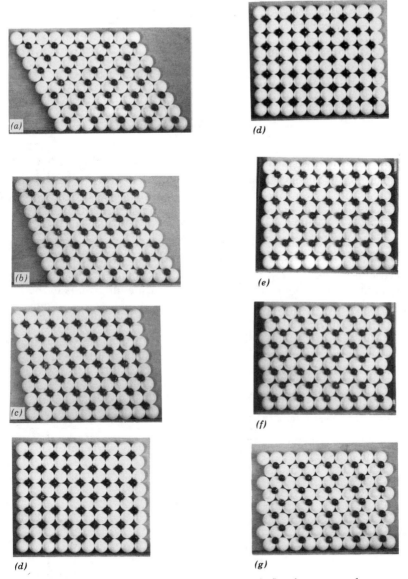

Figure 17. Transformation of (a) rutile type to (d) fluorite type under pressure, followed by the reversion to (g) α-PbO₂ type as the pressure is released. In each case the model is of a (100) layer of the structure. Large spheres are anions; small spheres are cations.

**TABLE 1   Axial Ratios of α-PbO$_2$, Fluorite, and
Related Types**

Structure Type	Axial Ratio c/b
Ideal (hcp) α-PbO$_2$	0.866
TiO$_2$(II)	0.898
Real α-PbO$_2$	0.924
Perfect fluorite	1

{101}$_{rutile}$. The double chains are connected by corner-sharing (so that, in contrast to rutile, the anions are rather closer to hcp—the (001) anion layers being fairly flat]. Neutron diffraction data show that O–H $\cdots$ O hydrogen bonding occurs: The hydrogen atoms lie in the shared edges in the [001] chains of empty octahedra, which explains the more nearly hcp array of anions (cf. marcasite above). Diaspore is the structure type of several minerals, including goethite FeO(OH), groutite MnO(OH), ramsdellite γ-MnO$_2$, montroseite (V,Fe)O(OH), paramontroseite VO$_2$, and Zn(OH)F. (The absence of O–H $\cdots$ O bonds in, for example, ramsdellite results in the empty tunnels again being closer to orthogonal in cross section, as in rutile.)

Geometrically, γ-MnO$_2$ can easily be converted to the rutile-type pyrolusite, β-MnO$_2$, again by small shifts of the cations or by slip. (These relations are more obvious using the HCP II or III projections, especially the latter, described in Chapter V.)

Data Table 14   Diaspore, α-AlO(OH) (35)

Orthorhombic, space group *Pbnm*, No. 62; $a = 4.396$, $b = 9.426$, $c = 2.844$ Å; $Z = 4$, $V = 117.85$ Å^3

Atomic Positions

All atoms are in 4(c):	$\pm(x,y,\frac{1}{4}; \frac{1}{2} - x, y + \frac{1}{2},\frac{1}{4})$
Al:	$x = -0.0451$, $y = 0.1446$
O(1):	$x = 0.2880$, $y = -0.1989$
O(2):	$x = -0.1970$, $y = -0.0532$
H:	$x = -0.4095$, $y = -0.0876$

Li$_2$ZrF$_6$

The hexagonal structure of Li$_2$ZrF$_6$ is projected down its c axis in Figure 19; the structural data are given in Data Table 15. (Li$_2$NbOF$_5$ is isostructural; see Data Table 16.) The parameters for the anions indicate that they are in almost perfect hcp. (Ideal $x = \frac{1}{3}$, $z = \frac{1}{4}$). The cations are in the centers of the octahedra. The only distortion makes the LiX$_6$ octahedra (necessarily) a little larger than the octahedra containing Zr or Nb. In Figure 19, the Li-

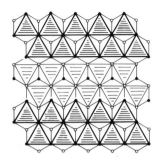

Figure 18. The structure of diaspore, α-AlO(OH).

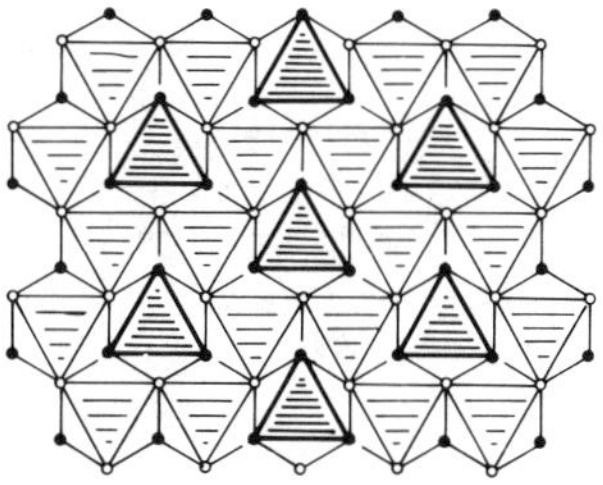

Figure 19. The structure of Li$_2$ZrF$_6$.

containing octahedra (lightly shaded) all share edges and are in the same plane. The Zr-, Nb-containing octahedra share all their corners with LiX$_6$ octahedra and are $\frac{1}{2}\mathbf{c}$ above (and below) them. (In Figure 19 they are indicated by the heavy lines.)

Data Table 15   Li$_2$ZrF$_6$ (36)

Hexagonal, space group $P\bar{3}1m$, No. 164; a = 4.98, c = 4.66, c/a = 0.936, Z = 1, V = 100.08 Å^3

Atomic Positions
Zr in 1(a):   0,0,0
F in 6(k): $\pm(x,0,z; 0,x,z; x,x,z)$; x = 0.33, z = 0.24
Li not determined.

Data Table 16   Li$_2$NbOF$_5$ (37)

Hexagonal, space group $P\bar{3}1m$, No. 164; a = 4.965, c = 4.572 Å, c/a = 0.921 (calc. = 0.943)

Atomic Positions (cf. Data Table 15)
Nb in 1(a)
O,F (disordered) in 6(k): x = 0.325, z = 0.24
Li in 2(d): $\pm(\frac{1}{3},\frac{2}{3},z)$; z = $\frac{1}{2}$
Atomic Distances
Li–O,F = 2.05 Å (6×)
Nb–O,F = 1.95 Å (6×)

This structure is yet another way of filling half the octahedra in an hcp anion array, this time by filling one-third of the octahedra in one layer and two-thirds in the next. It is easily transformed into other structures. Removing the Zr, Nb atoms produces layers like those present (though differently stacked) in Al(OH)$_3$ and BiI$_3$. If, instead of these atoms being removed, they are shifted by $\frac{1}{2}\mathbf{c}$, we get (ignoring the fact that Li ≠ Zr or Nb) the Cd(OH)$_2$ type. Similar shifts of half of each type of cation produce the (ternary) trirutile structure. If only one-fourth of the Li atoms and one-third of the Zr or Nb atoms are moved, the result is a superstructure of the α-PbO$_2$ type (but *not* that of columbite, which requires a more intricate set of cation movements) (38).

These relations (and others are easily imagined) are of greater interest and significance if we consider binary antitype structures, which are of considerable metallurgical importance. Again ignoring the difference between Li and Zr, ε-Fe$_2$N and γ-Co$_2$N are antitypes of Li$_2$ZrF$_6$. Figure 19 therefore also represents hcp Fe or Co, with the octahedra occupied by N. These antitypes are stable only when slightly (or largely, e.g., M$_3$N) nitrogen-deficient. At the exact M$_2$N stoichiometry, the transformations already described produce δ-Co$_2$N (antirutile) and ζ-Fe$_2$N (anti-α-PbO$_2$).

BaNiO$_3$

In the important structure type of BaNiO$_3$, shown in Figure 20, the large barium ions (the largest circles in the figure) occupy a proportion of the anion sites in an hcp array. (Crystal data are given in Data Table 17.) They lie between endless chains of face-sharing NiO$_6$ octahedra. This sort of substitution is a common principle in solid state chemistry; in the perovskite structure the Sr atoms complete a ccp array of anions (Chapter II); and there are many other examples, especially among the more complicated structures. It is a fundamental part of the fascinating and difficult structural chemistry of the ferrites (see Figure 40 of Chapter II). Of course, it is necessary that this cation have a radius approximately equal to that of the anion. (Compare the radii of O^{2-}, F$^-$, Sr^{2+}, and Ba^{2+}). Its coordination number is 12. In hcp, as here, its coordination polyhedron is a so-called twinned cuboctahedron (TCO). The BaNiO$_3$ structure therefore consists of a space-filling combination of BaO$_{12}$ TCOs and NiO$_6$ octahedra. In the ccp analogue, the corresponding structure is cubic perovskite (e.g., SrTiO$_3$), and the corresponding SrO$_{12}$ is a cuboctahedron. The present structure may be regarded as the hexagonal analogue of the cubic perovskite type. (There are many other mixed, **h** + **c**, analogues.)

Examples of compounds isostructural with BaNiO$_3$ are BaMnO$_3$, SrTiS$_3$, BaTiS$_3$, and CsNiCl$_3$.

Data Table 17   BaNiO$_3$ (39)

Hexagonal, space group $P6_3mc$, No. 186; $a = 5.58$, $c = 4.832$ Å; $Z = 2$, $V = 150.45$ Å^3

Atomic Positions
Ba in 2(b): $\pm(\frac{1}{3},\frac{2}{3},\frac{1}{4})$
Ni in 2(a): $0,0,0$; $0,0,\frac{1}{2}$
O in 6(c): $x,\bar{x},z$; $x,2x,z$; $2\bar{x},\bar{x},z$; $\bar{x},x,\frac{1}{2}+z$; $\bar{x},2\bar{x},\frac{1}{2}+z$; $2x,x,\frac{1}{2}+z$; $x = \frac{1}{6}$, $z = \frac{1}{4}$
Atomic Distance
Ni–Ni = 2.416 Å (cf. 2.49 Å in Ni metal)

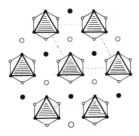

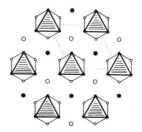

Figure 20. The structure of $BaNiO_3$. Large atoms are Ba. NiO_6 octahedra share faces.

Figure 21. The structure of β-$(NH_4)_2SiF_6$. Large circles are NH_4^+; SiF_6 octahedra are isolated.

β-$(NH_4)_2SiF_6$

The structure of β-$(NH_4)_2SiF_6$ (Figure 21) consists of isolated $[SiF_6]^{2-}$ anions held together by the large NH_4^+ cations. With a cation in $0,0,\frac{1}{2}$ as well as in $0,0,0$, it would be isostructural with $BaNiO_3$. (Compare the parameters and Figures 20 and 21.) The hcp anion net is completed by NH_4^+. (Compare the ionic radii of F^- and NH_4^+.) Crystal data are given in Data Table 18.

Some isostructural compounds are Rb_2GeF_6, Cs_2PuCl_6, Cs_2ThCl_6, Cs_2UCl_6, Cs_2ZrF_6, K_2PtF_6, Rb_2ZrF_6, Cs_2HfF_6, γ-K_2TiF_6, and Rb_2ReF_6.

Data Table 18 β-$(NH_4)_2SiF_6$ (40)

Hexagonal, space group $P\bar{3}m1$, No. 164; $a = 5.784$, $c = 4.796$ Å; $Z = 1$, $V = 139.0$ Å^3

Atomic Positions
Si in $1(a)$: $0,0,0$
(NH_4) in $2(d)$: $\pm(\frac{1}{3},\frac{2}{3},z)$; $z = 0.330$
F in $6(i)$: $\pm(x,\bar{x},z; 2\bar{x},\bar{x},z; x,2x,z)$; $x = 0.139$, $z = 0.799$

The isostructural K_2GeF_6 has also been studied (41). It has K in $2(d)$ with $z = 0.30$, and F in $6(i)$ with $x = 0.148$, $z = 0.780$.

β-TiCl₃

The structure of β-TiCl₃ is depicted in an idealized form ($x = \frac{1}{3}$) in Figure 22. It contains infinite linear chains of $TiCl_6$ octahedra sharing faces and lying parallel to **c**. These chains are held together only by van der Waals forces. Crystal data are given in Data Table 19.

Face-sharing between octahedra normally causes considerable distortion of the anion lattice. However, in this case the effect of metal–metal bonding as well as the large size of the (halogen) anions largely compensate the face-sharing effect, and the structure of β-TiCl$_3$ remains relatively undistorted, although the empty octahedra are a little larger than those containing cations. Hence $x = 0.315$ instead of the ideal value of $\frac{1}{3}$ (for perfect hcp), and c/a is slightly reduced also.

The structure has obvious similarities to that of BaNiO$_3$. It can be produced from the latter by removing the Ba and collapsing the remaining structure so that the chains of face-sharing octahedra come closer together and the hcp array is restored. This can be done by a simple geometrical (rotation) operation like the umbrella distortion used earlier.

Isostructural compounds are β-ZrCl$_3$, ZrBr$_3$, ZrI$_3$, HfI$_3$, and MoBr$_3$. Cs$_3$O is the antitype, but with the chains more widely spaced.

Data Table 19 β-TiCl$_3$ (42) [and β-ZrCl$_3$ (43, 44)]

Hexagonal, space group $P6_3/mcm$, No. 193; $a = 6.27$, $c = 5.82$ Å; $c/a = 0.928$, (ideal $= \sqrt{8}/3 = 0.943$); $Z = 2$, $V = 198.15$ Å^3

Atomic Positions
Ti in 2(b): $0,0,0; 0,0,\frac{1}{2}$
Cl in 6(g): $(x,0,\frac{1}{4}; 0,x,\frac{1}{4}; \bar{x},\bar{x},\frac{1}{4})$; $x = 0.315$ (42)*
Atomic Distances
β-TiCl$_3$:   Ti–Cl = 2.45 Å; Ti–Ti = 2.91 Å†
[β-ZrCl$_3$:   Zr–Cl = 2.55 Å; Zr–Zr = 3.07 Å†]

* β-ZrCl$_3$ would seem to be more accurately determined with x values of 0.319 and 0.320 by two different groups (43, 44).
† Cf. 2.86 Å in β-Ti, 2.90 Å in hcp Ti, 2.96 Å in rutile; 3.18 Å in β-Zr, 3.14 Å in bcc Zr. Orbitals on the metal atoms may overlap in the **c** direction, giving these short metal–metal distances.

ε-Fe$_3$N and PdF$_3$ and ReO$_3$

The same layers of octahedra (normal to **c**), joined by face-sharing in β-TiCl$_3$, are joined by corner-sharing in ε-Fe$_3$N; every corner of every octahedron is joined to a corner of another octahedron (see Figure 30). (Crystal data are given in Data Table 20.) This is also true of PdF$_3$, in which the

Figure 22. The structure of β-TiCl$_3$. TiCl$_6$ octahedra share faces.

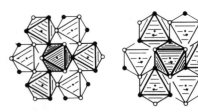

Figure 23. The transformation ReO$_3$ $\rightleftharpoons$ PdF$_3$.

anions are also in hcp, but it and ε-Fe₃N are topologically distinct. PdF₃ has a six-layer repeat along **c**, whereas ε-Fe₃N has the simpler two-layer repeat. They are simply different stacking variants related by slip on (0001) (see Chapter IV).

The ReO₃ type (also a six-layer structure) also consists of corner-connected octahedra but with the anions in incomplete ccp. There is a very important simple topological transformation relating PdF₃ to ReO₃, which is topologically identical: the "umbrella distortion" already mentioned, which is illustrated in Figure 23 (17, 45). No such topological change is possible between anti-ε-Fe₃N and ReO₃ or PdF₃.

Data Table 20 ε-**Fe₃N (46)**

Hexagonal, space group $P6_322$, No. 182; $a = 4.677$, $c = 4.371$ Å; $c/a = 0.935$; $Z = 2$, $V = 82.80$ Å³

Atomic Positions
Fe in 6(g): $x,0,0; 0,x,0; \bar{x},\bar{x},0; \bar{x},0,\frac{1}{2}; 0,\bar{x},\frac{1}{2}; x,\bar{x},\frac{1}{2}; x = \frac{1}{3}$
N in 2(c): $\frac{1}{3},\frac{2}{3},\frac{1}{4}; \frac{2}{3},\frac{1}{3},\frac{3}{4}$
Atomic Distances
Fe–N = 1.90 Å (6×)

ε-Fe₃N is said to have a homogeneity range extending almost to Fe₂N although *"throughout the range, the nitrogen atoms are completely ordered"* (47). In the resulting ε-Fe₂N$_{1-\delta}$, the octahedral sites at $0,0,\frac{1}{4}$ are also occupied by N; that is, the structure is the binary antitype of Li₂ZrF₆. However, close to the exact Fe₂N stoichiometry there is an abrupt transformation from ε-Fe₂N to ζ-Fe₂N with the anti-α-PbO₂ structure.

The anti-Fe₃N structure is closely related to that of BCl₃ (see above). It is produced when all the B atoms in the latter are shifted by **c**/4, which takes them from the centers of Cl₃ triangles to the centers of Cl₆ octahedra.

Isostructural compounds are V₃N, Co₃N, and Ni₃N. A polymorph of ReO₃ has been reported to have the anti-ε-Fe₃N structure. It is nonstoichiometric, and the structure determination was carried out on a single crystal of approximate composition Re$_{1.16}$O₃ (48). (Compare Fe₃N → Fe₂N above, with which it appears to be completely isostructural—extra N in the latter and extra Re in the former partly occupy the same sites in the same space group.)

MoOCl₃

The structure of MoOCl₃, idealized to perfect hcp, is shown in Figure 24, projected along **a**. Structural data are given in Data Table 21. Each octahedron, formed by five chlorines and one oxygen, is relatively undistorted, but the molybdenum atom is displaced from the center of the octahedron toward

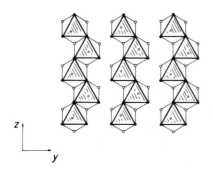

z

y

Figure 24. The idealized structure of $MoOCl_3$, but showing the displacements of the cations towards the O atoms.

the oxygen. This is indicated in the figure. That is why one Mo–Cl distance is as large as 2.82 Å. The structure can therefore also be approximated to $MoOCl_4$ square pyramids sharing (Cl) corners. But, idealized, the MX_4 stoichiometry is achieved by octahedral edge-sharing to form planar zigzag chains parallel to **c**. The same chains joined by additional corner-sharing yield the α-PbO_2 structure. ($MoOCl_3$ is α-PbO_2 type with all the cations omitted from alternate planes.) Note that the unit cells are comparable except that, because of the displacements of the Mo atoms, the b axis of $MoOCl_3$ is double that for α-PbO_2.

Data Table 21 $MoOCl_3$ (49)

Monoclinic, space group $P2_1/c$, No. 14; $a = 5.74$, $b = 13.51$, $c = 6.03$ Å, $\beta = 92.9°$; $Z = 4$, $V = 467.01$ Å^3

Atomic Positions
All atoms in 4(e): x,y,z; for details see ref. 49.
Atomic Distances
Mo–O = 1.60 Å
Mo–Cl = 2.28 Å; 2.37–2.82 Å (bridging)

$NbCl_5$ and UCl_6

At higher anion/cation ratios it becomes geometrically possible to obtain molecular structures. UCl_6 is one; $NbCl_5$ is another. In the latter we can identify isolated Nb_2Cl_{10} molecules. The structure of $NbCl_5$ is depicted (idealized) in Figure 25. (Structural data are given in Data Table 22.) We see that these molecules consist of two octahedra with a common edge. Two kinds of distortion of the close-packed anion array occur: one is that empty octahedra are larger than filled ones, and the second is the shortening of the anion–anion distance in the common octahedral edge due to metal–metal repulsion. (This is a general phenomenon; shared edges are usually shorter than unshared edges; see Chapter II.)

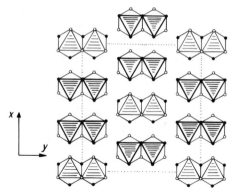

Figure 25. The idealized structure of $NbCl_5$.

Solid $NbCl_5$ vaporizes into monomer molecules: $NbCl_5$ with a trigonal bipyramid configuration.

$NbBr_5$ and $TaCl_5$ are isostructural; $MoCl_5$ is a very slightly distorted variant.

Data Table 22   NbCl₅ (50)

Monoclinic, space group $C2/m$, No. 12; $a = 18.30$, $b = 17.96$, $c = 5.888$ Å, $\beta = 90.6°$; $Z = 12$, $V = 1935.09$ Å³

Atomic Positions
Atoms occupy positions $4(g)$, $4(i)$, and $8(j)$; for details see ref. 50.
Atomic Distances
Nb–Cl = 2.25–2.30 Å (nonbridging); 2.56 Å (bridging)

The structure of UCl_6 is projected along **c** in Figure 26. Structural data are given in Data Table 23. The UCl_6 octahedra are isolated molecules held together by van der Waals forces. The hexagonal close-packing is only slightly distorted, the columns of empty octahedra being larger than the

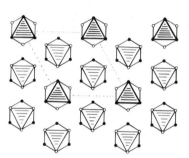

Figure 26. The idealized structure of UCl_6.

others. In the other columns, filled and empty octahedra alternate along the **c** axis and are identical in size. If they were filled with a cation, the β-$TiCl_3$ structure would result (cf. Figure 22). Note that at alternate levels, one-ninth and two-ninths of the octahedral sites are occupied by U.)

Data Table 23   UCl₆ (51)

Hexagonal, space group $P\bar{3}m1$, No. 164; $a = 10.97$, $c = 6.04$ Å, $c/a = 0.551$; $Z = 3$, $V = 629.28$ Å^3

Atomic Positions
U(1) in 1(*a*):   0,0,0
U(2) in 2(*d*): $\pm(\frac{1}{3},\frac{2}{3},z)$; $z = \frac{1}{2}$
Cl in 6(*i*): $\pm(x,\bar{x},z; x,2x,z; 2\bar{x},\bar{x},z)$
                          with Cl(1): $x = 0.10$, $z = 0.25$
                                 Cl(2): $x = 0.43$, $z = 0.25$
                                 Cl(3): $x = 0.77$, $z = 0.25$

Atomic Distances
U–Cl = 2.42 Å (six)

Distortion of Hexagonal Close-Packing and the Tricapped Triangular Prism (TCTP) Polyhedron

When octahedral and/or tetrahedral interstices are only partly filled, the close-packing of the anions may be perfect, or nearly so, but normally there are distortions. These may be due to the effect of ion size, chemical bonding, or stereochemically active lone pairs of electrons, or to the stoichiometry. Geometrically it is obvious that the hcp arrangement must be distorted if cations are accommodated in triangular bipyramids (TBPs) or tricapped triangular prisms (TCTPs). The structure of Na_2SiO_3 (Figure 9) provides an example of the distortions occurring when triangular bipyramidal polyhedra are filled.

A triangular bipyramid is formed by moving the cation (by $c_{hcp}/8$) from the center of an anion tetrahedron into the center of the triangular face at the "base" of the tetrahedron (regarded as a triangular pyramid with its three-fold axis parallel to **c**, cf. Figure 2 of Chapter IV and Figure 1 of Chapter V). Hypothetically, it would be possible to transform all the tetrahedrally coordinated cations in wurtzite into such fivefold coordination, but no compound with that structure is known.* Na_2SiO_3 is closely related, as described above. The similarity between the orthorhombic form of PbO and this hypothetical structure will be described in Chapter X.

Similarly, shifts of $c_{hcp}/4$ in hcp *octahedral* structures bring the cations

---

* The transformation under pressure is to *square* bipyramidal coordination (i.e., octahedral, CN = 6), the NaCl structure.

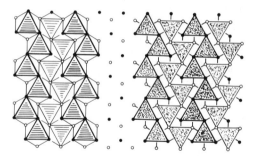

Figure 27. The idealized structures of α-PbO$_2$ and PbCl$_2$.

into positions corresponding to the centers of tricapped triangular prisms (tetrakaidecahedra).* These are common coordination polyhedra, occurring in many structures, and we will describe three different ways to distort hcp in order to produce them. Clearly, at least three hcp layers are necessary for TBP or TCTP formation. They are therefore *not* possible in ccp sequences (in which there is no face-sharing between octahedra or between tetrahedra).

First, it is worth emphasizing that such simple cation shifts relate, geometrically and very clearly, some of the most common and simplest octahedral structures to some of the most common and simplest TCTP structures. Figures 27–30 show the idealized structures of α-PbO$_2$, rutile, Li$_2$ZrF$_6$, and anti-ε-Fe$_3$N, together with the corresponding structures produced by simply translating the cations in these octahedral structures by $\pm c/4$, *all in the same direction* (i.e., $+c/4$ *or* $-c/4$). They are all TCTP structures: idealized PbCl$_2$, PbFCl, β_1-K$_2$ThF$_6$ = anti-Fe$_2$P, and Y(OH)$_3$ (= UCl$_3$), respectively. [Compare Y(OH)$_3$ in Figure 30 with BCl$_3$ in Figure 4!]

As stated above, the $c_{hcp}/4$ shifts of the cations into the cp anion plane at the centers of octahedral faces must cause these triangular faces to expand.†

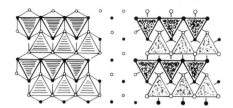

Figure 28. The idealized structures of rutile and PbFCl (BiOF).

---

* Note that this requires that the adjacent octahedron, toward which the cation moves, be *empty*. Two octahedra—one empty, one filled—with a common (0001) face → one filled TCTP. If all the TCTPs are filled, the stoichiometry is MX$_2$.

† Hence, there cannot be TCTP equivalents of structures with filled octahedral layers, such as Cd(OH)$_2$ unless cations move alternately $+c/4$ and $-c/4$; and certainly not for any octahedral structure with more than half the octahedra filled.

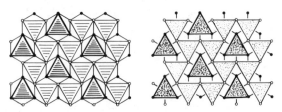

Figure 29.   The idealized structures of Li_2ZrF_6 and Fe_2P.

(The object of the exercise is to increase the CN of the cation, because the cation is too large for octahedral coordination: minimum radius ratio $\rho = r_c/r_a = 0.414$. That being so, it is very much too large for triangular coordination, $\rho = 0.155$. For TCTP coordination $\rho = 0.732$, CN = 9.) But there are other distortions also. Anion triangles immediately above and below the cation form the trigonal prism. (The one in the same plane forms the caps.) This prism is decreased in volume by the cation shift: its triangular faces decrease in size, as does the prism height. This last corresponds to a considerable decrease in c_{hcp} from that for the ideal, primitive hexagonal cell in hcp: c/a changes from 1.633 to 1.000 for ideal TCTPs (with all edges equal). Compare the values of this ratio: for ε-Fe_2N, $c/a = 1.599$; for Fe_2P, 1.021; for ε-Fe_3N, 1.619; for $Y(OH)_3$, 0.980.

In TCTP structures thus derived from hcp, the prisms may be joined in several ways: by sharing triangular faces along c_{hex} with similar prisms immediately above and/or below ($\Delta z = c$); by sharing triangular faces with adjacent prisms at the same height ($\Delta z = 0$); or by sharing edges with adjacent prisms at a different height ($\Delta z = c/2$). In the last case, atoms defining one prism serve as caps on adjacent prisms, and vice versa. It is clearly a geometrical necessity that all cp layers expand equally.* This is the case in the four structures already referred to: $PbCl_2$, $PbFCl$, Fe_2P, and $Y(OH)_3$.

In $Y(OH)_3$ the TCTPs are very regular, and (apart from the **c** axis compression) the deviations from hcp can be expressed by one distortion vector illustrated in the usual projection in Figure 31. Anion shifts are about 0.5 Å;

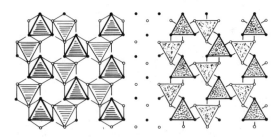

Figure 30.   The idealized structures of ε-Fe_3N or $Re_{1.16}O_3$ and $Y(OH)_3$.

---

* The fact that the original octahedral layers are not more than half-filled allows some triangles to increase in size and others to decrease, so that there is no net change in area of each "anion" layer.

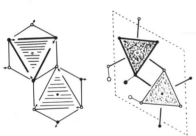
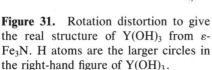
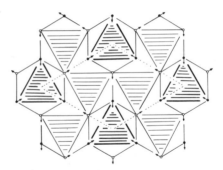

Figure 31. Rotation distortion to give the real structure of Y(OH)$_3$ from ε-Fe$_3$N. H atoms are the larger circles in the right-hand figure of Y(OH)$_3$.

Figure 32. Umbrella distortion to give Fe$_2$P from Li$_2$ZrF$_6$.

the distortion vector is in the octahedral edge and is repeated by threefold symmetry. The projected area of the new polyhedron is equal to that of the original octahedron, with approximately the same cation–anion distances, and the resulting structure is very regular—it is still hexagonal.

The simplest and most obvious way to reduce the triangular ends of the prism would be to move the atoms toward the center of each triangle. At the same time, the "caps" would move out away from the center of their triangle. This "umbrella distortion" is depicted in Figure 32. (It is the same as that used in describing the B$_2$O$_3$ structure.) The structure of Fe$_2$P (Figure 33) is thus derived from the ε-Fe$_2$N structure type. Again the atom shifts are about 0.5 Å. Note that the distortion vector in edge-shared pairs of octahedra is the vector sum of such distortions in each of these octahedra.

In the case of rutile → PbFCl type (Figure 28), the distortion vectors are shown in Figure 34. They may, of course, operate in either of the equivalent **a** or **b** directions of the rutile type. Their magnitude is one-twelfth of the rutile a axis—about 0.4 Å. There is an additional, associated small shift of the cations. The shifts all cooperate to provide space for the larger anions in

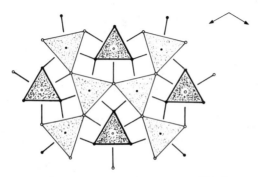

Figure 33. The real structure of Fe$_2$P.

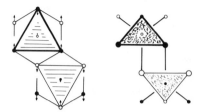

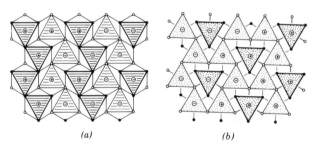

Figure 34. The parallel distortion to give PbFCl from rutile.

the structure of PbFCl or the stereochemically active lone pairs (of electrons) as in BiOF.

In every case the cation displacement vector is $\pm c/4$. It is not essential that all cations move in the same direction, $+c/4$ or $-c/4$. By judicious selection of positive and negative vectors, rutile may be transformed to Fe_2P or $PbCl_2$ (as well as PbFCl), α-PbO_2 to Fe_2P or PbFCl (as well as $PbCl_2$), and so on, and vice versa. Figure 35 gives just one example, α-PbO_2 to Fe_2P, which may be compared with Figures 27 and 29.

Indeed, we can also generate new structure types (for which no octahedral equivalent is yet known) such as that of NbCoB (52), shown in Figure 36, which is an intergrowth of $PbCl_2$ and Fe_2P. Similarly, shifting some cations in the trigonal prism structures by $2 \times (\pm c_{hcp}/4)$ [$= c/2$ for the trigonal prism structure] will transform them one into another; $PbCl_2 \rightleftarrows Fe_2P$, and so on (as, indeed, similar cation displacements transform one octahedral structure into another).

The real structure of $PbCl_2$ (Figure 37, cf. Figure 27) is built up from puckered (100) layers, each composed of TCTPs sharing triangular faces along c and b. Layers, displaced $\frac{1}{2}c$, are joined by common edges. Structural data are in Data Table 24.

It is a common structure, some isostructural compounds being α-PbF_2, $PbBr_2$, ThS_2, $BaBr_2$, $BaCl_2$, $EuCl_2$, $SmCl_2$, β-US_2, ZrF_2, and also Pb(OH)Cl and BiSCl. (Note that, crystallographically, it is a ternary structure.) The structures of the alkaline earth hydrides (BaH_2, etc.) and EuD_2 and YbD_2 are also similar. Some antitypes are Co_2P, Ru_2P, Re_2P, Ca_2Si, Co_2Si, and Ni_2Si. In some cases a small topological distortion increases the number of caps on each TP to four (Figure 32 of Chapter VIII). A somewhat larger topological distortion of the same kind increases the number of caps to

(a)   (b)

Figure 35. α-$PbO_2 \rightarrow Fe_2P$.

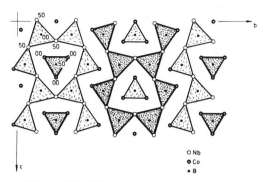

Figure 36. The real structure of NbCoB.

five, straightens the zigzag chains, and yields a higher symmetry (hexagonal) structure—that of Ni$_2$In (Figure 14 of Chapter VI). PbCl$_2$- or Ni$_2$In-$^{-1}$ transformations are therefore simple topological distortions, sometimes called "displacive transformations." This type has been studied in detail in one or two instances; an example is, MnCoGe (53), in which it is accompanied by a volume decrease of 3.9%. It also occurs with Ni$_2$Si, Co$_2$Si, NiTiSi, and CoTiSi.

The PbCl$_2$ structure is also described later in terms of twinning (Chapters VI and VIII).

Data Table 24   PbCl$_2$ (54)

Orthorhombic, space group *Pbnm* (very common, nonstandard setting of *Pnma*), No. 62; $a = 9.05$, $b = 7.62$, $c = 4.535$ Å; $Z = 4$, $V = 312.74$ Å^3

Atomic Positions

All atoms in 4(*c*): $\pm(x, y, \frac{1}{4}; \frac{1}{2} - x, \frac{1}{2} + y, \frac{1}{4})$
Pb: $x = 0.096$, $y = 0.262$
Cl(1): $x = 0.426$, $y = 0.359$
Cl(2): $x = 0.663$, $y = 0.975$

Atomic Distances

Pb–Cl = 2.86–3.08 Å (7×), 3.64 Å (2×)

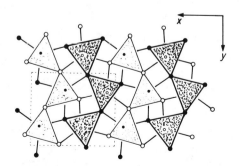

Figure 37. The real structure of PbCl$_2$.

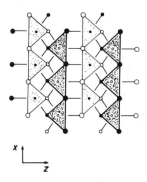

Figure 38. The real structure of BiOF.

The PbFCl structure type is very common for substances containing two different-sized anions. Reference 55 lists almost 200 examples. Only two have both anions of the same size: UTe_2 and BiOF. The structure of the latter is accurately known, and we give its data also.

Figure 38 depicts the structure of PbFCl-type BiOF projected along **b** ($\equiv$**a**). It consists of rows of TCTPs formed by sharing triangular prism faces parallel to **b**, joined into layers by triangular (cap) face-sharing in the **a** direction. These layers are joined by edge-sharing between TCTPs at different levels. (In PbFCl, the largest circles represent Cl and the smallest, Pb. In BiOF, the largest are F and the smallest, Bi.)

The structure can also be described as a layer structure. In BiOF, for example, $Bi_2O_2^{2+}$ layers of edge-sharing, O-centered, Bi_4 tetrahedra* (at $z = 0 \pm 0.2068$) are separated by $(F_2)^{2-}$ layers (two layers of F^- at $z = \frac{1}{2} \pm 0.176$). We assume that the lone pair on the Bi is stereochemically active, pushing the (more weakly bonded) fluorine atoms into separate (double) layers. High pressure could conceivably force the lone pair into a more spherical symmetry, and the structure could transform into the $PbCl_2$ type, which should be more stable at higher pressures; compare the rutile/α-PbO_2 transformation. In the other cases there is a simpler explanation: the geometry is a consequence only of the different anion sizes. (Anions are about the same size in BiOF.) Structural data for PbFCl and BiOF are given in Data Tables 25 and 26, respectively.

The $M_2O_2^{n+}$ layers in oxyhalides (which are of the type occurring alone in tetragonal PbO) can also be interleaved with anion layers other than the X_2 type observed here—for example, with a single anion layer, X, giving M_2O_2X, such as nadorite, $PbSbO_2Cl$, and perite, $PbBiO_2Cl$. In this way many other layer structures can be understood and related to the PbFCl type.†

---

* *Anion*-centered tetrahedra.

† Compare also the "Aurivillius phases," which are intergrowths of BiO and perovskite layers (56), cf. p. 34 of Chapter II; fluorite and the yttrium oxide fluorides, and so on, in Chapter XII; as well as La_2MoO_6, Bi_2MoO_6, and various layer-type sulfosalts (57).

Isostructural compounds include lanthanide oxide halides such as NdOBr, actinide oxide halides such as PuOCl, oxide chalcogenides such as UOSe, alkaline earth hydride halides such as CaHCl and BaHBr, and UP_2, UTe_2, $ThAs_2$, USb_2, etc.

Fe_2As or Cu_2Sb is the antitype structure. Other antitype compounds include Mn_2As and Cr_2As.

Data Table 25   PbFCl (58)

Tetragonal, space group $P4/nmm$, No. 129; $a = 4.106$, $c = 7.23$ Å*; $Z = 2$, $V = 121.89$ Å^3

Atomic Positions
Pb in 2(c): $0,\frac{1}{2},z$; $\frac{1}{2},0,\bar{z}$; $z = 0.20$
Cl in 2(c): $z = 0.65$
F in 2(a): $0,0,0$; $\frac{1}{2},\frac{1}{2},0$
Atomic Distances
Pb–F = 2.52 Å (4×)
Pb–Cl = 3.07–3.21 Å (5×)

* Note that b(PbFCl) $\equiv$ c(rutile) and vice versa. In rutile the a axis, normal to the plane of the diagram (Figure 15) equals the long (b) axis in the plane of the diagram. In PbFCl, a equals the *short* axis in the plane of the diagram (Figure 38). This is the result of the compression of the close-packed layers when the coordination changes from octahedral to TCTP.

Data Table 26   BiOF (59)

Unit cell dimensions: $a = 3.7469$, $c = 6.226$ Å

Atomic Positions
Bi in 2(c): $z = 0.2068 \pm 0.0003$
F in 2(c): $z = 0.676 \pm 0.009$
O in 2(a)
Atomic Distances
Bi–O = 2.27 Å (4×)
Bi–F = 2.92 Å, 2.75 Å (4×)

This is a convenient place to point out that any distortion of the tetrakaidecahedron may make it difficult to recognize it in a structure or, alternatively, make it easy to pick out some alternative coordination polyhedron. In the present instance, the distortion of the ideal TCTP (Figure 34) flattens the prism so that its crosssection is almost a 90° triangle. (In Fe_2As it is even flatter.) As a consequence, one can easily recognize a slightly distorted square antiprism coordination (with one cap) about the cation. Thus a bicapped triangular prism and a square antiprism are really very similar. The packing of lone pairs and anions in BiOF or PbFCl is similar to the packing of anions and potassium atoms in K_2MgF_4 and will be discussed below.

The structure of Fe_2P, shown in Figure 33, consists of TCTPs of Fe atoms that share triangular faces to form rings of six TCTPs on the same level, extended to columns by further face-sharing in the **c** direction. Each such ring contains one TCTP on the next level ($\Delta z = \frac{1}{2}\mathbf{c}$) by edge-sharing. Structural data are given in Data Table 27.

Mn_2P, Co_2P, and Ni_2P are isostructural.

Data Table 27 Fe₂P (60)

Hexagonal, space group $P\bar{6}2m$, No. 189; $a = 5.865$, $c = 3.456$ Å, $c/a = 0.589$; $Z = 3$, $V = 102.95$ Å^3

Atomic Positions
Fe in 3(f): $x,0,0$; $0,x,0$; $\bar{x},\bar{x},0$; $x = 0.256$
Fe in 3(g): $x,0,\frac{1}{2}$; $0,x,\frac{1}{2}$; $\bar{x},\bar{x},\frac{1}{2}$; $x = 0.594$
P in 2(c): $\pm(\frac{1}{3},\frac{2}{3},0)$
P in 1(b): $0,0,\frac{1}{2}$
Atomic Distances
P–Fe = 2.22–2.48 Å (9×)

Note that while these compounds are chemically binary, crystallographically they are quaternary. (There are four crystallographically distinct atomic positions.) In particular, there are two distinct P sites in the ratio $1:2$. Hence antitypes such as β_1-K_2ThF_6 need not be, and are not, disordered. (However, we know of no quaternary example.)

Gagarinite, $Na(Ca,Ln)_2F_6$ with Y = Ln, is an interesting variant: intermediate between Li_2ZrF_6 and Fe_2P, being derived from the former by shifting only the "Li" atoms (Ca,Ln) from their octahedral to trigonal prism sites; the "Zr" atom (Na in this case) remains in its octahedral site (compare Figure 29). Another is $Al_8FeMg_3Si_6$ (with a doubled c axis) in which the continuous net of trigonal prisms consists of alternate layers of Mg (at $\frac{1}{2}$) and Al (at 0); these are centered by Al at $\pm\frac{1}{4}$. The isolated TP columns are formed by Si (at $\pm\frac{1}{4}$) and centered by Fe at 0 and Al at $\frac{1}{2}$. So $Fe_3Fe_3PP_2$ becomes $Mg_{1.5}Al_{1.5}Si_3Fe_{0.5}Al_{0.5}Al_2 = Al_4Fe_{0.5}Mg_{1.5}Si_3$, or $Al_8FeMg_3Si_6$.

The structure of $Y(OH)_3$, projected along **c** in Figure 39 (and Figure 31), is

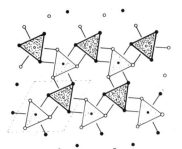

Figure 39. The real structure of $Y(OH)_3$. (Hydrogen positions in Figure 31.)

composed of very regular TCTPs sharing edges between different levels, and faces in the **c** direction. This arrangement leaves a column of empty, squat, face-sharing octahedra (or expanded TCTPs) at the origin. No H-bonding occurs. The hydrogen positions are shown in Figure 31; they line the tunnels at the origin. Data are given in Data Table 28.

Some isostructural compounds are $SmCl_3$, UCl_3, $AcBr_3$, $AcCl_3$, $AmCl_3$, and many lanthanide halides and hydroxides. There appear to be no anti-types.

Data Table 28 $Y(OH)_3$* (61)

Hexagonal, space group $P6_3/m$, No. 176; $a = 6.241$, $c = 3.539$ Å, $3c/a = 1.701$ (cf. 1.633 for ideal hcp); $Z = 2$, $V = 119.38$ Å^3

Atomic Positions
Y in 2(d): $\quad\quad\quad\quad\pm(\frac{2}{3},\frac{1}{3},\frac{1}{4})$
OH has O in 6(h): $\quad\pm(x,y,\frac{1}{4}; \bar{y},x - y,\frac{1}{4}; y - x,\bar{x},\frac{1}{4})$; $x = 0.396$, $y = 0.311$
Atomic Distances
Y—O = 2.437(3) Å (6×), 2.403(3) Å (3×)

* For $Y(OD)_3$ the parameters were refined with neutron diffraction data: this showed that D is also in 6(h), with $x = 0.279$, $y = 0.142$, so that l(O—D) = 0.94(2) Å.

The structure of $TlFe_3Te_3$ is very similar, with Tl in place of Y and Te in place of O, but Fe shifted by $c/2$ (or rotated by 60° about [000z]) from the positions of the D atoms in $Y(OD)_3$ (62). The structure of $TlSe_3Mo_3$ (63) is similar.

$La_7(OH)_{18}I_3$ is a superstructure of this type, in which one-seventh of the hydroxyl groups are substituted by iodine atoms in an ordered fashion so that $\mathbf{a} = 3\mathbf{a}_0 + \mathbf{b}_0$, $\mathbf{b} = -\mathbf{a}_0 + 2\mathbf{b}_0$, and $\mathbf{c} = \mathbf{c}_0$, where the subscript zero indicates the $Y(OH)_3$ structure; that is, $a = b = \sqrt{7}a_0$.

If the TCTP sites at the origin are filled, some topological distortion must occur, because both a and b triangles (at $z = 0$ and $\frac{1}{2}$) are otherwise the same size. The Fe_2P type is the most likely derivative because it involves only this distortion. Filling these sites alternately at $z = \frac{1}{4}, \frac{3}{4}$ could, with appropriate cation shuffles, produce the anti-$PbCl_2$ type.

The Umbrella Distortion

This fundamental distortion of a close-packed layer has been frequently used in this chapter and is quite common in crystal chemistry. It is very useful for relating structures, and so we now consider it in more detail (anticipating, to some extent, a fuller treatment of topological distortions in Chapter XI).

A close-packed (3^6) anion net (layer) can readily be transformed to the expanded net $3 \cdot 6 \cdot 3 \cdot 6$ (the so-called kagome net) as shown in Figure 40.

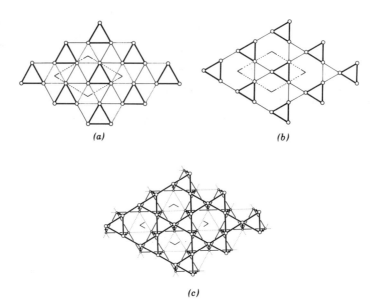

Figure 40. Transformation of a 3^6 net [in (a) and c)] to a $3 \cdot 6 \cdot 3 \cdot 6$ net [in (b) and (c)]. (c) emphasizes the atomic movements for (a) → (b) [the "jack" operation (17) or "umbrella" distortion].

Examination of Figure 40c shows that the atomic shifts can be described as an expansion of one-sixth of the triangles—the "umbrella distortion"—or as a rotation of the triangles in a corner-connected array, alternately clockwise and anticlockwise by 30°. This is exactly the mechanism of transformation from PdF_3 to ReO_3 shown in Figure 23. A similar distortion of half this size (15° rotation of triangles) is shown in Figure 41. This is closer to the umbrella distortion used here and is needed to discuss the relations between α-PbO_2 and $PbCl_2$, for example.

We start with the most symmetrical case: that for $Li_2ZrF_6 \rightarrow K_2ThF_6$ or ε-$Fe_2N \rightarrow Fe_2P$. This is exactly the umbrella distortion, with a rotation angle of $\sim12\frac{1}{2}°$ for Fe_2P, as can be verified by measurements in Figures 32 and 33.

The $Y(OH)_3$ structure can be formally obtained from that of Fe_2P by emptying alternate trigonal prisms in the edge-connected array in Figure 33; compare Figure 39 (and Figure 30 of Chapter IX). Of course, the empty prisms will then expand at their ends and contract at their caps, introducing an additional form (set) of distortion vectors. The net result is vectors in the edges of the rotation triangles (Figures 31 and 39) and octahedra instead of TCTPs at the cell origin.

In passing, we note again that often the "empty prisms" are not really empty; in $Y(OH)_3$ (Figure 31), each contains six H atoms (forming a column of fairly regular H_6 octahedra sharing faces as in hcp). That is, $Y(OH)_3$ is really $Y_2(H_6)O_6$ (cf. K_2ThF_6 of the Fe_2P type, with H_6 substituting for Th).

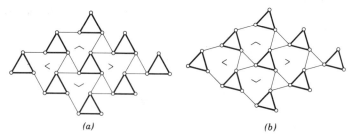

Figure 41. Half the transformation from 3^6 toward $3 \cdot 6 \cdot 3 \cdot 6$ (the "half jack" operation).

The structure of α-LiIO$_3$ (64) is an interesting variant. We write it as I$_2$Li$_2$O$_6$; I atoms are displaced along **z** toward one end face of their TCTPs (O$_9$) by virtue of their stereochemically active lone pairs of electrons, and the Li atoms occupy the (O$_6$) ocahedra, which form a face-shared column at the origin and parallel to **c**.

Furthermore, the full umbrella distortion (triangles rotated by 30°) transforms ε-Fe$_3$N or Y(OH)$_3$ to Ni$_3$Sn, the analogue in hcp of Cu$_3$Au in ccp, in which Ni + Sn together form a perfect hcp array (but with $c/a = 1.605$ instead of 1.633). Each Sn is coordinated by 12 Ni atoms: three above, three below, and six in the same plane [a regular hexagon, which is irregular in Y(OH)$_3$] together forming a "twinned" cuboctahedron. (In an exactly analogous way, PdF$_3$ can be transformed to Cu$_3$Au.)

We now turn to two groups of structures derived from some of the simple structures considered earlier, especially from the rutile type. Their derivation is, perhaps, a little more complex than we have previously encountered. The groups are (*i*) the "humite" or "chondrodite" and "leucophoenicite" series, related to olivine, and (*ii*) the CS derivatives of the rutile type, the "reduced rutiles."

The Humite Series of Minerals and the Related Leucophoenicite Family

The humite series is a group of related mineral structures (originally called the chondrodite series) determined as long ago as 1928 (65). It consists of

Norbergite	Mg$_2$SiO$_4 \cdot$ Mg(F,OH)$_2$
Chondrodite	2Mg$_2$SiO$_4 \cdot$ Mg(F,OH)$_2$
Humite	3Mg$_2$SiO$_4 \cdot$ Mg(F,OH)$_2$
Clinohumite	4Mg$_2$SiO$_4 \cdot$ Mg(F,OH)$_2$
Forsterite	Mg$_2$SiO$_4$ (olivine type)

TABLE 2   Unit Cell Parameters of Synthetic Fluoride End Members of the Humite Series
$(n\text{Mg}_2\text{SiO}_4 \cdot \text{MgF}_2; \; n = 1, 2, 3, 4, \infty)$

	Norbergite	Chondrodite	Humite	Clinohumite	Fosterite
Space group	$Pbnm$	$P2_1/b$	$Pbnm$	$P2_1/b$	$Pbnm$
a (Å)	4.71	4.73	4.74	4.74	4.76
b (Å)	10.27	10.25	10.24	10.23	10.20
c (Å)	8.72	7.79	20.72	13.58	5.98
β	90°	109.2°	90°	100.9°	90°
$c \sin \beta$ (Å)	8.72	7.35	20.72	13.34	5.98
$[= d(001)]$	$= 2 \times (2.99$ $+ 1.37)$	$= 2 \times 2.99$ $+ 1.37$	$= 2 \times (3 \times 2.99$ $+ 1.39)$	$= 4 \times 2.99$ $+1.38$	$= 2 \times 2.99$ $+ 0$

That they were related in structure was deduced even earlier (around 1900), from the similarity of their origins and physical properties and, particularly, from their chemical analyses. This was confirmed by a comparison of their unit-cell parameters (Table 2), especially their virtually identical parameters a and b. [One chooses $Pbnm$, a nonstandard (but very commonly used) setting of $Pnma$, so that a, b, and c are equivalent (corresponding) axes in both the orthorhombic and monoclinic space groups.]

The nature of the relationship between the structures can be expressed in several ways. Originally (65) they were described as consisting of "alternate slabs of the olivine structure and of magnesium hydroxide" [i.e., the Mg(OH)_2 structure]. The last line of Table 2 suggests that the unit slab of olivine is $\frac{1}{2} c_{ol}$ in thickness and that, in this series of structures, one, two, three, four, or an infinite number of these units alternate with a ~ 1.38 Å unit of the second structure type [$\equiv d(11\bar{2}0)$ of brucite, $\text{Mg(OH)}_2 = 1.57$ Å].

This description has been frequently repeated during the succeeding 60 years but is not quite correct. The thin (~ 1.38 Å) lamella is not a half-unit-cell element of Mg(OH)_2 but a {031} slice of sellaite, MgF_2, which has the rutile structure. Using the unit-cell parameters of MgF_2 ($P4_2/mnm$, with $a = 4.628$, $c = 3.045$ Å), we calculate the thickness of this slice as $d(031) = 1.376 \approx 1.38$ Å, a better agreement. The correct description is therefore that n (002) olivine units of Mg_2SiO_4 alternate with unit {031} slabs of sellaite, with $n = 1, 2, 3, 4,$ or ∞ for norbergite to forsterite,* and this suggests a more elegant description, as follows.

Figure 42a shows the rutile structure projected on (100) as in Figure 15 but now idealized to perfect hcp anions. It also shows the trace of an $(03\bar{1})$ plane (which is parallel to a and therefore projects edge-on). If this is used as a twin and composition plane, the result is that shown in Figure 42b. [The twinning operation is a rotation of one half of the crystal by 180° about the

---

* In the minerals, of course, there is some substitution for Mg, for example, by Fe, Mn, etc., and some substitution of OH for F.

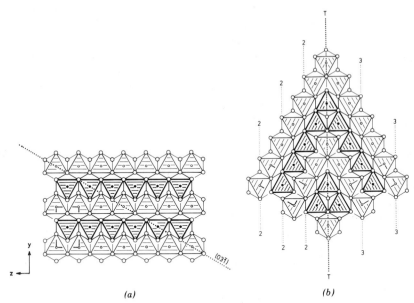

(a)   (b)

Figure 42. (*a*) The idealized structure of rutile (hcp anions) projected on (100). (Compare Figure 15.) The dotted line indicates a (03$\bar{1}$) plane (edge-on). Smaller circles are cations at $x = 0$ (open) and $\frac{1}{2}$ (filled). Larger circles are anions at $\frac{1}{4}$ or $\frac{3}{4}$. (*b*) The rutile structure as in (*a*), but now twinned on (03$\bar{1}$) = T$\cdots$T. The lighter dotted lines indicate parallel planes for mimetic twinning to give (*2*, on the left) the olivine structure (cf. Figure 13) or (*3*, on the right) the norbergite structure.

normal to (03$\bar{1}$), equivalent to reflection in (03$\bar{1}$).*] The structure in the region of the composition/mirror plane is exactly that of the octahedral component at the mirror planes in the olivine type (Figures 13*a,b*). The octahedral part of olivine may therefore be produced by mimetic (i.e., regularly repeated) twinning of the rutile type on {031}. The width of each twin band in olivine is $2 \times d\{031\}_r$ (subscript r indicates the rutile-type structure), as indicated by the lighter dotted lines marked "2" in Figure 42*b* and the symbol *2* in the following discussion.

At the twin plane (the heavier dotted line T$\cdots$T in Figure 42*b*, groups of three edge-sharing octahedra are generated; the structure of olivine is produced from unit-cell-twinned rutile, ..., *2,2,2*, ..., by inserting an additional cation into each anion tetrahedron that shares three of its edges (*but no faces*) with an octahedron in such a group of three octahedra. If r = rutile and tr = rutile twinned on {031} as described, one can imagine

$$2MgF_2 = Mg_2F_4(r) \rightarrow Mg_2F_4(tr) \rightarrow Mg_2O_4^{4-}(tr) \xrightarrow{+Si^{4+}} Mg_2SiO_4 \text{ (olivine)}$$

* It is relevant to point out that contact twins on {031}, although rare, are well known and documented for rutile.

Also shown in Figure 42*b*, but on the right, are the {031} planes on which additional mimetic twinning has to occur in order to produce twinned rutile, ..., *3,3,3,* ..., similar to the octahedral framework of olivine, but with twin-band width $3 \times d\{031\}_r$. This is the octahedral framework of norbergite, whose idealized structure (with anions in perfect hcp) is depicted in Figure 43. Again tetrahedral cations are inserted between the groups of three edge-sharing octahedra at the twin plane.

[Note that the {031} twinning plane is a mirror plane in the anion array of idealized rutile (= hcp). Hence the twinning operation leaves the anion array unaffected; it changes only the cation array. The same changes can clearly be achieved by shuffling some of the cations. The anion arrays are close to hcp in all structures of the humite series.]

The remaining structures in the series are similarly derived by mixing the two twin-band widths, *2* and *3*: chondrodite = ..., *3, 2, 3, 2,* ..., humite = ..., *3, 2, 2, 3, 2, 2,* ... = *3, 2^2*, clinohumite = ..., *3, 2, 2, 2, 3, 2, 2, 2,* ... = *3, 2^3*. A repeat unit in a unit-cell-twinned structure must contain an even number of lamellae; therefore, the series can be written in short form as

Norbergite	$(3)^2$
Chondrodite	$(3,2)$
Humite	$(3,2^2)^2$
Clinohumite	$(3,2^3)$
Olivine	$(2)^2$

The superscripts 2 and 3 indicate a doubling or tripling of the unit concerned: of *3* in norbergite, of *2* and $(3,2^2)$ in humite, and of *2* in clinohumite and in olivine. The general symbol is therefore $(3,2^{n-1})$ if *n* is even or $(3,2^{n-1})^2$ if *n* is odd.

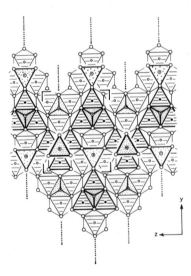

Figure 43. The idealized structure (with hcp anions) of norbergite, $Mg_2SiO_4 \cdot Mg(F,OH)_2$, projected on (100).

All the structures in the series can be described as lamellar intergrowths of olivine and rutile types (the earliest example of the intergrowth principle?), as lamellar intergrowths of norbergite and olivine, or as unit-cell-twinned rutile type (twinned on {031}) with additional (tetrahedral) cations in the composition planes. [An alternative description, related to the last, has been given elsewhere (66–68).]

Not surprisingly in view of the above description, these substances are commonly twinned on (001) and also tend to intergrow on the same plane. Both these phenomena correspond to infrequent "errors" or variation in the

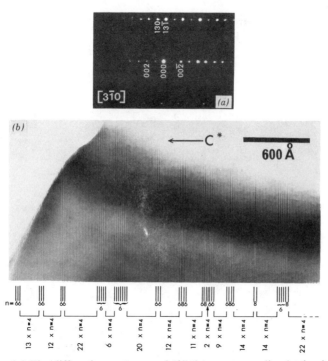

Figure 44. (*a*) The diffraction pattern and (*b*) the corresponding lattice image from a sample of clinohumite, $4Mg_2SiO_4 \cdot Mg(OH,F)_2$, i.e., $n = 4$ and $(3,2^3)$. The light fringes at the top of (*b*) become the dark fringes at the bottom (the contrast changes across the broad, horizontal dark band—an extinction contour—near the center of the image). These are the *3* lamellae of the norbergite type. The darker "fringes" at the top—lighter at the bottom—are, in fact, groups of closely spaced narrow fringes (well resolved in the original print) corresponding to the narrower *2* lamellae of the olivine type: groups of 3×2 in the regions $n = 4$ (clinohumite), and groups of 5×2 and 7×2 at $n = 6, 8$, respectively. The obvious spacing variation shows the specimen inhomogeneity. Most of the area is clinohumite, but careful examination and counting of the fringes indicates that the overall composition is $157[4Mg_2SiO_4 \cdot Mg(OH,F)_2] + 29[6Mg_2SiO_4 \cdot Mg(OH,F)_2] + 5[8Mg_2SiO_4 \cdot Mg(OH,F)_2]$ $= 842Mg_2SiO_4 \cdot 191Mg(OH,F)_2 = 191 \times [4.40_8Mg_2SiO_4 \cdot Mg(OH,F)_2]$ (rather than $4Mg_2SiO_4 \cdot Mg(OH,F)_2$).

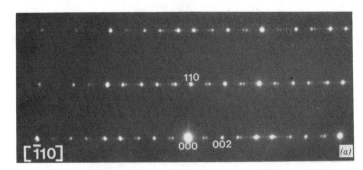

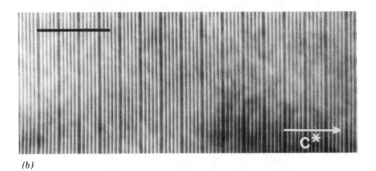

(b)

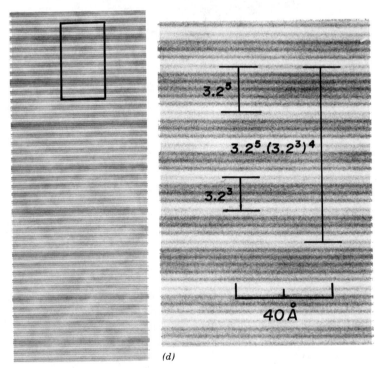

(c)

(d)

twinning frequency, and one might anticipate three further possibilities: (i) frequent errors of the same sort, that is, fine-scale disorder (and therefore nonstoichiometry); (ii) other members of the family $3,2^{n-1}$ with $n > 4$, or members such as $(3^m,2)$; and (iii) ordered mixed intergrowths of the sort $(3,2^p,3,2^q)$, $|p - q| = 1$, or more complex variations. All these have been looked for, and most have been found. (i) occurs, but not very frequently; an example is shown in Figure 44. (ii) occur as thin fault lamellae, but not in macroscopic amounts; see Figure 44 [but $(3^m,2)$ has never been observed]. Complex intergrowths (iii) with dimensions of up to a few thousand ångströms in the **c** direction have been seen, for example, $[3,2^5,(3,2^3)^4]$, which is shown in Figure 45. But the broad conclusion is that all such variations (outside the five members of the series) are nonequilibrium and tend to vanish if well enough annealed. It goes without saying that such fine-scale effects cannot be resolved by X-ray diffraction but only by electron microscopy (67, 68).

Another variation on this twinning theme is realized in the leucophoenicite family of structures, which utilizes twin band widths of *1* and *2* (instead of *2* and *3* as in the humites) (66, 69). Until 1983 the only known member of this family was leucophoenicite itself, $3Mn_2SiO_4 \cdot Mn(OH)_2$, a dimorph of manganhumite, but $(1,2^3)$ instead of $(3,2^2)^2$ and therefore monoclinic instead of orthorhombic. Its structure had been determined by Moore in 1970 (70). Since then two further members have been discovered: jerrygibbsite, $4Mn_2SiO_4 \cdot Mn(OH)_2$, a dimorph of sonolite = Mn clinohumite (71, 72), $(1,2^4)^2$ instead of $(3,2^3)$ and therefore orthorhombic instead of monoclinic; and, very recently (73), ribbeite, a dimorph of alleghanyite = Mn chondrodite, $2Mn_2SiO_4 \cdot Mn(OH)_2$, and therefore $(1,2^2)^2$ instead of $(3,2)$ and also orthorhombic instead of monoclinic. [The structure of jerrygibbsite has been solved by Kato (72), but that of ribbeite has not yet been reported.] Olivine, $(2)^2$, is also an end member of this structural family, of course; and so the only missing analogue in the leucophoenicite family of the humite series is the one corresponding to norbergite, $(1,2)$ instead of $(3)^2$. It will obviously be monoclinic.

In leucophoenicite family members, the *3* lamella of the humite analogue has been resolved into *1,2*; that is, an additional twin plane has been intro-

---

Figure 45. (*a*) The electron diffraction pattern from a mainly clinohumite specimen. The indexing is for clinohumite, $d(001)_{cl} = 13.3$ Å. (*b*) A region of ~ 1200 Å (part of a somewhat larger area) with a perfectly ordered complex superstructure corresponding to a metastable member of the humite series—$[3,2^5,(3,2^3)^4]$ with $d(001) \approx 73$ Å ($= 5 \times 4.36 + 17 \times 2.99$ Å or $19.32 + (4 \times 13.34) = 72.7$ Å). The black bar is ~ 250 Å. The broader dark bands are $(3,2^5)$, and the narrower ones $(3,2^3)$ ($4\times$). (*c*) An enlargement of part of (*b*) to show the fringe resolution clearly. (*d*) A further enlargement of the boxed area in (*c*); of the dark fringes, the lighter ones are *3* lamellae (norbergite type) and the darker ones are *2* lamellae (olivine type). The $[3,2^5,(3,2^3)^4]$ repeat is clearly visible.

duced. However, it should be pointed out, additional tetrahedral cations are *not* inserted into the additional twin/composition planes; to do so would involve pairs of SiO_4 tetrahedra sharing an edge—a high-energy situation.* It appears that one tetrahedron of the pair contains Si and the other two hydrogen atoms, 2H. The occupancy is statistically disordered in leucophoenicite (70, 74) but apparently ordered in jerrygibbsite (72).

Electron microscopic studies of leucophoenicite (74) and of jerrygibbsite (75, 76) reveal phenomena similar to those found in the humites—variation in twin-plane spacing, intergrowth of family members, and superperiodicities—but also occasional lamellae of humite family members.

It is fairly obvious that all these structures (in both the humite and leucophoenicite families) are related by crystallographic shear. This can transform *3 → 2* and *2 → 1*. The CS vector is, as in earlier examples, an $O \cdots O$ octahedral edge.

CS Derivatives of the Rutile Type, Mainly the "Reduced Rutiles"

These form as important a chapter in modern solid state chemistry as do their analogues derived from the ReO_3 type discussed in Chapter II. Unfortunately they are less readily depicted, the CS planes not being parallel to a principal axis of the rutile subcell. (Compare the $\{10l\}_R$ CS planes in ReO_3-derived structures that are all parallel to $\langle 010 \rangle_R$.) First discovered in 1957 (77a, 77b), their structures were resolved by a single-crystal X-ray diffraction study of Ti_5O_9 (78) and, later, by electron microscopy/diffraction (79). Similar structures occur in "reduced VO_2" and in $(Cr,Ti)O_{2-\delta}$ (80) and others, but we will concentrate mainly on the system TiO_x.

A large number of ordered phases intermediate between Ti_3O_5 and TiO_2 have (stoichiometric) compositions conforming to the general formula Ti_nO_{2n-p}; n and p being integers. They fall into three groups:

(i) $1.75 \leq x \leq 1.89$, with $p = 1$, $n = 4$ to 9 (Ti_4O_7 to Ti_9O_{17})
(ii) $1.93 \leq x \lesssim 1.98$, with $p = 1$, $n = 16$ to about 40–60†, probably with only even values of n
(iii) The intermediate region $1.89 < x < 1.93$, with $p > 1$ and $9 < n/p < 16$

Their structures all consist of (identical) rutile slabs of a given width joined on various CS planes:

---

* There are, for example, borate analogues of this sort, but then the B atoms are strongly repelled, almost into opposite triangular faces of the tetrahedral pair (69).

† This large uncertainty in n translates to a small uncertainty in stoichiometry: $Ti_{40}O_{79} = TiO_{1.975}$, $Ti_{60}O_{119} = TiO_{1.983}$, a difference of only 0.4% in oxygen content.

(i) $(121)_r$

(ii) $(132)_r$

(iii) The planes between $(121)_r$ and $(132)_r$

These last planes are $(hkl)_r = p \cdot (121)_r + q \cdot (011)_r$, with p and q integers, but varying across the composition range, as the plane "swings" from $(121)_r$, $p/q = \infty$ at lower x toward and eventually to $(132)_r$, $p/q = 1$, as x increases to 1.93. (x is the composition variable in TiO_x.)

The CS operators (plane plus displacement vector) are

(i) (121) $\frac{1}{2}[0\bar{1}1]$;

(ii) (132) $\frac{1}{2}[0\bar{1}1]$;

(iii) (hkl) $\frac{1}{2}[0\bar{1}1]$

(all based on the rutile subcell). That is, the displacement vector $\mathbf{R} = \frac{1}{2}[0\bar{1}1]_r$ is common to all three groups (and is again an $O \cdots O$ octahedral edge).

The structures can be quite simply described, but, because there is no convenient, low-index projection axis that shows the structures clearly (as does $[010]_R$ for ReO_3-derived CS structures; Chapter II), one considers first $(100)_r$ layers. All the structures are generated by stacking identical layers along $[1\bar{1}1]_r$ (at intervals of $\frac{1}{2}[1\bar{1}1]_r$), i.e., at 50° to the $[100]_r$ projection axis. In this way, Figure 46a (top left) represents rutile itself. Figure 46b (top center) shows the result of the "CS" operation (011) $\frac{1}{2}[0\bar{1}1]$, which, in fact, involves no oxygen loss. {The displacement vector $\mathbf{R} = \frac{1}{2}[0\bar{1}1]$, shown on the left, is parallel to the (011) planar boundary, which is therefore a slip plane producing an antiphase boundary (APB) rather than a CS plane.} Regular repetition of this (slip) operation on alternate $(011)_r$ planes yields Figure 46h (bottom center), which is readily recognizable as the α-PbO_2-type structure (cf. Figures 16 and 17g). Figure 46c (top right) shows operation (i) = (121) $\frac{1}{2}[0\bar{1}1]$, the production of a single $(121)_r$ CS plane. $\mathbf{R}$ is not now parallel to the fault plane, and if the right-hand half has been translated by $\mathbf{R}$ there is a consequent elimination of a (121) plane of oxygens. Regular repetition at every eighth, tenth, and so on anion-only (121) plane (i.e., the loss of one in $2n$ of the anion planes) produces Ti_4O_7, Ti_5O_9, $\cdots$, Ti_9O_{17}.

Figure 46d (left center) shows the production of (ii) a $(132)_r$ CS plane. It introduces something new; in b the displacement $\mathbf{R}$ introduces steps $\mathbf{A}$ into the rows of cations parallel to $\mathbf{c}$; in c, steps of type $\mathbf{C}$ are produced in these rows*; in d, $\mathbf{A}$ and $\mathbf{C}$ steps alternate along the CS plane. [Compare the resolution $(132)_r = (011)_r + (121)_r$.] The oxygen loss per CS plane is only one-half of that for $(121)_r$ CS. Again, regular repetition at intervals of $2n$

---

* Note that a $\mathbf{C}$ step is an $\mathbf{A}$ step *plus an additional* ("*interstitial*") *cation:* hence the reduction in the stoichiometric anion/cation ratio (= x in TiO_x) in the CS structures (but not in the α-PbO_2 type).

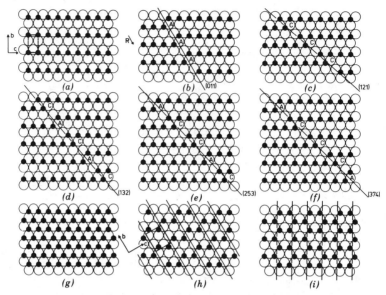

Figure 46. Single layers of cp anions (large open circles) and their overlying, octahedrally coordinated cations (smaller filled circles) in idealized rutile type and related structures. In all cases the complete structure is generated by stacking identical layers along $[1\bar{1}1]$ at intervals of $\frac{1}{2}[1\bar{1}1]_r$. Reading horizontally from top left to bottom right, the layers represent (a) rutile (note unit cell and axes); (b) an isolated $(011)_r$ $\frac{1}{2}[0\bar{1}1]_r$ antiphase boundary characteristic of the α-PbO$_2$ structure (or plastically deformed rutile) (note displacement vector $\mathbf{R} = \frac{1}{2}[0\bar{1}1]_r$); (c)–(f) isolated CS planes of type $(121)_r$, $(132)_r$, $(253)_r$, and $(374)_r$, all with the same displacement vector $\mathbf{R} = \frac{1}{2}[0\bar{1}1]_r$, and all belonging to the family of CS planes $(hkl)_r = p(121)_r + q(001)_r$. Note the **A**-type steps in (b), the **C**-type steps in c, and their simultaneous presence in (d)–(f). (g) A NiAs-type layer. (h) A layer of α-PbO$_2$ type generated from (a) by regular repetition of the operation shown in (b). [Note the α-PbO$_2$-type unit cell and axes in (h).] (i) A layer of corundum type that can easily be generated from rutile by regular repetition of the CS operation $(001)_r \frac{1}{2}[0\bar{1}1]_r$.

In all cases the full line represents the "fault" planes, APB or CS.

$(132)_r$ anion-only planes produces a member (Ti_nO_{2n-1}) of the $(132)_r$ family of CS structures.

Two examples of intermediate CS planes (iii), namely, $(253)_r$ and $(374)_r$, are shown in Figures 46e and f (center, and center right). These are $(hkl) = p \cdot (121) + q \cdot (011)$, with $q = 1$ and $p = 2$ and 3, respectively. Not surprisingly, they consist of p steps of type **C** alternating with single **A** steps. Regular repetition at an interval of $n \cdot d(hkl)$ will produce perfectly ordered Ti_nO_{2n-p} structures.*

---

* Notice the possibility of structural ambiguity if both p and q are nonzero: Step sequences $\cdots$**ACAC**$\cdots$ and $\cdots$**AACC**$\cdots$ correspond to the same CS plane but different structures, as do $\cdots$**ACCACC**$\cdots$ and $\cdots$**AACCCC**$\cdots$, and so on. In equilibrium structures, maximum dispersion appears to obtain—that is, the first rather than the second case in each pair.

It appears that the CS plane orientation changes *continuously* ("swings") as x varies between ~1.89 and ~1.93, from $(121)_r$ to $(132)_r$.

[Figure 46i (bottom right) is the product of a different CS operation, $(001)_r \frac{1}{2}[0\bar{1}1]_r$, repeated on every fourth anion-only (001) plane. It is the corundum-type structure of Ti_2O_3. Figure 46g (bottom left), the NiAs structure, can be produced in several ways—by CS on every second $(001)_r$ plane, CS on every second $(010)_r$ plane, and so on.]

Determining these CS structures is by no means trivial. All have triclinic cells, and only the smallest of these can be readily solved by single-crystal X-ray diffraction. Others were unraveled by electron microscopy/diffraction (79), but even that fails at very high n values $(TiO_{\sim 1.97})$.

Figure 47 (see color plates) shows a model of the structure of Ti_5O_9, $n = 5(121)$, idealized to regular octahedra. Alternate rutile blocks—on either side of the $(121)_r$ CS planes and $n = 5$ octahedra wide—are differently colored.

Figure 48 shows just a few examples of $[1\bar{1}1]_r$ zone axis, electron diffraction patterns from crystals with a range of CS planes. The orientation of the rows of superlattice spots (relative to the strong spots corresponding to the rutile-type subcell) reveals the direction of the CS planes, and the ratio p/q in $(hkl) = p \cdot (121) + q \cdot (011); h = p, k = 2p + q, l = p + q$. The spacing in these rows gives the CS plane spacing $n \cdot d \ (hkl)_r$, the value of n in M_nO_{2n-p}. (Not all cases have $q = 1$).

In regions (i) and (ii), compositions intermediate between adjacent integral n values (i.e., "nonstoichiometry") appear to be accommodated by coherent intergrowth of the two. An example is shown in Figure 49; a disordered, lamellar intergrowth of $n = 6$ and 7 (121), that is, of M_6O_{11} and M_7O_{13}. It is an interesting (but futile) question whether such a situation corresponds to a diphasic or to a monophasic crystal. It must also be remembered that these electron microscope studies are made at room temperature on samples prepared at high temperature and then cooled or "quenched" (cooled rapidly). [Note also that the experimentally observed thermodynamics of the system $TiO_x(s) + O_2(g)$ never corresponds to the phase rule for two solid phases in equilibrium, for $1.75 \le x < 2.00$ (79, 81). This is to be expected; it is an inevitable consequence of strain at the interfaces between adjacent lamellae of different n—the size and shape of the rutile subcell vary slightly as n varies.]

On the other hand, in the intermediate composition range—(iii) $1.89 < x < 1.93$—it appears that there is a *continuous* sequence of ordered structures, with no intermediate diphasic regions. As the composition changes, so n, p, and q adjust to accommodate it. [It also appears that the CS plane spacing is constant and the thermodynamic behavior reversible, or very nearly so, across this composition range (81). This too is as expected: the absence of interfaces means the absence of strain in the crystals.]

The composition (or composition range) in which a given CS plane orientation is stable depends on the system, that is, on the cations present, and on the temperature. The detailed description given above (but not all the fig-

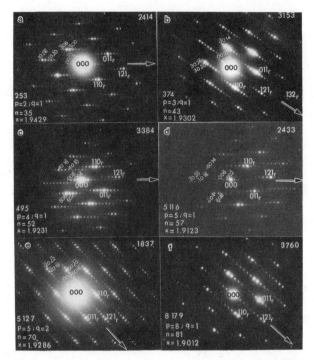

Figure 48. A few examples of indexed selected area diffraction patterns from various ordered CS structures $(Ti,V)_nO_{2n-p}$ with a range of CS plane orientations intermediate between $(121)_r$ and $(132)_r$. In the bottom left corner of each pattern are given the CS plane indices followed by the values of p and q in $(hkl) = p(121) + q(011)$, and then n in M_nO_{2n-p} and x in MO_x. On the patterns the larger indices are for the rutile subcell reflections (the stronger ones), and the smaller ones are for the unit cell of the CS structure. (The figures in each top right-hand corner are simply photographic plate numbers.) (These pictures were kindly provided by Dr. Carlos Otero-Diaz.)

ures) is for Ti_nO_{2n-p} (i.e., $pTi_2O_3 \cdot (n - 2p)TiO_2$). ($V_nO_{2n-p}$ is similar.) If Ti^{3+} is replaced by Cr^{3+}, the resulting $(Cr,Ti)O_x$ show $(121)_r$ CS (but only up to $n = 6$ in $Cr_2Ti_4O_{11}$) and "swinging CS," but no $(132)_r$ CS structures. Substitution of Fe^{3+} for Ti^{3+} and/or Zr^4 for Ti^{4+} leads to additional complexities of the same sort. Substitution of Ga^{3+} for Ti^{3+} leads to different fault planes, $(210)_r$. Samples quenched from very high temperature ($\geq 1450°C$), especially those with Fe^{3+} substitution, also exhibit different CS systems, with CS planes parallel to $(0kl)_r = p \cdot (010)_r + q \cdot (011)_r = (0, p + q, q)_r$; all in the $[100]_r$ zone (82). The APB component and the displacement vector $\mathbf{R} = \frac{1}{2}[0\bar{1}1]_r$ are the same as before, but the nonstoichiometry component, $(010)_r$ is different. The observed CS planes again swing continuously from $(013)_r$ through $(041)_r$, $(051)_r$, etc., to $(010)_r$. There are many possibilities, but we will not consider them here.

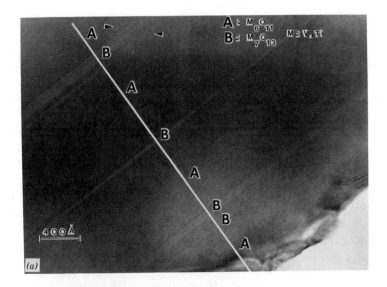

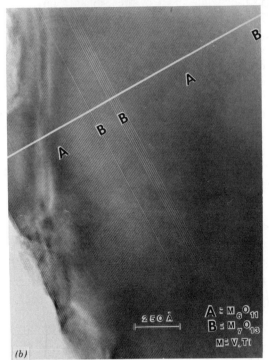

Figure 49. An example of a "nonstoichiometric" TiO_x crystal that is, in fact, a lamellar intergrowth of the $n = 6(121)$ phase with a smaller amount of $n = 7(121)$: M_6O_{11} ($= \mathbf{A}$) $+ M_7O_{13}$ ($= \mathbf{B}$), with M = (Ti,V). (*a*) and (*b*) are, respectively, lower- and higher-magnification images of the same area of crystal. (*c*) is the corresponding selected area diffraction pattern showing strong reflections (arrowheads) at 121*/6 from the major phase M_6O_{11}, and weaker ones (arrows) (streaked because the $\mathbf{B}$ lamellae are thin) at 121*/7 from the smaller amount of M_7O_{13}. (All pictures were kindly supplied by Dr. Carlos Otero-Diaz.)

In $(121)_r$ CS planes there are rods of corundum-type structure. In $(132)_r$ CS planes there are alternate rods of corundum and α-PbO$_2$ types. [Higher index $(hkl)_r$ CS planes also contain these two elements.] In Ga$_2$O$_3$ + TiO$_2$, the $(210)_r$ "fault" planes contain rods of the β-Ga$_2$O$_3$ structure (83). Thus, many of these structures may be regarded as being constructed of translationally related slabs of rutile type. All may be regarded as intergrowths of rutile type (TiO$_2$) with rods of a different structure type appropriate to the added component, e.g., corundum-type Ti$_2$O$_3$ or Cr$_2$O$_3$, or β-Ga$_2$O$_3$ type (the high-temperature form of Ga$_2$O$_3$).

It seems likely (84) that "nonstoichiometry" (the oxygen deficiency in TiO$_{2-\delta}$, WO$_{3-\delta}$, etc.) is accommodated by CS planes only when the material involved has a very high dielectric constant (is almost ferroelectric) and/or, perhaps, is compliant, that is, able to accommodate local elastic strains without large stresses. Compare rutile [dielectric constant, $\varepsilon = 173(\|)$, $89(\perp)$] with SnO$_2$ [$\varepsilon = 24.0(\|)$, $23.4(\perp)$], which does not exhibit CS.

We conclude by remarking that while most of these phases have been produced in only gram quantities in the laboratory—sufficient to study their interesting structural properties by X-ray diffraction or electron microscopy/diffraction—this is not so for the "swinging" and $(132)_r$ CS phases of reduced rutile. One of the methods for upgrading ilmenite (i.e., removing the Fe from FeTiO$_3$ to produce rutile) actually produces many thousands of tons of these complex oxides (TiO$_{\sim 1.90}$–TiO$_{\sim 1.95}$) annually at a plant that processes beach sands in Western Australia. Rutile, prepared by this and other routes, is widely used to produce white pigment (for paints), welding rods, and titanium metal.

References

1. K. Aurivillius, *Acta Chem. Scand.* **18,** 1305 (1964).

2. E. T. Lance, J. M. Haschke, and D. R. Peacor, *Inorg. Chem.* **15,** 780 (1976).

3. M. Atoji and W. N. Lipscomb, *J. Chem. Phys.* **27,** 195 (1955).

4. T. M. Sabine and Suzanne Hogg, *Acta Cryst. B* **25,** 2254 (1968).

5. S. C. Abrahams and J. L. Bernstein, *Acta Cryst. B* **25,** 1233 (1969).

6. M. O'Keeffe and B. G. Hyde, *Acta Cryst. B* **34,** 3519 (1978).

7a. G. Hägg, *Ark. Kemi Mineral. Geol.* **16B,** 1 (1943).

7b. T. J. McLarnan and P. B. Moore, in *Structure and Bonding in Crystals,* Vol. 2, M. O'Keeffe and A. Navrotsky, Eds. Academic Press, New York, 1981, p. 133.

8. L. V. Azaroff and M. J. Buerger, *Am. Mineral.* **40,** 213 (1955).

9. F. Jellinek, *Acta Cryst.* **10,** 620 (1957).

10. A. Kjekshus and W. B. Pearson, *Prog. Solid State Chem.* **1,** 83 (1964).

11. R. W. G. Wyckoff, *Crystal Structures,* 2nd ed., Vol. 1. Wiley, New York, 1963, p. 125.

12. I. Idrestedt and C. Brosset, *Acta Chem. Scand.* **18,** 1879 (1964).

13. L. Solov'eva and N. V. Belov, *Sov. Phys. Crystallogr.* **9,** 458 (1964).
14. C. T. Prewitt and R. D. Shannon, *Acta Cryst. B* **24,** 869 (1968).
15. W. S. McDonald and K. W. J. Cruickshank, *Acta Cryst.* **22,** 37 (1967).
16. K.-F. Hesse, *Acta Cryst. B* **33,** 901 (1977).
17. M. O'Keeffe and B. G. Hyde, *Phil. Trans. Roy. Soc. Lond. A* **295,** 553 (1980).
18. M. Marezio, *Acta Cryst.* **18,** 481 (1965).
19. C. Keffer, A. Mighell, F. Mauer, H. Swanson, and S. Block, *Inorg. Chem.* **6,** 119 (1967).
20. E. F. Farrell, J. H. Fang, and R. E. Newnham, *Amer. Mineral.* **48,** 804 (1963).
21. A. D. Wadsley, A. F. Reid, and A. E. Ringwood, *Acta Cryst. B* **24,** 740 (1968).
22. B. G. Hyde, T. J. White, M. O'Keeffe, and A. W. S. Johnson, *Z. Kristallogr.* **160,** 53 (1982).
23. W. R. Busing and H. A. Levy, *J. Chem. Phys.* **26,** 563 (1957).
24. S. C. Abrahams and J. L. Bernstein, *J. Chem. Phys.* **55,** 3206 (1971).
25. B. G. Hyde, A. N. Bagshaw, S. Andersson, and M. O'Keeffe, *Ann. Rev. Mater. Sci.* **4,** 43 (1974).
26. B. G. Brandt and A. C. Skapski, *Acta Chem. Scand.* **21,** 661 (1967).
27. H. Schäfer and H. G. von Schnering, *Angew. Chem.* **76,** 883 (1964).
28. A. Simon, *Angew. Chem. Int. Ed. (Engl.)* **20,** 1 (1981).
29. D. B. McWhan, F. M. Rice, and J. P. Remeika, *Phys. Rev. Lett.* **23,** 384 (1969).
30. P. Y. Simons and F. Dachille, *Acta Cryst.* **23,** 334 (1967).
31. S. Andersson and J. Galy, *Bull. Soc. Chim. France* **4,** 1065 (1969).
32. L. A. Bursill, B. G. Hyde, and D. K. Philp, *Phil. Mag.* **23,** 1501 (1971).
33. B. G. Hyde, L. A. Bursill, M. O'Keeffe, and S. Andersson, *Nature Phys. Sci.* **237,** 35 (1972).
34. B. G. Hyde, in *Reactivity of Solids,* J. S. Anderson et al., Eds. Chapman & Hall, London, 1972, p. 23.
35. W. R. Busing and H. A. Levy, *Acta Cryst.* **11,** 798 (1958).
36. R. Hoppe and W. Dähne, *Naturwiss.* **47,** 397 (1960).
37. S. Andersson and J. Galy, *Acta Chem. Scand.* **23,** 2949 (1969).
38. J. Galy and S. Andersson, *J. Solid State Chem.* **3,** 525 (1971).
39. J. J. Lander, *Acta Cryst.* **4,** 148 (1951).
40. E. O. Schlemper and W. C. Hamilton, *J. Chem. Phys.* **45,** 408 (1966).
41. H. Bode and R. Brockmann, *Z. Anorg. Chem.* **269,** 173 (1952).
42. C. Natta, P. Corradini, I. W. Bassi, and L. Porri, *Atti. Accad. Naz. Lincei. Rend. Class Sci. Fis. Mat. Nat.* **24,** 121 (1958).
43. L. F. Dahl, T.-I. Chiang, P. W. Seabaugh, and E. M. Larsen, *Inorg. Chem.* **9,** 1236 (1964).
44. J. A. Watts, *Inorg. Chem.* **5,** 281 (1966).
45. M. A. Hepworth, K. H. Jack, R. D. Peacock, and G. J. Westland, *Acta Cryst.* **10,** 63 (1957).
46. K. H. Jack, *Acta Cryst.* **5,** 404 (1952).
47. D. H. Jack and K. H. Jack, *Mater. Sci. Eng.* **11,** 1 (1973).

48. W. Jeitschko and A. N. Sleight, *J. Solid State Chem.* **4,** 324 (1972).

49. G. Ferguson, M. Mercer, and W. A. Sharp, *J. Chem. Soc. (London)* A **1969,** 2415.

50. A. Zalkin and D. E. Sands, *Acta Cryst.* **11,** 615 (1958).

51. W. H. Zachariasen, *Acta Cryst.* **1,** 285 (1948).

52. P. I. Krypyakevich, Yu. B. Kuz'ma, Yu. V. Voroshilov, C. B. Shoemaker, and D. Shoemaker, *Acta Cryst.* B **27,** 257 (1971).

53. W. Jeitschko, *Acta Cryst.* B **31,** 1187 (1975).

54. K. Sahl, *Beitr. Mineral. Petrogr.* **9,** 111 (1963).

55. F. Hulliger, *Structural Chemistry of Layer-Type Phases,* F. Levy, Ed. Reidel, Dordrecht, 1976, pp. 260–263.

56. B. Aurivillius, *Arkiv Kemi* **1,** 463, 499 (1949); **2,** 519 (1950); **5,** 39 (1952).

57. E. Makovicky and B. G. Hyde, *Struct. Bonding* **46,** 101 (1981).

58. R. W. G. Wyckoff, *Crystal Structures,* 2nd ed., Vol. 1. Wiley, New York, 1963, p. 295.

59. B. Aurivillius, *Acta Chem. Scand.* **18,** 1823 (1964).

60. S. Rundqvist and F. Jellinek, *Acta Chem. Scand.* **13,** 425 (1959).

61. A. N. Christensen, R. G. Hazell, and Å. Nilsson, *Acta Chem. Scand.* **21,** 481 (1967).

62. K. Klepp and H. Boller, *Monatsh. Chem.* **110,** 677 (1979).

63. M. Potel, R. Chevrel, and M. Sergent, *Acta Cryst.* B **36,** 1545 (1980).

64. *Structure Reports* **31A,** 214 (1966); **39A,** 328 (1973).

65. L. Bragg and G. F. Claringbull, *Crystal Structures of Minerals,* Bell and Sons, London, 1965.

66. M. O'Keeffe and B. G. Hyde, *Struct. Bonding* **61,** 77 (1985).

67. T. J. White and B. G. Hyde, *Phys. Chem. Miner.* **8,** 55 (1982).

68. T. J. White and B. G. Hyde, *Phys. Chem. Miner.* **8,** 167 (1982).

69. T. J. White and B. G. Hyde, *Acta Cryst.* B **39,** 10 (1983).

70. P. B. Moore, *Am. Mineral.* **55,** 1146 (1970).

71. P. J. Dunn, D. R. Peacor, W. B. Simmons, and E. J. Essene, *Am. Mineral.* **69,** 546 (1984).

72. T. Kato, private communication, 1983.

73. D. R. Peacor, P. J. Dunn, S.-C. Su, and J. Innes, *Am. Mineral.* **72,** 213 (1987).

74. T. J. White and B. G. Hyde, *Am. Mineral.* **68,** 1009 (1983).

75. L. Stenberg and B. G. Hyde, unpublished work, 1984.

76. Y.-C. Yau and D. R. Peacor, *Am. Mineral.* **71,** 985 (1986).

77a. S. Andersson, B. Collén, U. Kuylenstierna, and A. Magnéli, *Acta Chem. Scand.* **11,** 1641 (1957).

77b. S. Andersson, B. Collén, G. Kruuse, U. Kuylenstierna, A. Magnéli, H. Pestmalis, and S. Åsbrink, *Acta Chem. Scand.* **11,** 1653 (1957).

78. S. Andersson, *Acta Chem. Scand.* **14,** 1161 (1960).

79. L. A. Bursill and B. G. Hyde, *Prog. Solid State Chem.* **7,** 177 (1972).

80. S. Andersson and L. Jahnberg, *Arkiv Kemi* **21,** 413 (1964).

81. R. R. Merritt and B. G. Hyde, *Phil. Trans. Roy. Soc. Lond. A* **274,** 627 (1973).

82. L. A. Bursill, I. E. Grey, and D. J. Lloyd, in *Diffraction Studies of Real Atoms and Real Crystals.* Australian Academy of Science, Melbourne, 1974, p. 265.

83. D. J. Lloyd, I. E. Grey, and L. A. Bursill, *Acta Cryst. B* **32,** 1756 (1976).

84. L. A. Bursill, B. G. Hyde, and M. O'Keeffe, in *Solid State Chemistry* (Proceedings of the 5th Materials Research Symposium, NBS, 1971), R. S. Roth and S. J. Schneider, Eds. U.S. Government Printing Office, Washington, DC, 1972, p. 197.

CHAPTER IV

HCP II:
Hexagonal-Close-Packed Arrays Projected along an Axis Parallel to a cp Plane but Normal to a cp Row in That Plane, $\langle 10\bar{1}0 \rangle$ (= $\sqrt{3}a \approx$ 4.5–5 Å for Many Oxides)

The following structures are described in this chapter: wurtzite (ZnS), NiAs, $Cd(OH)_2$, rutile (TiO_2), α-PbO_2, Li_2ZrF_6, β-$TiCl_3$, ε-Fe_3N, corundum (α-Al_2O_3), ilmenite, V_2O_3, B-Nb_2O_5, PdF_3, BiI_3, $AlBr_3$, WCl_6, cementite (Fe_3C), YF_3, Pd_5B_2 (= Fe_5C_2), Sr_7Pt_3, baddeleyite (monoclinic ZrO_2 = ScOF = TaON), tetragonal YbO(OH), orthorhombic SrI_2, FeB, $LuFeO_3$, wittichenite ($BiCu_3S_3$), U_2FeS_5, Ru_7B_3, Fe_7C_3, $Ba_3Fe_3S_7$, $La_3A_nBX_7$.

Introduction

In Figure 1*a* of Chapter III, the arrow marked (2) shows the vector along which the various structures are projected in the present chapter. This projection is called HCP II. It and equivalent vectors $\langle 1\bar{1}00 \rangle$ are shown in Figure 1, in which the four-index Miller–Bravais system of notation for planes and vectors is used. For oxides or fluorides this projection axis is about 4.5–5 Å long, normally a few percent larger than the projection axis in Chapter III. In Figure 2 we see the projected hcp array and the sequence of the close-

108

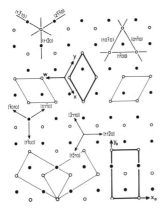

Figure 1. HCP I (referred to the heavily outlined hexagonal cell at the center of the diagram) = hcp projected along **c**. The diagram shows three equivalent orientations for the hexagonal unit cell and for the orthohexagonal cell (o) of twice the volume, as well as various low-index planes and directions. Four-index Miller–Bravais notation is used: $\{hkil\}$ for planes and $\langle uvtw \rangle$ for vectors. (Note $i = -(h + k)$, $t = -(u + v)$.]

packed layers, $\cdots abab \cdots$, as well as the projected hexagonal unit cell (outlined with broken lines) and the axes $\mathbf{x} = \mathbf{a}_0$ and $\mathbf{z}$ above. The projection axis, $\sqrt{3}a$ in length, we define as $\mathbf{b}_0$. It is the $\mathbf{b}$ axis of the orthohexagonal unit cell. Atoms in the same layer differ in height by $\frac{1}{2}\mathbf{b}_0$, and their various heights are given (as fractions of $\mathbf{b}_0$) in the right-hand part of the figure. Tetrahedra on different levels are shown with one corner pointing into and

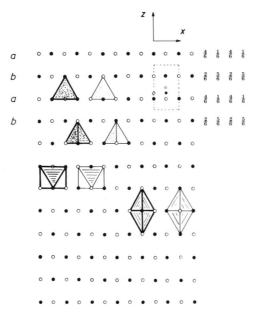

Figure 2. HCP II. Tetrahedra, octahedra, and trigonal bipyramids, from top to bottom. Close-packed planes, $a\ b\ a\ b$, are perpendicular to the plane of paper. The square is a tetrahedral position with $y_0 = \frac{5}{6}$ or $\frac{2}{6}$, $z = \frac{3}{8}$. The origin is in an octahedral position (octahedron center, 000); another octahedral position is marked with an open circle, $y_0 = \frac{1}{2}$, $z = \frac{1}{2}$. Atomic heights (y_0) for close-packed atoms (large open and filled circles) are given in the upper right corner.

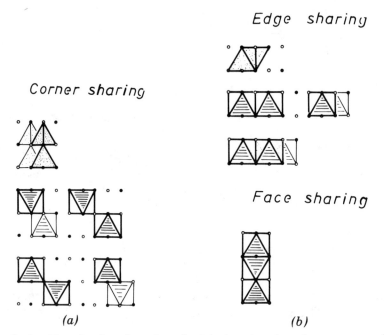

Figure 3a,b. Keys for the orientation of polyhedra when they share corners, edges, or faces, in this projection.

one corner pointing out of the paper, respectively. Two octahedra are shown, also on different levels, and below the octahedra, two trigonal bipyramids.

In Figures 3a and b, a key is given for the relative positions of polyhedra when they share corners, edges, or faces in this projection. (Pairs of face-sharing tetrahedra are shown in Figure 2 as trigonal bipyramids.)

This projection is not easy to see or use, but it is very important because it is the most convenient way to describe some of the most common structures.* Some structures described in Chapter III are repeated here; one reason is to demonstrate relationships.

Simple Structures

In Figure 4 the wurtzite structure is shown. Tetrahedra share corners only, each tetrahedron sharing every corner; if this structure is compared with the idealized nickel arsenide structure given in Figure 5, it is easy to see the cation shifts that transform one structure into the other. As shown in Figure 5, the NiAs structure can formally be said to consist of two identical layers. One layer is shown with heavy lines and consists of octahedra sharing faces to form columns, these columns being joined by having edges in common. This layer is repeated by the $\frac{1}{2}\mathbf{a}_o + \frac{1}{2}\mathbf{b}_o$ translation ($= a_h\langle 1\bar{2}10\rangle/3$) into an

---

* It was first systematically used to describe crystallographic shear in hcp (1).

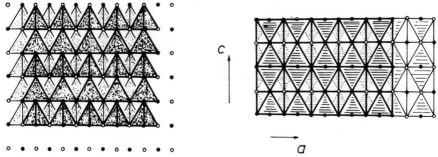

Figure 4. The idealized structure of wurtzite projected on (1100).

Figure 5. The idealized structure of NiAs.

identical layer almost hidden in the figure. This layer is indicated by dotted lines and is also extended with light lines on the right-hand side of the figure.

In this description, nickel atoms are in the centers of octahedra that have arsenic atoms at their corners. The nickel array itself is primitive hexagonal and will be discussed here. The related WC structure type will be considered in Chapter IX.*

It is very easy to relate the NiAs structure to the $Cd(OH)_2$, rutile, α-PbO_2, Li_2ZrF_6, β-$TiCl_3$, and ε-Fe_3N structure types. These structures are given in Figures 6–11, and they are obtained from the NiAs structure by taking out rows of cations in different directions or different ways. (Their structural data are given in Chapter III.) It is important to show these structures in this projection because we can then understand some more complicated structures and some important transformations. Note also the very simple rela-

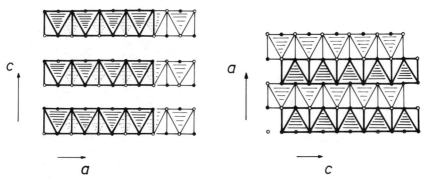

Figure 6. The idealized structure of $Cd(OH)_2$.

Figure 7. The idealized structure of rutile.

* Stacking sequences of the close-packed layers of atoms are:

Wurtzite:	a	β	b	α	a	β $b \cdots$,
NiAs:	a	γ	b	γ	a	γ $b \cdots$,
WC:	a	γ	a	γ	a	γ $a \cdots$.

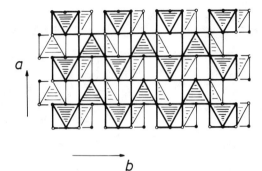

Figure 8. The idealized structure of α-PbO$_2$.

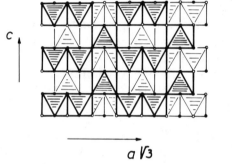

Figure 9. The idealized structure of Li$_2$ZrF$_6$.

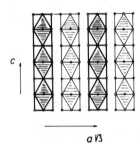

Figure 10. The idealized structure of β-TiCl$_3$.

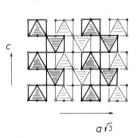

Figure 11. The idealized structure of ε-Fe$_3$N.

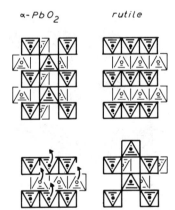

Figure 12. Cation shifts indicated by arrows to transform rutile to α-PbO$_2$.

tion between the β-TiCl$_3$ and ε-Fe$_3$N structures: identical (0001) layers are stacked differently. Each can be transformed to the other by cation "hopping" or by slip on (0001). The transformation rutile $\rightarrow$ α-PbO$_2$ is shown by cation jumps in Figure 12.

The complicated polytypism that occurs for several compounds of the Cd(OH)$_2$ structure type can clearly be demonstrated using this projection axis. [It involves mixed hexagonal (**h**) and cubic (**c**) close-packing of the anions.*]

The structure of corundum or α-Al$_2$O$_3$ is rhombohedrally centered hexagonal, $a = 4.76$ Å and $c = 13.0$ Å, given idealized in projection along **a** in Figure 13. It can be seen to consist of pairs of octahedra sharing faces along **c**. Such pairs form chains by edge-sharing in the plane of the paper. These chains join up in three dimensions to other identical chains, also by edge-sharing. The essential deviation from ideal hexagonal close-packing is a consequence of the face-sharing. (It can also be described as a compromise between regular octahedral coordination of cations by anions and regular tetrahedral coordination of anions by cations.) Structural data are in Data Table 1.

Data Table 1   Corundum, Al$_2$O$_3$ (2)

Hexagonal, space group $R\bar{3}c$, No. 167; $a = 4.759$, $c = 12.991$ Å, $c/a = 2.730$ (ideal hcp = 2.828); $Z = 6$, $V = 254.8$ Å^3

Atomic Positions: $(0,0,0; \frac{1}{3},\frac{2}{3},\frac{2}{3}; \frac{2}{3},\frac{1}{3},\frac{1}{3})+$
Al in 12(c): $\pm(0,0,z; 0,0,z + \frac{1}{2})$; $z = 0.352$
O in 18(e): $\pm(x,0,\frac{1}{4}; 0,x,\frac{1}{4}; \bar{x},\bar{x},\frac{1}{4})$; $x = 0.306$
Atomic Distances
Al–O = 1.85 Å (3×), 1.97 Å (3×)

The corundum structure is easily derived from that of rutile by means of simple parallel translation or crystallographic shear (CS). Figure 14 shows how rutile slabs join up by face-sharing, as indicated by dotted lines, to form the corundum structure. If these slabs are joined by edge-sharing, the rutile or α-PbO$_2$ structures are formed. Figure 15 (see color plates) shows a polyhedral model of corundum in which alternate rutile slabs are yellow and blue.

Most oxides of the corundum type are materials of great hardness and have high melting points. Al$_2$O$_3$ doped with Cr$_2$O$_3$ is the red gemstone ruby. Al$_2$O$_3$ doped with Ti$_2$O$_3$ is sapphire.

Some isostructural compounds are α-Fe$_2$O$_3$ (hematite), Ti$_2$O$_3$, V$_2$O$_3$, Cr$_2$O$_3$, α-Ga$_2$O$_3$ (low-temperature form), and Rh$_2$O$_3$. Compounds belonging

---

* Similar polytypism—variation of layer stacking between h, c, and hc—relates α-PbO$_2$, anatase, and brookite, respectively. (These are all TiO$_2$ polymorphs.)

c

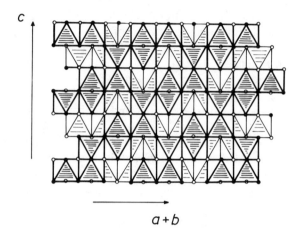

a + b

Figure 13.　The idealized structure of corundum, $\overset{\text{``}}{\alpha}$-$Al_2O_3$.

to the ilmenite ($FeTiO_3$) group have the same sort of structure, but the Fe and Ti atoms are ordered on distinct sets of cation sites: they segregate onto *alternate* planes parallel to (0001). Data are given in Data Table 2.

Data Table 2　Ilmenite, $FeTiO_3$ (3)

Rhombohedrally centered hexagonal, space group $R\bar{3}$, No. 148; a = 5.087, c = 14.042 Å, c/a = 2.760, Z = 6, V = 314.7 Å³

Atomic Positions: $(0,0,0; \frac{1}{3},\frac{2}{3},\frac{2}{3}; \frac{2}{3},\frac{1}{3},\frac{1}{3})+$
Fe in 6(*c*):　　$\pm(0,0,z);\ z$ = 0.1446
Ti in 6(*c*):　　z = 0.3536
O in 18(*f*):　　$\pm(x,y,z;\ \bar{y},x-y,z;\ y-x,\bar{x},z)$; x = 0.295, y = −0.022, z = 0.2548
Atomic Distances
Fe–O = 2.07_0 Å (3×), 2.19_7 Å (3×)
Ti–O = 1.87_5 Å (3×), 2.08_7 Å (3×)

When V_2O_3 is cooled, it transforms at ~155 K from the corundum structure, when it is metallic, to a monoclinic structure (4), when it is an insulator and is also antiferromagnetic. The main difference between the two structures is due to very small shifts in the metal atom positions. The structure of monoclinic V_2O_3 is shown in Figure 16; it is close to the real structure of Al_2O_3. This transition is similar in nature to the Mott transition observed by McWhan in the Cr_2O_3–V_2O_3 system, which, however, is a metal–insulator transition accompanied only by a volume change and no change in crystal symmetry (5, 6).

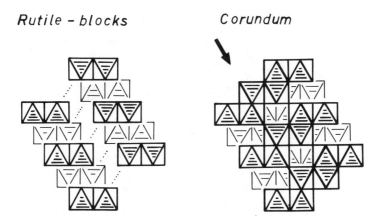

Figure 14. Rutile blocks on the left are joined by face-sharing to form the structure of corundum, on the right. Note that when such rutile blocks are joined by edges, they form α-PbO_2 (or rutile, of course), but when joined by corners, they form B-Nb_2O_5.

B-Nb_2O_5 is the densest of all the Nb_2O_5 modifications ($d = 5.29$ g/cm³). For this reason and because ordinary H-Nb_2O_5 ($d = 4.55$ g/cm³) (7) transforms to B-Nb_2O_5 under high pressure (8), it is also believed to be the high-temperature–high-pressure modification. The oxygens are in a slightly distorted hcp array, and the structure is shown in Figure 17. It is easy to recognize the same rutile slabs that formed corundum by octahedral face-sharing, but in the B-Nb_2O_5 structure they share corners. (In rutile itself they share edges.) Crystal data are given in Data Table 3. Sb_2O_5 (9) is isostructural.

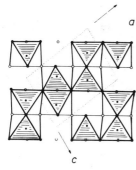

Figure 16. The real structure of a layer of monoclinic and antiferromagnetic V_2O_3, a very small distortion of Al_2O_3.

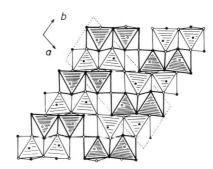

Figure 17. The real structure of a layer of B-Nb_2O_5. Rutile blocks as in Figure 14 are joined by corners. The structure can also be described as a crystallographic shear derivative of the hcp PdF_3 structure (1).

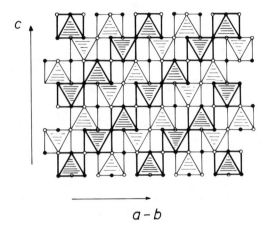

Figure 18.   The idealized structure of PdF$_3$.

Data Table 3   B-Nb$_2$O$_5$ (10)

Monoclinic, space group $B2/b$, No. 15; $a = 12.73$, $b = 5.56$, $c = 4.88$ Å,
$\gamma = 105.05°$; $Z = 4$, $V = 333.55$ Å^3

Atomic Positions: $(0,0,0; \frac{1}{2},0,\frac{1}{2})+$
Nb in 8(f): $\pm(x,y,z; \bar{x},\frac{1}{2} - y,z)$; $x = 0.140$, $y = 0.249$, $z = 0.238$
O(1) in 4(e): $\pm(0,\frac{1}{4},z)$; $z = 0.099$
O(2) in 8(f): $x = 0.389$, $y = 0.031$, $z = 0.054$
O(3) in 8(f): $x = 0.295$, $y = 0.375$, $z = 0.426$
Atomic Distances
Nb–O: 1.81, 1.91, 1.94, 2.06, 2.12, and 2.19 Å

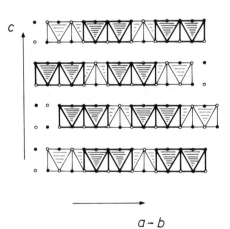

Figure 19.   The idealized structure of BiI$_3$.

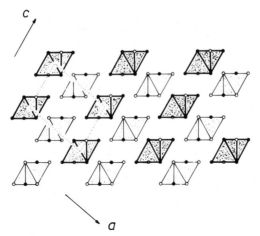

Figure 20. The idealized structure of AlBr$_3$. Two tetrahedra share edges and form the separate molecules Al$_2$Br$_6$ in the structure.

Figures 18–21 show, in idealized form, a number of interesting structures with anions in hcp. They all have the same repeat along the hexagonal **c** axis as corundum. The last two, AlBr$_3$ and WCl$_6$, are examples of molecular compounds (which often fit this general scheme of description). Crystallographic data for the four compounds are in Data Tables 4–7.

Data Table 4   PdF$_3$ (11)

Rhombohedral, space group $R\bar{3}c$, No. 167; $a_h = 5.009$, $c_h = 14.118$ Å, $c_h/a_h = 2.819$; $Z = 6$, $V_h = 306.7$ Å^3

Atomic Positions: $(0,0,0, \frac{1}{3},\frac{2}{3},\frac{2}{3}; \frac{2}{3},\frac{1}{3},\frac{1}{3})+$
Pd in 6(b): $(0,0,0; 0,0,\frac{1}{2})$
F in 18(c): $\pm(x,0,\frac{1}{4}; 0,x,\frac{1}{4}; x,x,\frac{3}{4})$; $x = -0.333$
Atomic Distances
Pd–F = 2.04 Å (6×)

Data Table 5   BiI$_3$ (12)

Rhombohedral space group $R\bar{3}$, No. 148; $a_h = 7.516$, $c_h = 20.718$ Å, $c_h/a_h = 2.757$; $Z = 6$, $V = 1013.6$ Å^3

Atomic Positions*: $(0,0,0; \frac{1}{3},\frac{2}{3},\frac{2}{3}; \frac{2}{3},\frac{1}{3},\frac{1}{3})+$
Bi in 6(c): $\pm(0,0,z)$; $z \approx 0.1667$
I in 18(f): $\pm(x,y,z; \bar{y},x-y,z; y-x,\bar{x},z)$; $x = 0.3415$, $y = 0.3395$, $z = 0.0805$

* Powder diffraction data only.

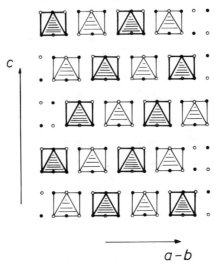

c

a−b

Figure 21. The idealized structure of WCl₆.

Data Table 6   AlBr₃ (13)

Monoclinic, space group $P2_1/a$ (nonstandard setting of $P2_1/c$; a and c interchanged), No. 14; $a = 10.22$, $b = 7.10$, $c = 7.49$ Å, $\beta = 96°$; $Z = 4$, $V = 541$ Å³

Atomic Positions
All atoms in 4(*e*): $\pm(x,y,z; \frac{1}{2} + x, \frac{1}{2} - y, z)$, with

	Al	Br(1)	Br(2)	Br(3)
x	0.050	0.150	0.169	0.008
y	0.095	0.075	−0.078	0.392
z	0.183	−0.083	0.411	0.252

Atomic Distances
Al–Br = 2.23–2.42 Å

Data Table 7   WCl₆* (14)

Rhombohedral, space group $R\bar{3}$, No. 148; $a_h = 6.10$, $c_h = 16.71$ Å, $c_h/a_h = 2.73_9$; $Z = 3$, $V = 538$ Å³

Atomic Positions: $(0,0,0; \frac{1}{3},\frac{2}{3},\frac{2}{3}; \frac{2}{3},\frac{1}{3},\frac{1}{3})+$
W in 3(*a*):   0,0,0
Cl in 18(*f*): $\pm(x,y,z; \bar{y},x - y,z; y - x,\bar{x},z)$; $x = y = 0.295$, $z = 0.080$
Atomic Distances
W–Cl = 2.24 Å

* This is the α form. There is also a β polymorph (15), a two-layer structure ($a_h = 10.493$, $c_h = 5.725$ Å, $V = 545.89$ Å³, $Z = 3$; space group $P\bar{3}m1$) isostructural with UCl₆ (Chapter III, Figure 32). It too consists of WCl₆ octahedra, packed so that the anions form a slightly distorted hcp array.

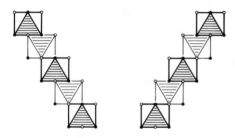

Figure 22. The diagonal string of corner-sharing octahedra common to the corundum, rutile, B-Nb_2O_5, and PdF_3 structures: two alternative equivalent orientations. [Corner-connected across the mirror plane, these units also appear in the structures of Fe_3C etc. (below).]

Note how PdF_3 (Figure 18) and ε-Fe_3N (Figure 11) are topologically distinct. They are composed of identical *pairs* of close-packed octahedral layers, differently stacked. There is a simple slip relation between them. [ε-Fe_3N is PdF_3 regularly twinned on (0001).]

Note also that while AsI_3 and SbI_3 are isostructural with BiI_3, their cations are off-center in the octahedra due to their lone pairs of electrons. (The effect is As > Sb > Bi $\approx$ 0.)

The structures of corundum, rutile, B-Nb_2O_5, and PdF_3 merit further consideration. Each contains the structure element shown in Figure 22; compare Figures 13, 7, 17, and 18. The difference between them is only in the way in which these lamellar elements are connected:

Corner connection, giving stoichiometry MX_3; cf. Figure 18

Corner connection of edge-shared pairs, giving stoichiometry $MX_{2.5}$; cf. Figure 17

Edge connection of edge-shared pairs, i.e., all edge connection, giving stoichiometry MX_2; cf. Figure 7;

Face connection of edge-shared pairs, giving stoichiometry $MX_{1.5}$; cf. Figure 13

Starting with PdF_3, all the others are derivable by CS on $(1\bar{1}01)$ planes of the hexagonal unit cell of PdF_3 (the plane of the unit lamellae). The displacement vector $\frac{1}{3}[10\bar{1}0]_{PdF_3}$ is the same in each case. [As is usual for CS in octahedral structures, it is an octahedral edge, $d(X\cdots X)$.]

Exactly the same slabs of hcp atoms (with empty octahedra) are utilized in the next section—joined in a very different mode of corner connection.

Structures Described by Parallel Reflection Operations (16–18)

If we cut the hcp array of Figure 2 on a $(11\bar{2}2)$ or $(11\bar{2}\bar{2})$ plane and rotate the block on one side of this plane by 180° about the normal to the plane* (use

* The normal to $(11\bar{2}2)$ is not a rational vector for perfect hcp, $c/a = \sqrt{8/3} = 1.633$.

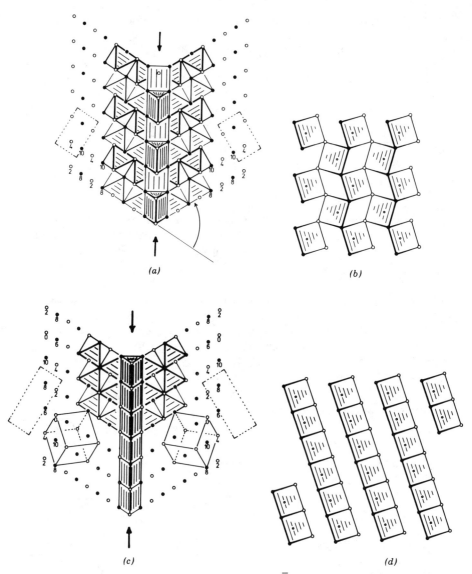

(a)   *(b)*

(c)   *(d)*

Figure 23. (*a*) An hcp atom array projected on $\{1\bar{1}00\}$ and twinned by reflection in $\{11\bar{2}2\}$, which is also the composition plane and is arrowed. Heights are in units of one-twelfth of the repeat distance normal to the plane of the projection $a\langle1\bar{1}00\rangle$ (cf. Figure 2). Sites at the centers of the trigonal prisms are at heights of 11 (prisms projecting forward) and 7 (prisms projecting backward). Trigonal prisms and some octahedra are shown, and hexagonal unit cells are outlined for both twins. The curved arrow represents the homogeneous shear that will produce the structure by mechanical twinning of PdF_3. (Compare Figure 22.) (*b*) The $\{11\bar{2}2\}$ sheet of trigonal prisms in (*a*) projected along the normal to the sheet. (*c*) Analogous to (*a*), but ccp instead of hcp. Atomic heights are again in units of one-twelfth of the projection axis. Face-centered cubic unit cells are outlined by full lines, and projected hexagonal cells by broken lines. The twin/composition plane is $\{113\}_{fcc} = \{11\bar{2}3\}_{hex}$. Prism centers are at heights of $\frac{1}{12}$, $\frac{3}{12}$, $\frac{5}{12}$, $\frac{7}{12}$, $\frac{9}{12}$, and $\frac{11}{12}$. (*d*) The prism layer from (*c*), analogous to (*b*).

120

transparent paper), we have carried out a twin operation on the hcp array of Figure 2, which gives us Figure 23a (cf. Fig. 22). The operation is equivalent to reflection in $\{11\bar{2}2\}$. The composition plane (on which the twin crystals are joined) is the same as this mirror (or twin) plane. The operation creates trigonal prisms in the composition plane, and these are shown in Figure 23b. (The corresponding twinning operation on ccp is shown in Figure 23c. It creates the topologically different trigonal prism array shown in Figure 23d. Because the former has octahedra at three different levels, and the latter at six levels, this projection is less useful for ccp than for hcp.)

The same result may also be attained by a homogeneous shear of ~60° on $(11\bar{2}2)$, as indicated in Figure 23a. (The angle of shear is 62.96° for perfect hcp.) Note that when an octahedron is sheared in this way it transforms to an octahedron with a different (twinned) orientation. Note also that shear of only *parts* of octahedra (at the mirror plane) transforms them to trigonal prisms. We shall return to this point later.

The best refinement of the cementite structure type, Fe_3C, seems to have been done on the isostructural $Fe_{2.7}Mn_{0.3}C$, for which we give the data in Data Table 8. (Cementite itself, Fe_3C, is metastable under all conditions. Hence, although of great importance in steels, large single crystals suitable for X-ray diffraction studies cannot be prepared at atmospheric pressure.)

Data Table 8   Cementite-Type $Fe_{2.7}Mn_{0.3}C$ (19)

Orthorhombic, space group *Pnma*, No. 62; $a = 5.0598$, $b = 6.7462$, $c = 4.5074$ Å; $Z = 4$, $V = 147.02$ Å^3

Atomic Positions

(Fe, Mn) in 4(c): $\pm(x,\frac{1}{4},z; \frac{1}{2} + x,\frac{1}{4},\frac{1}{2} - z)$; $x = 0.0367$, $z = 0.8402$

(Fe, Mn) in 8(d): $\pm(x,y,z; \frac{1}{2} + x,\frac{1}{2} - y,\bar{z}; x,\frac{1}{2} - y,z; \frac{1}{2} + x,y,\frac{1}{2} - z)$;
$x = 0.1816$, $y = 0.066$, $z = 0.3374$

C in 4(c): $x = 0.877$, $z = 0.444$

Atomic Distances

C-Fe = 1.96 Å, 1.98 Å, 2.03 Å (4×), 2.37 Å (2×), and 2.39 Å. Average of the first six values = 2.01 Å.

The structure is almost ideal twinned hcp, with reflection block size *3*, as shown in Figure 24a,b (the real structure). It is easier to accommodate the carbon atoms in the larger trigonal prisms (Figure 24c) than in the octahedral interstices, and this is the reason the structure exists. (Carbon dissolved in ccp or bcc Fe causes substantial strain because it is too large for even an octahedral interstice, Fe_6.) For detailed discussion of this structure and its relationship with the ε-carbide, see refs. 16–18. Among the many isostructural compounds we mention only Co_3C, Ni_3C, Co_3B, Ni_3B, Pd_3P, and Pd_3Si. Among the antitypes we shall discuss YF_3, SbF_3, $SbCl_3$, AsF_3, BiF_3, and XeO_3 later. The first of these (YF_3) is shown in Figures 24d, e and f, with crystallographic data in Data Table 9.

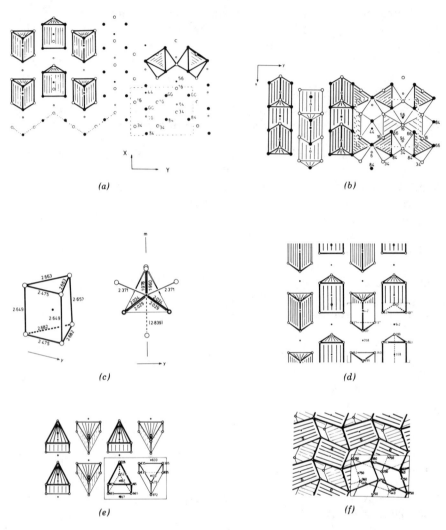

Figure 24. (*a*) The real structure of Fe$_3$C, cementite. Small circles represent carbon; larger circles, iron. The dotted zigzag line indicates sizes of the hcp twin units, $3d(11\bar{2}2)_{hcp}$. Trigonal prisms as created in the twin planes contain carbon. (Note: Only half the prisms are drawn.) Empty octahedra in the hcp blocks are almost undistorted, as can be seen in the right part of the drawing. The shortening of one pair of the trigonal prism edges, the "east–west" one, is a consequence of the twin operation (cf. Figure 23). (*b*) Similar to (*a*), but showing the topology of the trigonal prisms and of the octahedra. (*c*) The dimensions in Å of the CFe$_6$ trigonal prism edges and C–Fe bond lengths in Fe$_3$C. (*d, e, f*). The three axial projections of YF$_3$: (001), (100), and (010). (Space group setting *Pnma*.)

Data Table 9   YF$_3$ (20)

Orthorhombic, space group *Pnma*, No. 62; $a = 6.353$, $b = 6.850$, $c = 4.393$ Å; $Z = 4$, $V = 191.17$ Å^3

Atomic Positions
Y in 4(c): $\pm(x,\frac{1}{4},z; \frac{1}{2} + x,\frac{1}{4},\frac{1}{2} - z)$; $x = 0.367$, $z = 0.058$
F(1) in 4(c): $x = 0.528$, $z = 0.601$
F(2) in 8(d): $\pm(x,y,z; \frac{1}{2} + x,\frac{1}{2} - y,\frac{1}{2} - z; x,\frac{1}{2} - y,z; \frac{1}{2} + x,y,\frac{1}{2} - z)$; $x = 0.165$,
                  $y = 0.060$, $z = 0.363$
Atomic Distances
Y–F = 2.25–2.32 Å (8×), 2.60 Å

The block size can be changed from *3,3,3** in Fe$_3$C to *2,3,2,3*, which is the structure of Pd$_5$B$_2$ (21) given in Figure 25 (with the crystallographic data given in Data Table 10). Fe$_5$C$_2$ and Mn$_5$C$_2$ are isostructural. The possibilities for disorder, nonstoichiometry, and a sequence of intermediate structures are obvious. A recent and new example of the last is the structure of Sr$_7$Pt$_3$ (22), which is twinned hcp . . . , *2,2,3,2,2,3*, ... (Fe$_7$C$_3$ is *not* isostructural; cf. below with Ru$_7$B$_3$.) Even more recent evidence appears to confirm the existence of all three possibilities in the Fe–C system (23).

Data Table 10   Pd$_5$B$_2$ (21)

Monoclinic, space group *C2/c*, No. 15; $a = 12.786$, $b = 4.955$, $c = 5.472$ Å, $\beta = 97°2'$; $Z = 4$, $V = 344.1$ Å^3

Atomic Positions: $(0,0,0; \frac{1}{2},\frac{1}{2},0)+$
Pd(1) in 8(f): $\pm(x,y,z; x,\bar{y},\frac{1}{2} + z)$; $x = 0.0958$, $y = 0.0952$, $z = 0.4213$
Pd(2) in 8(f): $x = 0.2127$, $y = 0.5726$, $z = 0.3138$
Pd(3) in 4(e): $\pm(0,y,\frac{1}{4})$; $y = 0.5727$
B in 8(f): $x = 0.106$, $y = 0.311$, $z = 0.077$
Atomic Distances
B–Pd = 2.18 Å (4×), 2.19 Å (2×), 2.60 Å, 2.80 Å, 3.06 Å

A drawing of the corresponding structure with block size *2,2,2* is given in Figure 26. No example of this ideal structure is known. But, slightly distorted, it describes several structure types: monoclinic ZrO$_2$ (baddeleyite), tetragonal YbO(OH), and (orthorhombic) SrI$_2$. The refinement of baddeleyite, monoclinic ZrO$_2$, was on a synthetic single crystal, free of HfO$_2$. The crystal data are given in Data Table 11.

---

* Each number denotes the spacing of the twin planes in units of $d(11\bar{2}2)_{hcp}$; see the dotted zigzag line at the bottom of Figure 24*a*.

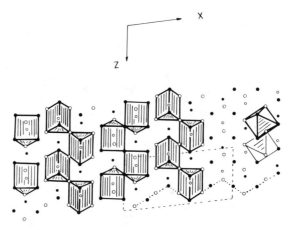

Figure 25. The real structure of Pd_5B_2. Fe_5C_2 is isostructural. The block size is . . . , 2,3,2,3, . . . , as indicated by the dotted lines. The mixing of twin block widths leads to monoclinic symmetry and a different composition.

Data Table 11   Baddeleyite, ZrO_2 (24)

Monoclinic, space group/ $P2_1/c$, No. 14; $a = 5.1454$, $b = 5.2075$, $c = 5.3107$ Å, $\beta = 99.23°$; $Z = 4$, $V = 140.46$ Å^3

Atomic Positions
All atoms in 4(e): $\pm(x,y,z; x,\frac{1}{2} - y,\frac{1}{2} + z)$, with

	x	y	z
Zr	0.2758	0.0411	0.2082
O(1)	0.0703	0.3359	0.3406
O(2)	0.4423	0.7549	0.4789

In Figures 27a–c, a monocapped trigonal prism has been selected as the ZrO_7 coordination polyhedron. In Figure 27a the structure can be compared with that of Fe_3C and especially Figure 26: it is the analogous projection. The polyhedra are twisted, apparently to improve the bonding to the seventh

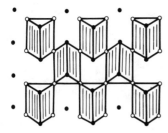

Figure 26. Idealized structure of twinned blocks of hcp: . . . , 2,2,2, ...

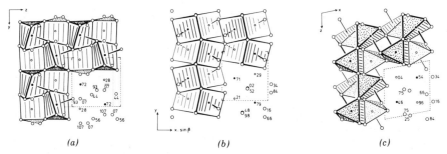

(a)   *(b)*   *(c)*

Figure 27. The monoclinic baddeleyite structure (ZrO$_2$) projected (*a*) on (100), (*b*) along [001], and (*c*) on (010). Large circles are oxygen, small circles are zirconium atoms; heights are in hundredths of the (unit cell) projection axis.

oxygen, and this is also shown in the other two projections of Figures 27*b* and *c*. ScOF, TaON, and NbON are isostructural. (Note that, crystallographically, it is a ternary structure, not a binary; in ZrO$_2$, oxygen occupies *two* distinct sets of sites.)

In the FeB structure the hcp twin bands have the minimum width . . . , 1,1,1, ... It is shown in Figure 28. Structural data are given in Data Table 12.

Data Table 12   FeB (25)

Orthorhombic, space group *Pbnm*, No. 62; $a = 4.053$, $b = 5.495$, $c = 2.946$ Å; $V = 65.61$ Å^3

Atomic Positions

Fe in 4(*c*): $\pm(x,y,\tfrac{1}{4}; \tfrac{1}{2} - x, \tfrac{1}{2} + y, \tfrac{1}{4})$; $x = 0.125$, $y = 0.180$

B in 4(*c*): $x = 0.61$, $y = 0.04$

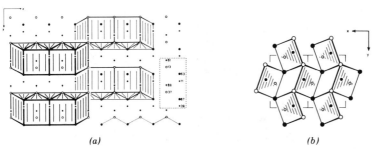

(a)   *(b)*

Figure 28. (*a*) The structure of FeB, twinned hcp . . . , (*1,1*), . . . projected on (100). Large circles are Fe, small circles are B atoms. Only about one-half of the BFe$_6$ trigonal prisms are drawn. (Unit height = $a/100$.) (*b*) FeB projected on (001), atoms at $z/c = \tfrac{1}{4}$ (open circles) and $\tfrac{3}{4}$ (filled circles). (Space group setting *Pbnm*.)

Filling of Twinned hcp Arrays

Polyhedra like the octahedra in the cementite type, previously unoccupied, are filled in, for example, the gadolinium orthoferrite type ("orthorhombic perovskite"), especially $LuFeO_3$; see Figure 29 and Data Table 13. $UCrS_3$, $CaUS_3$, $CaZrS_3$, and a high pressure form of $MgSiO_3$ (to mention only a few) are isostructural. They will be considered from a different viewpoint (26) later (Chapter XI). $BiCu_3S_3$ and $SbCu_3S_3$ are also cementite-related, but the copper atoms are now in triangular coordination (in the *faces* of the S_6 octahedra in the hcp twin lamellae), whereas in $LuFeO_3$ the Fe atoms center the O_6 octahedra.

U_2FeS_5 (Figure 30 and Data Table 14), is the corresponding filled antitype of Pd_5B_2, with Fe filling all the octahedral sites. Dy_2LiCl_5 is isostructural.

The structure of α-Sb_2O_4 is an example of large twinned blocks of hcp, but it will be described among the lone-pair structures (Chapter X).

Data Table 13 $LuFeO_3$ (27)

Orthorhombic, space group *Pnma*, No. 62; $a = 5.547$, $b = 7.565$, $c = 5.213$ Å, $Z = 4$, $V = 218.75$ Å^3

Atomic Positions
Lu in 4(*c*): $\pm(x,\frac{1}{4},z; \frac{1}{2} + x,\frac{1}{4},\frac{1}{2} - z)$; $x = 0.07149$, $z = 0.01997$
Fe in 4(*b*): $(0,0,\frac{1}{2}; 0,\frac{1}{2},\frac{1}{2}; \frac{1}{2},0,0; \frac{1}{2},\frac{1}{2},0)$
O(1) in 4(*c*): $x = 0.454$, $z = 0.120$
O(2) in 8(*d*): $\pm(x,y,z; \frac{1}{2} + x,\frac{1}{2} - y,\frac{1}{2} - z; x,\frac{1}{2} - y,z; \frac{1}{2} + x,y,\frac{1}{2} - z)$; $x = 0.3071$,
         $y = 0.0621$, $z = 0.6893$

Atomic Distances
Lu–O = 2.185 Å, 2.225 Å (2×), 2.243 Å, 2.455 Å (2×), 2.687 Å (2×), and then
  $\geqq 3.195$ Å
Fe–O = 1.997 Å (2×), 2.008 Å (2×), 2.024 Å (2×)

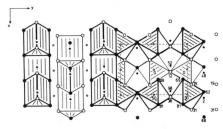

Figure 29. The real structure of $LuFeO_3$. Lu in trigonal prisms, Fe in octahedra. Octahedra as well as trigonal prisms are filled in a "Fe_3C" structure type. (*Pnma*.)

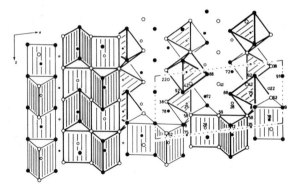

Figure 30. U_2FeS_5 projected on (010); compare Figures 29 and 25. (U in trigonal prisms, Fe in octahedra.)

Data Table 14　U_2FeS_5 (28)

Monoclinic, space group $C2/c$, No. 15; $a = 14.697$, $b = 6.326$, $c = 7.024$ Å, $\beta = 96.50°$; $Z = 4$, $V = 648.85$ Å^3

Atomic Positions: $(0,0,0; \frac{1}{2},\frac{1}{2},0)+$

U in 8(f):	$\pm(x,y,z; x,\bar{y},\frac{1}{2} + z)$; $x = 0.10555$, $y = 0.21972$, $z = 0.08147$
Fe in 4(d):	$(\frac{1}{4},\frac{1}{4},\frac{1}{2}; \frac{3}{4},\frac{1}{4},0)$
S(1) in 4(e):	$\pm(0,y,\frac{1}{4})$; $y = 0.9116$
S(2) in 8(f):	$x = 0.0891$, $y = 0.3798$, $z = 0.4282$
S(3) in 8(f):	$x = 0.2804$, $y = 0.4200$, $z = 0.1880$

Cyclic Reflection Operation

In Figure 31 we have excised a unit from the hcp array of Figure 2. The angle for perfect hcp is 121.5°, but with a small distortion (to make the angle exactly 120°)* this column serves as a trilling (= triple twinning) unit to build Figure 32. This is triply twinned hcp, with lamellae of Fe_3C type along the composition planes and a large unit in the center, which, if repeated regularly, forms the structure of Ru_7B_3 shown in Figure 33. Structural data are given in Data Table 15. Th_7Fe_3 is isostructural.

---

* This changes $(c/a)_{hcp}$ from the ideal $\sqrt{8/3} = 1.633$ to $\sqrt{3} = 1.732$, a change of 6%. See the introduction to the cementite structure above.

Trilling unit of hcp

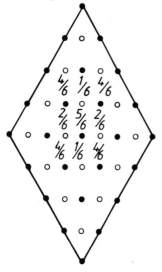

Figure 31. A column cut out of hcp.

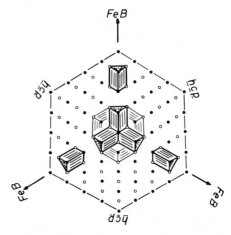

Figure 32. The column of Figure 31 used to construct a trilling of hcp. Fe₃C type is formed in the twin planes. The central unit is a piece of the Ru_7B_3 structure, which is shown in Figure 33.

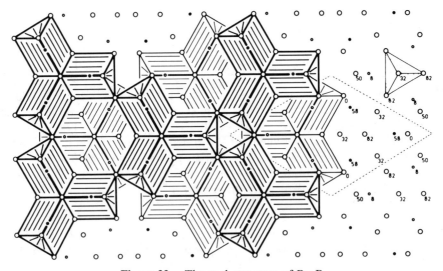

Figure 33. The real structure of Ru_7B_3.

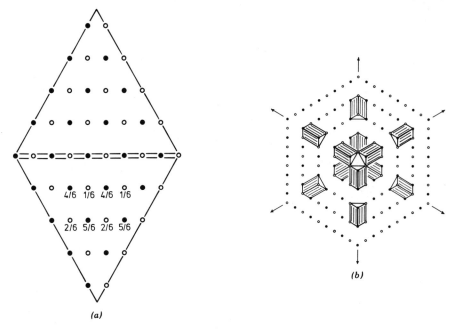

$\frac{4}{6}$ $\frac{1}{6}$ $\frac{4}{6}$ $\frac{1}{6}$

$\frac{2}{6}$ $\frac{5}{6}$ $\frac{2}{6}$ $\frac{5}{6}$

(a)

(b)

Figure 34. (*a*) A column of hcp similar to that in Figure 31 but with a different boundary face. (*b*) A sixling formed from the blocks in (*a*). The twin planes and blocks have the same structures as in Figure 32, but at the center an octahedron is generated instead of a tetrahedron.

Data Table 15 Ru$_7$B$_3$ (29)

Hexagonal, space group $P6_3mc$, No. 186, $a = 7.47$, $c = 4.71$ Å (orthohexagonal $a = 7.47$, $b = 12.94$, $c = 4.71$ Å, $c/a = 0.6305$); $Z = 2$, $V = 247.61$ Å^3

Atomic Positions

Ru(1) in 6(c): $x,\bar{x},z;\ x,2x,z;\ 2\bar{x},\bar{x},z;\ \bar{x},x,\frac{1}{2} + z;\ \bar{x},2\bar{x},\frac{1}{2} + z;\ 2x,x,\frac{1}{2} + z;\ x = 0.456,$
$z = 0.318$

Ru(2) in 6(c): $x = 0.122$, $z = 0$

Ru(3) in 2(b): $\frac{1}{3},\frac{2}{3},z;\ \frac{2}{3},\frac{1}{3},\frac{1}{2} + z;\ z = 0.818$

B in 6(c): $x = 0.19$, $z = 0.58$

Atomic Distances

Ru–B = 2.15–2.20 Å

The structure of Ru$_7$B$_3$ is not only triply twinned hcp but also triply twinned Fe$_3$C type (repeated mimetically in both cases). We note that in this structure there are two ways in which the twin planes (in hcp) intersect; one, at a three-fold axis, is shown in Figure 32. The other, at a $\bar{3}$ axis, may be formed in a similar way but with a shift in the boundaries of the blocks in Figure 31 and the formation of a sixling instead of a trilling. This is shown in Figure 34, the center of which is an octahedron rather than a tetrahedron (as in Figure 32).

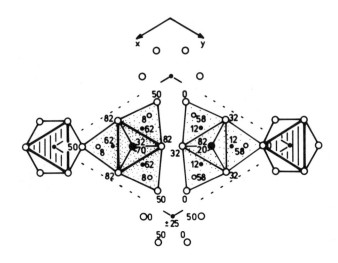

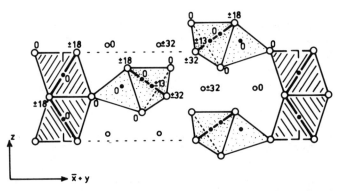

Figure 35. The available A (octahedral) and B (tetrahedral) interstices in the Ru$_7$B$_3$-type structure (Figure 33) are shown as filled circles. The atoms in the trigonal prisms are omitted, and the trigonal prisms are not shown (although they are readily delineated).

The structure of Fe$_7$C$_3$ (and the isostructural Mn$_7$C$_3$) is closely related to that of Ru$_7$B$_3$; it is formed by inverting the **c** axis of half the octahedral columns in Figure 33 [which inverts the associated trigonal prisms by reflection in (0001)]. Its orthorhombic unit cell is identical to the orthohexagonal cell of Ru$_7$B$_3$. In a third type, only one-fourth of these columns are inverted, so it is hexagonal, with **a** $\equiv$ 2 × **a**(Ru$_7$B$_3$) (30). This inversion of **c** for trigonal columns of trigonal prisms corresponds only to a shift of central octahedra by **c**/3. This is equivalent to introducing a **c** layer into the **h** layers—a stacking fault [so that, in this projection, atomic heights in cp layers are, e.g., 0/6, 3/6 + 1/6, 4/6 instead of 2/6, 5/6 + 1/6, 4/6 (compare Figures 2, 31, 34a, and 23a,c)]. These structures may therefore be regarded as twinned again, by rotating the appropriate columns by 180°!

$Ba_3Fe_3Se_7$ (31) and $Ln_3A_nBX_7$ with $n = 1, \frac{2}{3}, \frac{1}{2}$, or $\frac{1}{4}$ for various A and B (at least 250 compounds) are filled forms of the antitype of Ru_7B_3, with a proportion of the octahedra and half the tetrahedra occupied (by A and B, respectively; Ln in S_6 trigonal prisms). Various atoms may occupy some or all of the A and B positions: Cu, Ag, Cr, Mn, Fe, Al, etc. in A; and Si, Ge, Al, etc. in B. These positions are shown in Figure 35.

References

1. S. Andersson and J. Galy, *J. Solid State Chem.* **1,** 576 (1970).

2. R. E. Newnham and Y. M. de Haan, *Z. Kristallogr.* **117,** 235 (1962).

3. B. Morosin, R. J. Baughman, D. S. Ginley, and M. A. Butler, *J. Appl. Cryst.* **11,** 121 (1978).

4. P. D. Dernier and M. Marezio, *Phys. Rev. B* **2,** 3771 (1970).

5. D. B. McWhan, T. M. Rice, and J. P. Remeika, *Phys. Rev. Lett.* **23,** 1384 (1969).

6. D. B. McWhan and J. P. Remeika, *Phys. Rev.* **132,** 3734 (1970).

7. B. M. Gatehouse and A. D. Wadsley, *Acta Cryst.* **17,** 1545 (1964).

8. A. D. Wadsley and S. Andersson, *Perspect. Struct. Chem.* **1970,** 1.

9. M. Jansen, *Acta Cryst. B* **35,** 539 (1979).

10. F. Laves, H. Wulf, and W. Petter, *Naturwiss,* **51,** 633 (1964).

11. M. A. Hepworth, K. H. Jack, R. D. Peacock, and G. J. Westland, *Acta Cryst.* **10,** 63 (1957).

12. J. Trotter and T. Zobel, *Z. Kristallogr.* **123,** 67 (1966).

13. P. A. Renes and C. H. MacGillavry, *Rec. Trav. Chim. Pays-Bas* **64,** 275 (1945).

14. J. A. A. Ketelaar and G. W. van Oosterhout, *Rec. Trav. Chim. Pays-Bas* **62,** 197 (1943).

15. J. C. Taylor and P. W. Wilson, *Acta Cryst. B* **30,** 1216 (1974).

16. S. Andersson and B. G. Hyde, *J. Solid State Chem.* **9,** 92 (1974).

17. B. G. Hyde, A. N. Bagshaw, S. Andersson, and M. O'Keeffe, *Ann. Rev. Mater. Sci.* **4,** 43 (1974).

18. B. G. Hyde, S. Andersson, M. Bakker, C. M. Plug, and M. O'Keeffe, *Prog. Solid State Chem.* **12,** 273 (1979).

19. E. J. Fasiska and G. A. Jeffrey, *Acta Cryst.* **19,** 463 (1965).

20. A. Zalkin and D. H. Templeton, *J. Am. Chem. Soc.* **75,** 2453 (1953).

21. E. Stenberg, *Acta Chem. Scand.* **15,** 861 (1961).

22. M. L. Fornasini and A. Palenzona, *J. Solid State Chem.* **47,** 30 (1983).

23. E. Bauer-Grosse and G. Le Caër, *J. Phys. F.: Met. Phys.* **16,** 399 (1986).

24. D. K. Smith and H. W. Newkirk, *Acta Cryst.* **18,** 983 (1965).

25. T. Bjurström, *Arkiv Kemi Mineral. Geol.* **11**, 12 (1933).

26. M. O'Keeffe and B. G. Hyde, *Acta Cryst. B* **33**, 3802 (1977).

27. M. Marezio, J. P. Remeika, and P. D. Dernier, *Acta Cryst. B* **26**, 2008 (1970).

28. H. Noël, M. Potel, and J. Padiou, *Acta Cryst. B* **32**, 605 (1976).

29. B. Aronsson, *Acta Chem. Scand.* **13**, 109 (1959).

30. R. Fruchart et al., *Struct. Rep.* **30A,** 36 (1965); **38A,** 61 (1972).

31. J. Flahaut and P. Laruelle, in *The Chemistry of Extended Defects in Non-metal-lic Solids,* L. Eyring and M. O'Keeffe, Eds., North-Holland, Amsterdam, 1970, p. 109.

CHAPTER V

HCP III:
Hexagonal-Close-Packed Arrays Projected along a cp Row of Atoms, i.e., along $\langle 11\bar{2}0 \rangle$ onto $\{11\bar{2}0\}$; Repeat Distance $= a \approx 3.5$ Å for Many Oxides

The following structures are described in this chapter: β-BeO, rutile (TiO_2), wurtzite (ZnS), NiAs, marcasite (FeS_2), $CaCl_2$, α-$Cd(OH)_2$ ($\equiv CdI_2$), γ-$Cd(OH)_2$, A-La_2O_3, ω-Ti (and bcc, AlB_2, Ni_2In), $CuGeO_3$, hollandite ($\sim BaMn_4O_8$), ramsdellite (MnO_2), $CaFe_2O_4$, Si_3N_4, Nb_3Te_4, fluoborite $(Mg_3BO_3[OH,F]_3)$, $Y(OH)_3$.

We choose a projection of hcp (HCP III) at right angles to that in Chapter IV. It has the shortest possible repeat distance—an octahedral or tetrahedral edge. It is shown in Figure 1, where, on the left, we see tetrahedra in the upper part, octahedra in the middle, and trigonal bipyramids in the lower part. Figure 2 gives a comparison between the (idealized) structures of rutile, NiAs, and wurtzite, all in this same hcp net.

Simple Structures

Figure 3 gives the real structures of β-BeO and rutile and reveals the obvious similarities in the distortions of their "hcp" arrays. Structural data for the former are given in Data Table 1. (Data for the latter are in Data Table 12 of

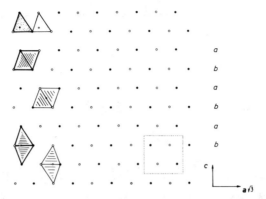

Figure 1. HCP III. All atoms on 0 or $\frac{1}{2}$. Small circles mark the centers of the polyhedra, which are, from the top: tetrahedra, octahedra, and trigonal bipyramids. Closepacked planes are perpendicular to the plane of paper and are marked a, b, a, b.

Chapter III.) The edge-sharing of tetrahedra in β-BeO (the high-temperature form, $\geq 2100°C$) results, of course, in a short Be–Be distance, 2.24 Å, which is the same as in the metal. [The Ti–Ti distance in rutile across the shared octahedral edge is also approximately the same as in the metal. The shortest distances $d(\text{Ti} \cdots \text{Ti})$ are, in hexagonal α-Ti metal 2.90 Å; in bcc β-Ti metal 2.86 Å; in rutile 2.96 Å; in β-TiCl$_3$ 2.91 Å (Chapter III).] α-BeO (the low-temperature form, $\leq 2100°C$) is isostructural with wurtzite, with $a = 2.76$, $c = 4.47$ Å, $c/a = 1.620$, at 2000°C ($V = 14.7$ Å^3 per formula unit).

Data Table 1 β-BeO (1)

Tetragonal, space group $P4_2/mnm$, No. 136; $a = 4.75$, $c = 2.74$ Å at 2100°C; $Z = 4$, $V = 61.82$ Å^3 (15.5 Å^3 per formula unit)

Atomic Positions
Be in 4(g): $\pm(x,\bar{x},0; \frac{1}{2} + x,\frac{1}{2} + x,\frac{1}{2})$; $x = 0.336$
O in 4(f): $\pm(x,x,0; \frac{1}{2} + x,\frac{1}{2} - x,\frac{1}{2})$; $x = 0.310$

The marcasite polymorph of FeS$_2$ is a rutile type deformed by S–S bonding (Figure 4a). There are several isostructural selenides, tellurides, phos-

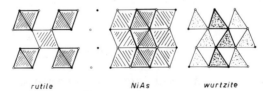

rutile   NiAs   wurtzite

Figure 2. The idealized structures of (left to right) rutile, NiAs, and wurtzite.

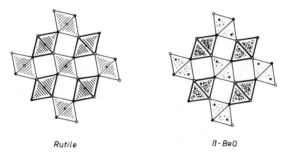

Rutile   β-BeO

Figure 3. Rutile (left) and β-BeO (right). Small circles are Ti^{4+} or Be^{2+}. In β-BeO, tetrahedra share edges; in TiO_2, octahedra share edges.

phides, arsenides, and antimonides, and alkali metal superoxides such as NaO_2 (III). $CaCl_2$ and $CaBr_2$ have a similar but smaller orthorhombic distortion of the rutile type (but, of course, no anion–anion bonds).

The three topologically equivalent structures of rutile (TiO_2),* $CaBr_2$ ($CaCl_2$ type), and marcasite are compared in Figures 4b–d.

Two forms of $Cd(OH)_2$ are described in Figures 5 and 6. The first one is the well-known $Mg(OH)_2$ (= brucite) type, and has already been described in Chapters III and IV, while the γ-$Cd(OH)_2$ in Figure 6 is not. In the latter, octahedra share faces, edges, and corners. From the figure it is obvious how the structure can be transformed into the rutile structure by simple translations (slip) on planes parallel to (001). [Note that it can also be described as rutile twinned by mirror planes at (001) and (002).] Structural data are given in Data Table 2.

Data Table 2 γ-$Cd(OH)_2$ (3)

Monoclinic, space group Im (nonstandard setting of Cm), No. 8; a = 5.67, b = 10.25, c = 3.41 Å, β = 91.4°; Z = 4, V = 198.1 Å^3

Atomic Positions: $(0,0,0; \frac{1}{2},\frac{1}{2},\frac{1}{2})+$
Cd in 4(b): $x,y,z; x,\bar{y},z; x$ = 0, y = 0.1522, z = 0
OH(1) in 2(a): $x,0,z; x$ = 0.105, z = 0.508
OH(2) in 2(a): x = 0.714, z = 0.006
OH(3) in 4(b): x = 0.879, y = 0.879, z = 0.583
Atomic Distances
Cd–OH = 2.16–2.39 Å

* It is perhaps worth pointing out that while the geometry of the anion array in rutile is usually described as distorted hcp, in fact the distortion is such that the array is midway between hcp and ccp. [However, ccp anions are not possible with this topology of the octahedra (Figures 4e–g) (2).] Its anion array is a regular sphere packing with a coordination number of 11, compared with 12 in close-packing.

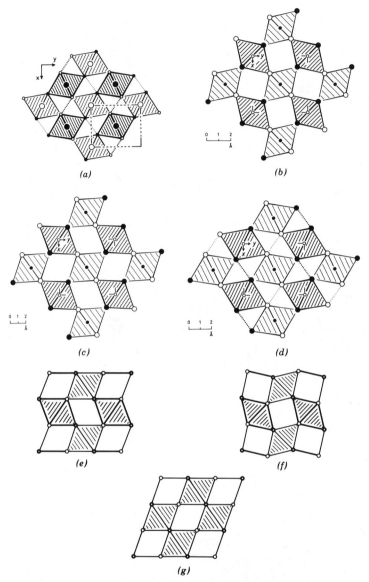

Figure 4. (*a*) The marcasite structure of FeS$_2$ projected on (001). (*b*)–(*d*) (001) projections of the structures of (*b*) rutile, (*c*) CaBr$_2$ (CaCl$_2$ type), and (*d*) marcasite. In order to make the similarities more obvious, the unit-cell axes are scaled so that $a \times b$ = constant. (*e*)–(*g*) Connected ribbons of edge-shared octahedra: (*e*) hcp, (*f*) rutile, and (*g*) ccp. Compare their topologies.

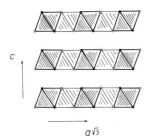

c

$a\sqrt{3}$

Figure 5. The idealized structure of ordinary α-Cd(OH)$_2$.

The La$_2$O$_3$-type structure of the A-type rare earth sesquioxides (also La$_2$O$_2$S, etc.) is not very well described in terms of cation-centered polyhedra of anions or its anion array (although we will so describe it in Chapter VIII). The simplest description seems to be as an antistructure (4): The cation array is virtually perfect hcp, and alternate layers of tetrahedral and octahedral sites are then filled with the anions. The structure is depicted this way in Figure 7, and the crystallographic data are given in Data Table 3.

Data Table 3   La$_2$O$_3$ (5)

Trigonal, space group $P\bar{3}m1$, No. 164; a = 3.938, c = 6.136 Å, c/a = 1.558; Z = 1, V = 82.41 Å^3

Atomic Positions
La in 2(d): $\pm(\frac{1}{3},\frac{2}{3},z)$; z = 0.2467
O(1) in 2(d): z = 0.6470
O(2) in 1(a):   0,0,0
Atomic Distances
La–O = 2.365 Å (3×), 2.456 Å (1×), 2.731 Å (3×)

[This closely resembles the brucite type. Mg(OH)$_2$ has the same space group, with c/a = 1.517; Mg is in the 1(a) position of O(2), O in the 2(d)

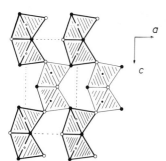

a

c

Figure 6. The real structure of γ-Cd(OH)$_2$. Octahedra share faces, edges (along **b**), and corners.

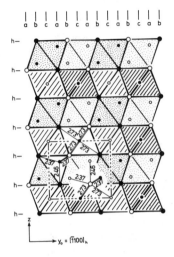

Figure 7. The A-type structure of La_2O_3 projected on (11̄20). Large circles are La, small circles are O (open at 0, filled at $\frac{1}{2}$). Bond lengths are shown, and a unit cell is outlined. The polyhedra are of *cations* and are *anion*-centered.

position of La (with $z = 0.2216$), and H in the 2(d) position of O(1) (with $z = 0.4303$; i.e., displaced from the center of the O_4 tetrahedron to give one O–H bond).]

True antistructures of La_2O_3 include $ZrLi_2N_2$, Li_2CeN_2, and $Na(LiZn)O_2$.

Cyclic Reflection and Translation Operations

The ω-phase structure is formed when hcp titanium (or zirconium) is subjected to high pressure at room temperature (6). It is also formed in binary systems like Ti–V and Ti–Cr, and isostructural compounds are δ-TiO_x ($x = 0.65$) (7), ε-TaN (8), and (slightly distorted) $CuGeO_3$ (9). The relations between the ω-phase structure and other structures like bcc and hcp have been given elsewhere (10). Here we describe it as a triple twinning of hcp in Figure 8; the structure is shown in Figure 9. Structural data are given in Data Table 4.

Data Table 4 ω-Ti(O) (6)

Hexagonal, space group $P6/mmm$, No. 191; $a = 4.625$, $c = 2.813$ Å, $c/a = 0.608$; $Z = 1$, $V = 52.42$ Å^{3}*

Atomic Positions
Ti(1) in 1(a):   0,0,0
Ti(2) in 2(d): $\pm(\frac{1}{3},\frac{2}{3},\frac{1}{2})$

CN: 14 for Ti(1), 11 for Ti(2)

* Only 3.3% less than that for 3 Ti atoms in bcc Ti.

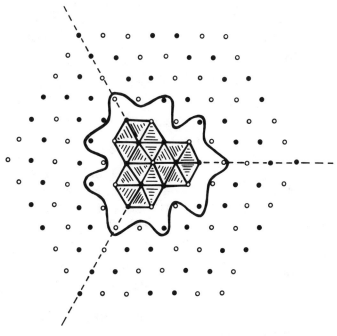

Figure 8. Triply twinned hcp to give the ω-phase structure of Ti. The wavy line delineates how much ω structure is formed by the operation.

Ti(1) is in a hexagonal prism of Ti(2) atoms, capped on the hexagonal faces by other Ti(1) atoms with 12 Ti–Ti = 3.018 and 2 Ti–Ti = 2.813 Å. Ti(2) is in an Edshammar polyhedron derived from two cube halves as shown in Figure 10 (11). In order to fit the ω-Ti structure, this 11-polyhedron has to shrink along its trigonal axis by about 12%.

The transformation of ω to bcc is simple: The Ti(2) atoms at $\frac{2}{3},\frac{1}{3},\frac{1}{2}$, and $\frac{1}{3},\frac{2}{3},\frac{1}{2}$ are shifted by $\pm\mathbf{c}/6$ so that the coordinates become $\frac{2}{3},\frac{1}{3},\frac{2}{3}$ and $\frac{1}{3},\frac{2}{3},\frac{1}{3}$. The axial

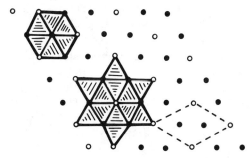

Figure 9. The real structure of ω-Ti. All atoms on 0 or $\frac{1}{2}$. The larger unit shown is six slightly distorted (elongated) octahedra sharing faces. The smaller unit of three octahedra sharing faces is identical with the well-known cluster in Nb_3Te_4.

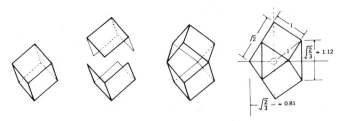

Figure 10. The Edshammar CN = 11 polyhedron. The polyhedron around Ti(2) in ω-Ti is thus derived from a cube.

ratio c/a changes from 0.608 for ω to 2 × 0.612 for bcc (the hexagonal cell of which has a doubled c axis). The ω-phase structure is thus a high-pressure form. It is as rich in octahedra as hcp or ccp, although the octahedra in the ω-structure cannot be regular.* The so-called tetrahedrally close-packed structures (for example, Cr_3Si, σ and μ phases) are surely high-pressure phases but contain no octahedra. Instead, they are very rich in tetrahedra, which again cannot be regular. It is also interesting that these structures are also easily derived from bcc, as we shall discuss later (in Chapter XIV).

The "cluster" of three face-sharing $NbTe_6$ octahedra in Nb_3Te_4 is also easily traced in the ω structure in Figure 9.

It is easy to confuse the ω-phase structure with that of AlB_2 (which we will consider in Chapter IX). In the projection used here, the drawings are identical, with Al in the Ti(1) sites and B in the Ti(2) sites. (Compare Figure 9 with Figure 1 of Chapter IX.) The two structures have the same space group. The difference between them is in their c/a ratio, which is about 0.6 for ω but 1.084 for AlB_2; a difference of degree so large (~2×) as to constitute a difference of kind. The rectangle of atoms at the "waist" of the octahedron (parallel to **c**) is 2.67 × 2.81 Å^2 in ω, but 1.74 × 3.25 Å^2 in AlB_2. The octahedra that are elongated in ω can hardly be described as octahedra at all in AlB_2.

The Ni_2In structure is also related (although we shall describe it differently in Chapter VI). It is an ordered superstructure with Ti(1) sites occupied by Ni and Ti(2) sites by Ni and In alternating along **c** and thus doubling the c axis. Its ratio $c/a = 0.614 × 2$, which is similar to ω.

A detailed discussion of these relations and possible underlying causes has been given by Pearson (12).

An example of a stuffed ω-Ti structure is found in the compound $CuGeO_3$, the structure of which is shown in Figure 11. The ω-structure type provides elongated octahedra which are very suitable for Cu^{2+}, with its preference for square coordination. Ge is in O_4 tetrahedra. Structural data are given in Data Table 5.

---

* The ratio of trans to cis distances is in one case $\sqrt{3}$ instead of $\sqrt{2}$.

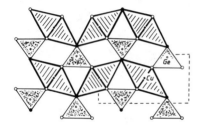

Figure 11. The real CuGeO$_3$ structure. The oxygen that is corner-shared between two octahedra corresponds to Ti(l) in ω-Ti. There are six octahedra around this oxygen.

Data Table 5   CuGeO$_3$ (9)

Orthorhombic, space group, *Pbmm*, No. 51; $a = 4.81$, $b = 8.47$, $c = 2.94$ Å; $Z = 2$, $V = 119.8$ Å^3

Atomic Positions
For coordinates see ref. 9.
Atomic Distances
Cu–O = 1.94 Å (4×); 2.77 Å (2×)
Ge–O = 1.72–1.77 Å

Cyclic translation (13) of the rutile structure (or double slip to give orthogonal APBs) produces the hollandite framework as shown in Figure 12. In the twin planes, lamellae of the ramsdellite structure are formed. Rotation of columns in the rutile structure also produces the hollandite structure, Figure 13 (14, 15). Structural data for hollandite are given in Data Table 6.

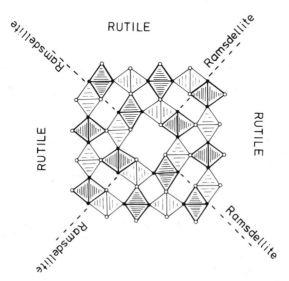

Figure 12. A fourling construction of rutile to obtain the hollandite structure (by cyclic translation).

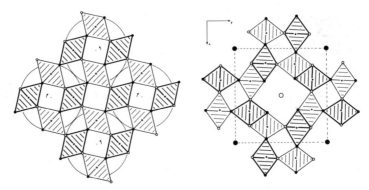

Figure 13. Rutile → hollandite (framework) by a rotation operation. The large circles are the A sites at 0, ½.

Data Table 6   Hollandite, $A_2B_8[O(OH)]_{16}$ (16)

Tetragonal, space group $I4/m$, No. 87; $a = 9.96$, $c = 2.86$ Å; $Z = 1$, $V = 284$ Å³

Atomic Positions: $(0,0,0; \frac{1}{2},\frac{1}{2},\frac{1}{2})+$
A in 2(b): $0,0,\frac{1}{2}$; A = Ba, Pb, Na or K
B in 8(h): $\pm(x,y,0; \bar{x},y,0)$; $x = 0.348$, $y = 0.167$, B = Fe or Mn
2(O,OH) in 8(h): $x_1 = 0.15$, $y_1 = 0.18$; $x_2 = 0.54$, $y_2 = 0.17$

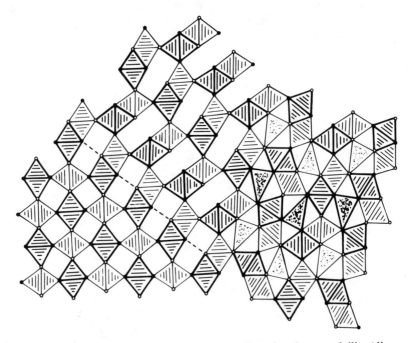

Figure 14. Starting from the left, rutile is twinned to give the ramsdellite (diaspore) structure, which is again twinned to give the $CaFe_2O_4$ structure.

There is a great deal of variation in the degree of filling of the A sites (in the tunnels) and in the exact positions of these A atoms, and even in the symmetry of the structure (some examples are monoclinic). Even today, all the problems are not resolved, and many people are still examining such structures—by X-ray methods and by electron microscopy. However, the basic features of the structure type (such as its topology) are clear.

If ramsdellite (hcp) is twinned by glide reflection, the important structure of $CaFe_2O_4$ is formed, as shown in Figure 14 (compare Figure 15). Many compounds are isostructural; examples are CaV_2O_4, $SrEu_2O_4$, Eu_3O_4, and $SrIn_2O_4$. Structural data for CaV_2O_4 are given in Data Table 7. (It can also be regarded as a twinned hc anion array (2).]

Data Table 7 CaV_2O_4 (17)

Orthorhombic, space group *Pbnm*, nonstandard setting of No. 62 (*Pnma*); $a = 10.66$, $b = 9.20$, $c = 3.01$ Å; $Z = 4$, $V = 295$ Å^3

Atomic Positions
All atoms in 4(*c*): $\pm(x,y,\frac{1}{4}; \frac{1}{2} - x,\frac{1}{2} + y,\frac{1}{4})$, with

	V(1)	V(2)	Ca	O(1)	O(2)	O(3)	O(4)
x	0.109	0.606	0.350	0.648	−0.016	0.225	−0.078
y	0.068	0.077	0.242	0.289	0.387	0.470	0.071

Atomic Distances
V–O: 1.98–2.04 Å (6×)
Ca–O: 2.39–2.64 Å (8×)

The structure of Nb_3Te_4 is shown in Figure 16 (compare Figure 9). It can be described as triply twinned hcp Te or NiAs type, the twin unit being just the unit cell content shown in the lower part of the figure. The structure is somewhat distorted due to Nb–Nb bonds across the octahedral edges within

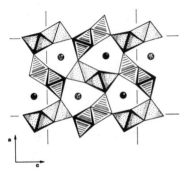

Figure 15.   The real structure of $NaFeTiO_4$ (isostructural with $CaFe_2O_4$). Setting *Pnma*.

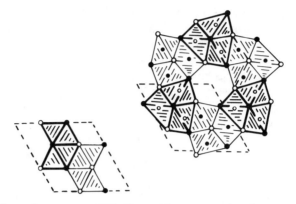

Figure 16. The real structure of Nb₃Te₄, with atoms on $\frac{1}{4}$ or $\frac{3}{4}$. The trilling unit (an element of NiAs) is in the left-hand corner.

the Nb_3Te_{11} clusters, which thus form zigzag chains of Nb atoms along **c**. Structural data are given in Data Table 8.

In Figure 17, another origin is chosen in triply twinned hcp and, in the center, it is shown how a large unit of the Nb_3Te_4 structure is formed.

Data Table 8   Nb₃Te₄ (18)

Hexagonal, space group $P6_3/m$, No. 176; $a = 10.671$, $c = 3.647$ Å, $c/a = 0.342$; $Z = 2$, $V = 359.65$ Å³

Atomic Positions
Nb in 6(*h*): $\pm(x,y,\frac{1}{4};\ \bar{y},x - y,\frac{1}{4};\ y - x,\bar{x},\frac{1}{4})$; $x = 0.489$, $y = 0.104$
Te(1) in 2(*c*): $\pm(\frac{2}{3},\frac{1}{3},\frac{3}{4})$
Te(2) in 6(*h*): $x = 0.339$, $y = 0.273$
Atomic Distances
Nb–Nb = 2.97 Å (2×)
Nb–Te = 2.76–2.95 Å (6×)

$K_{0.3}Ti_3S_4$ is a partly filled Nb_3Te_4 type, with $Ti_3S_4 \equiv Nb_3Te_4$. The K atoms are in the empty tunnels in Figure 16 (19). For this compound, $c/a = 0.359$.

The structure of β-Si_3N_4 is shown in Figure 18. (That of α-Si_3N_4 is only a small distortion of the β form.) It can also be described as triply twinned hcp, the twin unit being just the unit-cell content shown in the lower left part of the figure (the same as used for Nb_3Te_4). The exact structure of a twin unit is very near the ideal hcp arrangement, but now tetrahedra are occupied instead of octahedra. Structural data are given in Data Table 9.

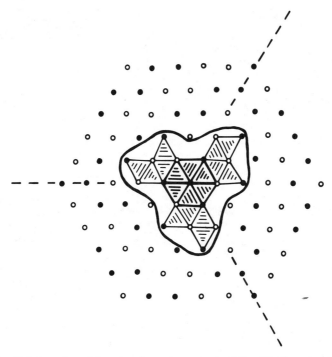

Figure 17.   Triple-twinned hcp to give a large part of the Nb_3Te_4 structure in the center.

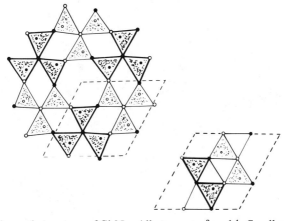

Figure 18.   The real structure of Si_3N_4. All atoms on $\frac{3}{4}$ and $\frac{1}{4}$. Smaller circles are Si. In phenacite, Be_2SiO_4, the c axis is three times as large, due to ordering of cations. Otherwise the structures are identical.

Data Table 9   β-Si₃N₄ (20)

Hexagonal, space group $P6_3/m$, No. 176; $a = 7.607$, $c = 2.911$ Å, $c/a = 0.383$;
$Z = 2$, $V = 145.8$ Å³

Atomic Positions
N in 2(c): $\pm(\frac{1}{3},\frac{2}{3},\frac{1}{4})$
N in 6(h): $\pm(x,y,\frac{1}{4}; \bar{y},x - y,\frac{1}{4}; y - x,\bar{x},\frac{1}{4})$; $x = 0.321$, $y = 0.025$
Si in 6(h): $x = 0.174$, $y = 0.766$
Atomic Distances
Si–N = 1.71–1.75 Å
N–N = 2.76–2.91 Å

It has recently been asserted that the space group is $P6_3$ rather than $P6_3/m$
(21), but the disagreement is still unresolved. The lower symmetry space
group ($P6_3$) implies small displacements of both types of N atoms along **c**, by

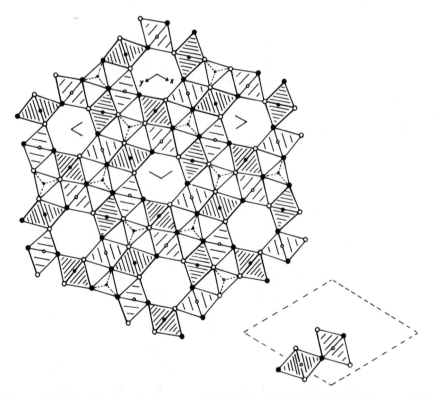

Figure 19. The idealized structure of fluoborite, $Mg_3BO_3[(OH)_{1-x}F_x]_3$, $x \approx 0.3$.
Atoms on $\frac{1}{4}$ or $\frac{3}{4}$. Twin-unit in right corner.

0.03–0.04 Å, thus transforming the NSi_3 triangle to a (very low) triangular pyramid.

If the twin unit is smaller, two octahedra instead of the four in Nb_3Te_4, the fluoborite structure is derived, as shown in Figure 19. Mg are in O_6 octahedra, and boron in O_3 triangles. Structural data are given in Data Table 10. (See also Chapter VI.)

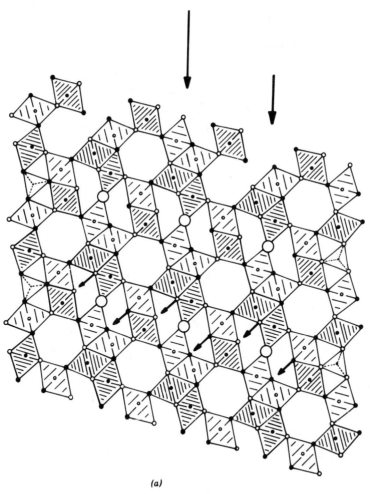

(a)

Figure 20.   Shear in fluoborite to give the Nb_3Te_4 structure. For further explanation, see text.

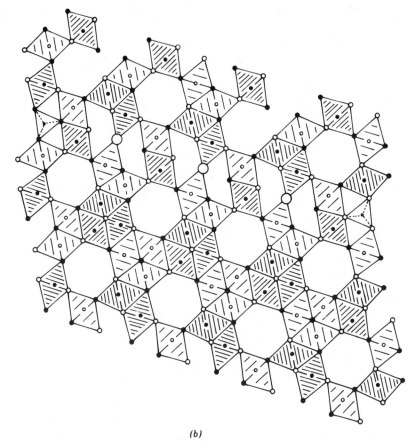

(b)

Figure 20. *(Continued)*

Data Table 10   Fluoborite, Mg₃BO₃[(OH)₁₋ₓFₓ]₃, x ≈ 0.3 (22, 23)

Hexagonal, space group $P6_3/m$, No. 176; $a = 9.06$, $c = 3.06$ Å; $Z = 2$, $V = 218$ Å³

Atomic Positions

B in 2(c):	$\pm(\frac{1}{3}, \frac{2}{3}, \frac{1}{4})$
Mg in 6(h):	$\pm(x, y, \frac{1}{4}; \bar{y}, x - y, \frac{1}{4}; y - x, \bar{x}, \frac{1}{4})$; $x = 0.381$, $y = 0.038$
O in 6(h):	$x = 0.381$, $y = 0.537$
[(OH),F] in 6(h):	$x = 0.310$, $y = 0.218$

Atomic Distances

B–O = 1.44 Å (3×)
O–O = 2.49 Å (in triangle) (3×)

It is interesting that the structure type of Nb₃Te₄ can also be formally derived by a double crystallographic shear mechanism on fluoborite. This is shown in the sequence of Figures 20*a* and *b*. Figure 20*a* is a single shear, and some atoms are ringed by large circles. They are at impossibly short distances, and one of them has to be omitted. Hence one of the two adjacent octahedra has to be transformed into a trigonal prism. However, this is just a "transition state" in the production of the Nb_3Te_4 structure. The second shear eliminates this difficulty.

Formally, the trilling unit described above can be reduced until it contains only two atoms. The structure of $Y(OH)_3$ is then derived. It was described in Chapter III.

References

1. D. K. Smith, C. F. Cline, and S. B. Austerman, *Acta Cryst.* **18**, 393 (1965).
2. B. G. Hyde, A. N. Bagshaw, S. Andersson, and M. O'Keeffe, *Ann. Rev. Mater. Sci.* **4**, 43 (1974).
3. P. M. de Wolff, *Acta Cryst.* **21**, 432 (1966).
4. M. O'Keeffe and B. G. Hyde, *Struct. Bonding* **61**, 77 (1984).
5. P. Aldebert and J. P. Traverse, *Mater. Res. Bull.* **14**, 303 (1979).
6. J. C. Jamieson, *Science* **139**, 762 (1963).
7. S. Andersson, *Acta Chem. Scand.* **13**, 415 (1959).
8. G. Brauer and K. H. Zapp, *Naturwiss.* **40**, 604 (1953).
9. H. Vollenkle, A. Wittman, and H. Nowotny, *Monatsh. Chem.* **98**, 1352 (1967).
10. S. Andersson, *Arkiv Kemi* **15**, 247 (1960).
11. L-E. Edshammar, X-ray studies on binary alloys of aluminium with platinum metals, thesis, Univ. Stockholm, 1969, 51 pp.
12. W. B. Pearson, *Proc. Roy. Soc. Lond. A* **365**, 523 (1979).
13. L. Stenberg and S. Andersson, *Z. Kristallogr.* **158**, 133 (1982).
14. B. G. Hyde, in *Reactivity of Solids,* J. S. Anderson *et al.,* Eds. Chapman & Hall, London, 1972 p. 23.
15. L. A. Bursill and B. G. Hyde, *Nature Phys. Sci.* **240**, 122 (1972).
16. A. Byström and A.-M. Byström, *Acta Cryst.* **3**, 146 (1950).
17. E. F. Bertaut, P. Blum, and G. Magnano, *Bull. Soc. Franc. Miner. Crist.* **129**, 536 (1956).
18. K. Selte and A. Kjekshus, *Acta Cryst.* **17**, 1568 (1964).
19. R. Schöllhorn, W. Schramm, and D. Fenske, *Angew. Chem. Int. Ed. Engl.* **19**, 492 (1980).
20. O. Borgen and H. M. Seip, *Acta Chem. Scand.* **15**, 1789 (1961).
21. R. Grün, *Acta Cryst. B* **35**, 800 (1979).
22. Y. Takeuchi, *Acta Cryst.* **3**, 208 (1950).
23. A. Dal Negro and C. Tadini, *Miner. (Tschermaks) Petrogr. Mitt.* **21**, 94 (1974).

CHAPTER VI

CCP II:
Projection along the Shortest Axis,
$\langle 110 \rangle$ of the fcc Unit Cell, a cp Row
of Atoms

The following structures are described in this chapter: NaCl, zinc blende (ZnS), $CdCl_2$, α-Ga_2O_3 and β-Ga_2O_3, spinel ($MgAl_2O_4$) and the β-phase, CrB, Ni_2In, Re_3B and $PuBr_3$, W_2CoB_2, $NbAs_2$, Dy_3Ni_2, $CaIrO_3$ (Cr_3GeC), SnI_2, $CaTi_2O_4$, Y_5S_7, lillianite ($Pb_3Bi_2S_6$), $CaFe_3O_5$, $CaFe_4O_6$, $CaFe_5O_7$, warwickite [$(Mg,Fe,Ti)_2BO_4$] and $CoFeBO_4$, Yb_3S_4, cuspidine [$Ca_4Si_2O_7(F,OH)_2$] etc., Si_3N_4, Nb_3Se_4 (Nb_3Te_4), phenacite (Be_2SiO_4), fluoborite, wightmanite [$Mg_5BO_4(OH)_5 \cdot nH_2O$], $BaBi_2S_4$, NH_4CdCl_3, $PbSnS_3$, Sn_2S_3, $PbCl_2$, La_2SnS_5.

Here we project ccp along the shortest possible axis, a polyhedral edge. Figure 1 shows the net of atoms, with octahedra and tetrahedra indicated. Note how all octahedra "point the same way." In hcp they point in opposite directions in adjacent layers parallel to the cp planes of anions (cf. Chapter V). Another way of describing this is to say that in hcp, octahedra (and tetrahedra) share faces and edges, whereas in ccp they share only edges. These are simple diagnostic distinctions between hcp and ccp, especially in complex structures.

Simple Structures

In Figure 2 we see the previously described relationship between the zinc blende and NaCl structures. The jump of a central atom that transforms the

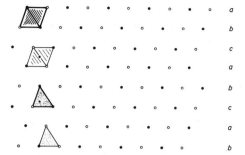

Figure 1. Cubic close packing, ccp. Smaller circles are at the centers of the poly-hedra. All atoms are on 0 or $\frac{1}{2}$.

former structure into the latter is indicated by an arrow. (Note that the vector represented by this arrow has a component normal to the projection plane. Note also that it does not matter whether the "central atom" is a cation or an anion: both NaCl and zinc blende are their own antitypes.)

The idealized $CdCl_2$ structure is shown in Figure 3, and it is easy to see how this structure is related to others like NaCl or $Cd(OH)_2$ (Figure 5 of Chapter V).

α-Ga_2O_3 has the corundum structure (Chapter IV), but the (high-tempera-ture) β-form has a new structure, shown in several ways in Figure 4. In it, half the Ga^{3+} ions are in tetrahedra, and the other half are in octahedra.

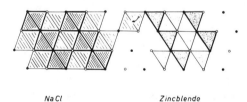

NaCl   Zincblende

Figure 2. The relationship between NaCl and zinc blende.

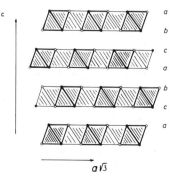

Figure 3. The structure of $CdCl_2$.

Structural data are given in Data Table 1.

This structure, of course, has similarities to the NaCl, zinc blende, rutile, and $CdCl_2$ types and, as we shall see, to the spinel structure also.

Data Table 1   **β-Ga$_2$O$_3$ (1)**

Monoclinic, space group $C2/m$, No. 12; $a = 12.23$, $b = 3.04$, $c = 5.80$ Å, $\beta = 103.7°$; $Z = 4$, $V = 209$ Å^3

Atomic Positions
All atoms in 4(i): $(0,0,0; \frac{1}{2},\frac{1}{2},0) \pm (x,0,z)$, with

	x	z
Ga(1)	0.0904	0.7948
Ga(2)	0.3414	0.6857
O(1)	0.1674	0.1011
O(2)	0.4957	0.2553
O(3)	0.8279	0.4365

Average Atomic Distances
Ga(1)–O = 1.83 Å (4×), O · · · O = 3.02 Å
Ga(2)–O = 2.00 Å (6×), O · · · O = 2.84 Å

We may note that there is an orthogonal unit cell (useful for drawing the structure and for comparing it with other structures) with $\mathbf{a}_o = 2\mathbf{a}_m + \mathbf{c}_m$, $\mathbf{b}_o = \mathbf{b}_m$, $\mathbf{c}_o = \mathbf{c}_m$. (Subscript o = orthogonal, m = monoclinic.) This is indicated in Figures 4*b* and *c*.

Furthermore, in Chapter II the spinel structure was depicted in projection down its 8-Å cubic unit-cell edge. Such a long projection axis necessitated two bounded projections to show the whole structure clearly. Using the projection of this chapter, the same structure can be depicted in a single diagram. Figure 5 shows that for a face-centered cubic structure there is always an equivalent, alternative unit cell of half the volume: body-centered tetragonal (t) instead of face-centered cubic (c). The new cell has

$$\mathbf{a}_t = \tfrac{1}{2}\mathbf{a}_c - \tfrac{1}{2}\mathbf{b}_c, \qquad \mathbf{b}_t = \tfrac{1}{2}\mathbf{a}_c + \tfrac{1}{2}\mathbf{b}_c, \qquad \mathbf{c}_t = \mathbf{c}_c$$

In other words, $a_t = b_t = a_c/\sqrt{2}$, and $c_t = a_c$.

With the shorter $\mathbf{a}_t$ axis, the complete spinel structure can be depicted by projection along $\mathbf{a}_t$, as is done for the (high-pressure) spinel form of γ-Fe$_2$SiO$_4$ in Figure 6. (The strings of edge-sharing octahedra are now either parallel to the projection plane or normal to it.)

Figure 7 shows the same drawing but with all the anions omitted, which emphasizes that the complete structure consists of identical blocks, each being half the volume of the tetragonal unit cell, arranged in zigzag fashion.

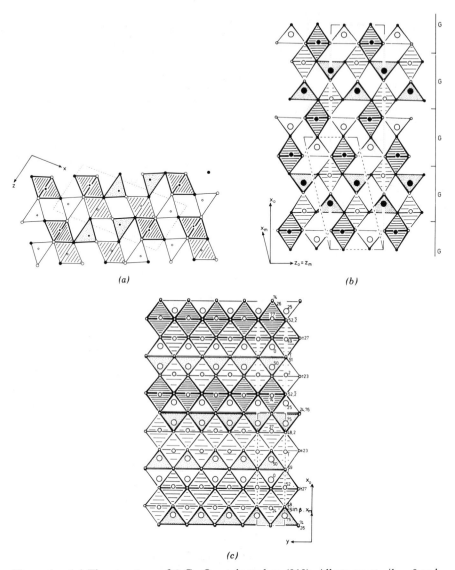

Figure 4. (*a*) The structure of β-Ga_2O_3 projected on (010). All atoms at $y/b = 0$ or $\frac{1}{2}$. (*b*) The same projection showing the orthogonal cell (o) of twice the volume of the monoclinic cell (m). The structure can be constructed of (100) slabs G. (*c*) The structure projected along [001]; compare Figure 6.

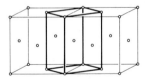

Figure 5. Relation between the face-centered cubic and body-centered tetragonal unit cells.

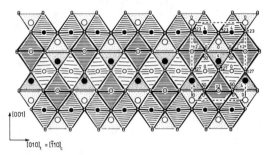

Figure 6. (High-pressure) spinel form of γ-Fe$_2$SiO$_4$ projected on $(110)_c \equiv (100)_t$. Large circles are tetrahedrally coordinated cations (Si); medium circles are octahedrally coordinated cations (Fe); small circles are anions (O). (Atom heights are in units of $a_t/8$ for cations, $a_t/100$ for anions.)

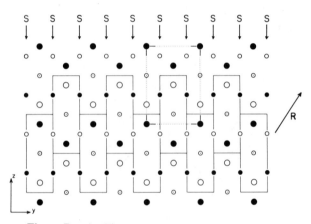

Figure 7. As Figure 6, but with anions omitted.

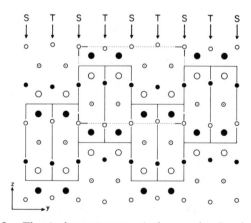

Figure 8. The β-phase structure (anions omitted); cf. Figure 7.

The shift in origin from the position of one block to the next is the vector $\mathbf{R} = \frac{1}{2}[111]_t$ (see Figure 7). (The blocks repeat in the $\mathbf{x}$ and $\mathbf{z}$ directions, forming infinite slabs.)

There is a whole series of structures that seem to be significant as possible intermediates in the geologically important spinel $\rightleftarrows$ olivine transformation referred to in Chapter III. They all contain the same blocks stacked together, with origin shifts of either $\mathbf{R} = \frac{1}{2}[111]_t$ (spinel, S) or $\mathbf{R} = \frac{1}{2}[010]_t$ (a pseudomirror plane, T) (2). We show only one such structure—that of the so-called β-phase (phase III) in Figure 8. Structural data are given in Data Table 2.

Data Table 2   **"Phase III,"** $Ni_{1.563}Al_{0.875}Si_{0.563}O_4$ **(3)***

Orthorhombic, space group *Imma*, No. 74; $a = 5.6646$, $b = 11.455$, $c = 8.1007$ Å, $Z = 8$, $V = 525.64$ Å^3

Atomic Positions $(0,0,0; \frac{1}{2},\frac{1}{2},\frac{1}{2})+$

M(1) in 4(a):	$0,0,0; 0,\frac{1}{2},0$	($\sim$3.2Ni + 0.8Al)
M(2) in 4(e):	$0,\frac{1}{4},z; 0,\frac{3}{4},\bar{z}; z = 0.9758$	($\sim$3.6Ni + 0.4Al)
O(1) in 4(e):	$z = 0.220$	
O(2) in 4(e):	$z = 0.722$	
M(3) in 8(g):	$\pm(\frac{1}{4},y,\frac{1}{4}; \frac{3}{4},y,\frac{1}{4}); y = 0.1246$	($\sim$5.7Ni + 2.3Al)
T in 8(h):	$\pm(0,y,z; 0,\frac{1}{2} - y,z); y = 0.1202, z = 0.6186$	($\sim$3.5Al + 4.5Si)
O(3) in 8(h):	$y = 0.9997, z = 0.253$	
O(4) in 16(j):	$\pm(x,y,z; x,\bar{y},\bar{z}; x,\frac{1}{2} + y,\bar{z}; x,\frac{1}{2} - y,z); x = 0.252, y = 0.127,$	
	$\quad z = 0.9982$	

* This is a high-temperature/high-pressure structure of the β-phase type, $0.563Ni_2SiO_4 \cdot 0.437NiAl_2O_4$.

A phase with this structure sometimes (but not always) appears between the olivine and spinel regions in $p-T$ phase diagrams of A_2BX_4 compounds, for example, in the cases of Mg_2SiO_4 and Co_2SiO_4 (4). It can be thought of as spinel with anti-phase boundaries (APBs) in every second $(020)_t$ plane (the mirror planes, T in Figure 8); or spinel can be thought of as β phase with APBs on every second S boundary (Figure 7). [The introduction of an APB does not affect the anion array, since $\frac{1}{2}[101]_t$, the APB displacement vector, is an anion–anion vector (see Figure 6). This array is approximately ccp in all these related structures.]

The structure of the unit block just referred to is interesting in another respect. In Figure 9 it is depicted in projection on $(100)_t \equiv (010)_t$ (i.e., half the projection axis of Figure 6). Figure 4b shows the analogous projection of β-Ga_2O_3 (cf. Figure 4a). (Structural data for the latter have already been given in Data Table 1.) They have a great deal in common: both contain identical bands G (cf. Figures 9 and 4b)—the building element of β-Ga_2O_3. In the spinel-like element, these alternate with narrow bands B, which are strips of edge-sharing octahedra, i.e., of NaCl (B1) type. Elimination of

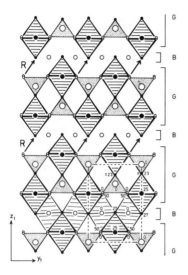

Figure 9. The structure of a "unit block" of spinel, etc., projected on $(010)_t = (100)_t$. Compare Figure 4*b*: the (001) G slabs are the same in the two figures.

these strips by CS on (001) planes with vector **R** yields the β-Ga$_2$O$_3$ structure: i.e.,

$$A_2BX_4 \xrightarrow[-AX]{CS} ABX_3 = \beta\text{-Ga}_2O_3$$

where A is an octahedral cation and B is a tetrahedral cation.

This means that the spinel structure consists of columns (one octahedron thick) of NaCl type in a β-Ga$_2$O$_3$ type matrix (Figure 10). That these two structures intergrow so readily may explain the rather strange behavior in the high-temperature part of the spinel phase diagram (Figure 11). The solubility of Al$_2$O$_3$ in spinel is enormous at temperatures above $\sim$1400°C; up to $\sim$84% Al$_2$O$_3$/16% MgAl$_2$O$_4$ at $\sim$1800°C. It is possible that this is related to a reported high-temperature form of "Al$_2$O$_3$," θ-Al$_2$O$_3$, that has the β-Ga$_2$O$_3$

Figure 10. Resolution of the spinel structure into ⟨100⟩ columns of β-Ga$_2$O$_3$ and NaCl types, heavily outlined. The former are V-shaped; the latter are just one octahedron in cross section.

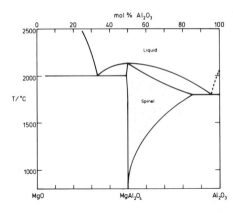

Figure 11. Phase diagram of MgO + Al_2O_3.

type structure (probably stabilized by impurities?). The solubility is simply due to an increase in the ratio of β-Ga_2O_3 to NaCl elements in the structure of the "spinel" phase.

Complex Structures

Figure 12 shows ccp twinned (by reflection) on {113} of the fcc unit cell (5). The trigonal prisms created in the twin plane by the twin operation are easily recognized. The simplest (although not the most obvious) structure derived by *periodic* twinning of this sort is that of CrB, shown (in this projection) in Figure 13. Boron atoms are in the centers of Cr_6 trigonal prisms, and the twin block size is the minimum possible [$1 \times d(113)_{fcc}$], and therefore written as . . . , *1,1,1,* . . . . Structural data are given in Data Table 3.

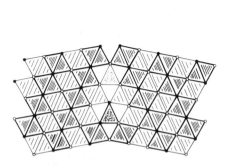

Figure 12. Twinning (reflection) in ccp. Trigonal prisms are created in the twin plane, which is {113}$_{fcc}$. (Cf. Figures 23c and d of Chapter IV.)

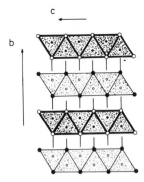

Figure 13. The structure of CrB, twinned ccp Cr . . . , *1,1,1,* . . . projected on (100). BCr$_6$ prisms are shown. (This is the analogue in ccp of FeB in hcp.)

Data Table 3 CrB

Orthorhombic, space group *Cmcm*, No. 63; $a = 2.696$, $b = 7.858$, $c = 2.932$ Å; $Z = 4$, $V = 68.40$ Å^3

Atomic Positions
Cr in 4(*c*): $\pm(0,y,\frac{1}{4}; \frac{1}{2},\frac{1}{2} + y,\frac{1}{4})$; $y = 0.146$
B in 4(*c*): $y = 0.440$

Isostructural compounds include LnSi (with Ln = Eu to Lu), all the rare earth monogallides and many monogermanides; and HoO(OH) is simply related (only alternate prisms are occupied by Ho). The simple topological relations between the CrB structure and the NaCl, SnS, and TlI structures were recently described (5) and will be discussed later (Chapter IX). The last is very similar to CrB: Tl occupy trigonal prisms of I but are off-center due to the stereochemically active lone pair of electrons on Tl$^+$.

A formal description of the hexagonal Ni$_2$In structure puts it in this family of twinned ccp structures. The size is . . . , 2,2,2, . . . , and it is shown in Figure 14. However, the trigonal prisms (as we might expect since they are pentacapped) are considerably elongated (cf. Chapter VIII, Figure 35), and it is really isostructural with the ω-phase structure of titanium (if the distinction between Ni and In is ignored) as was discussed in Chapter V. There are many isostructural compounds, often with somewhat different *c/a* ratios, e.g., 1.228 for Ni$_2$In, 1.301 for MnCoGe, 1.36 for UPt$_2$ (which is also *very* slightly deformed to orthorhombic). See also ref. 5. The 11-coordination

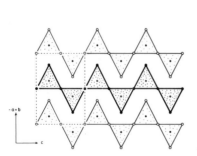

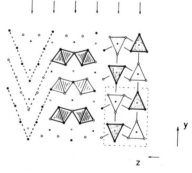

Figure 14. The hexagonal structure of Ni$_2$In projected on (1120): twinned ccp . . . , 2,2,2, . . . Ni. InNi$_6$ trigonal prisms are shown; note that they are pentacapped. Ni$_2$In is isostructural with ω-Ti; the doubling of *c* is due to ordering of Ni and In.

Figure 15. The structure of Re$_3$B projected on (100): twinned ccp . . . , 3,3,3, ... All atoms are at $x = 0$ (open circles) or at $x = \frac{1}{2}$ (filled circles). On the left, the herringbone pattern of {110} fcc planes is shown; in the center, the (empty) Re$_6$ octahedra; and on the right, the BRe$_6$ trigonal prisms.

polyhedron around In is the Edshammar polyhedron described by him for IrAl$_3$, which has already been mentioned in Chapter V and will be discussed again in Chapter VIII. Ni$_2$In can also be derived by slip from CaZn$_5$ and is very similar to CrO$_3$ (Figure 24 of Chapter XIV).

The structure of Re$_3$B, shown in Figure 15, is twinned ccp with block size . . . , *3,3,3*, . . . (5, 6). It is the ccp analogue of Fe$_3$C in twinned hcp (Chapter IV). As expected, the prism edge across the twin plane is again short: 2.73 Å compared with 3.04 Å for the other two triangular edges. The other edges of the trigonal prism are 2.81 Å long. The octahedra are only slightly distorted.

The structure of PuBr$_3$ is the antiform of Re$_3$B. It is the structure of most of the triiodides and some of the tribromides of the lanthanide and actinide metals.

Structural data are given in Data Tables 4 and 5.

Data Table 4   Re$_3$B (7)

Orthorhombic, space group *Cmcm*, No. 63; a = 2.890, b = 9.313, c = 7.258 Å; Z = 4, V = 195.3 Å^3

Atomic Positions
B in 4(c): $\pm(0,y,\frac{1}{4}; \frac{1}{2},\frac{1}{2} + y,\frac{1}{4})$; y = 0.744
Re in 4(c): y = 0.4262
Re in 8(f): $\pm(0,y,z; 0,y,\frac{1}{2} - z; \frac{1}{2},\frac{1}{2} + y,z; \frac{1}{2},\frac{1}{2} + y,\frac{1}{2} - z)$; y = 0.1345, z = 0.0620

Data Table 5   PuBr$_3$ (8)

Orthorhombic, space group *Cmcm*, No. 63; a = 4.10, b = 12.65, c = 9.15 Å; Z = 4, V = 474.6 Å^3

Atomic Positions
Pu in 4(c): y = 0.75
Br in 4(c): y = 0.43
Br in 8(f): y = 0.14, z = 0.05

There are several examples of asymmetrical twinning: the structure of W$_2$CoB$_2$ is twinned ccp . . . *1,2,1,2*, . . . ; NbAs$_2$ is slightly deformed . . . , *1,3,1,3*, . . . ; Dy$_3$Ni$_2$ is deformed . . . , *1,1,1,3,1,1,1,3*, ...

Filled Twinned ccp Structures

Cr$_3$GeC is twinned ccp Cr . . . , *3,3,3*, . . . , similar to Re$_3$B, with Ge in the trigonal prisms but also C in the octahedra. CaIrO$_3$ is an antiform with Ca in the bicapped trigonal prisms of oxygen and Ir in the O$_6$ octahedra. Its struc-

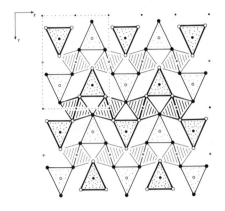

Figure 16. The real structure of $CaIrO_3$. Atoms are at 0 or $\frac{1}{2}$.

ture is shown in Figure 16, and structural data are given in Data Table 6. Isostructural compounds include $UFeS_3$, $UScS_3$, $TlPbI_3$, and $NdYbS_3$.

Data Table 6 $CaIrO_3$ (9)

Orthorhombic, space group *Cmcm*, No. 63; $a = 3.145$, $b = 9.855$, $c = 7.293$ Å; $Z = 4$, $V = 226.0$ Å^3

Atomic Positions

Ir in 4(b): $0,\frac{1}{2},0; 0,\frac{1}{2},\frac{1}{2}; \frac{1}{2},0,0; \frac{1}{2},0,\frac{1}{2}$

Ca in 4(c): $\pm(0,y,\frac{1}{4}; \frac{1}{2},\frac{1}{2} + y,\frac{1}{4}); y = 0.7498$

O(1) in 4(c): $y = 0.4331$

O(2) in 8(f): $\pm(0,y,z; 0,y,\frac{1}{2} - z; \frac{1}{2},\frac{1}{2} + y,z; \frac{1}{2},\frac{1}{2} + y,\frac{1}{2} - z); y = 0.1296,$
                        $z = 0.0553$

* The [IrO$_6$] octahedra are fairly regular: edge lengths 2.67_9–2.87_0 Å. The [CaO$_6$] trigonal prisms are bicapped, as expected: d(Ca–O) $= 2.39_4$ Å (2×), 2.42_8 Å (4×), 2.52_3 Å (2×), and 3.12_3 Å (one).

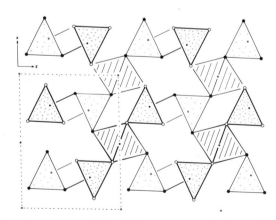

Figure 17. The real structure of SnI_2. If the empty octahedra are filled, the structure becomes identical with $CaIrO_3$.

SnI_2 is remarkably similar, with a structure in which all the trigonal prisms of I and the octahedra in alternate twin bands are occupied by Sn (the remainder being empty). Its monoclinic structure is shown in Figure 17, and structural data are given in Data Table 7.

Data Table 7 SnI₂ (10)

Monoclinic, space group $C2/m$, No. 12; $a = 14.17$, $b = 4.535$, $c = 10.87$ Å, $\beta = 92.0°$; $Z = 6$, $V = 698.09$ Å^3

Atomic Parameters: $(0,0,0; \frac{1}{2},\frac{1}{2},0)+$
Sn(1) in 2(a): 0,0,0
Sn(2) in 4(i): $\pm(x,0,z)$; $x = 0.266$; $z = 0.306$
I(1) in 4(i): $x = 0.078$; $z = 0.739$
I(2) in 4(i): $x = 0.360$; $z = 0.561$
I(3) in 4(i): $x = 0.352$; $z = 0.924$
Atomic Distances
Sn(1)–I = 3.15–3.17 Å (octahedra)
Sn(2)–I = 3.08–3.25 Å (trigonal prisms)

In the important $CaTi_2O_4$ structure, the ccp blocks are . . . , *4,4,4*, . . . (5, 6) with Ca atoms in bicapped trigonal prisms. Many compounds are isostructural, among them a number of ternary lanthanide sulfides such as MnY_2S_4. The structure of $CaTi_2O_4$ is shown in Figure 18, and the data are given in Data Table 8.

Data Table 8 CaTi₂O₄ (11)

Orthorhombic, space group $Cmcm$, No. 63; $a = 3.136$, $b = 9.727$, $c = 9.976$ Å; $Z = 4$, $V = 304.31$ Å^3

Atomic Positions
O(1) in 4(b): $0,\frac{1}{2},0$; $0,\frac{1}{2},\frac{1}{2}$; $\frac{1}{2},0,0$; $\frac{1}{2},0,\frac{1}{2}$
Ca in 4(c): $\pm(0,y,\frac{1}{4}; \frac{1}{2},\frac{1}{2} + y,\frac{1}{4})$; $y = 0.384$
O(2) in 4(c): $y = 0.060$
O(3) in 8(f): $\pm(0,y,z; 0,y,\frac{1}{2} - z; \frac{1}{2},\frac{1}{2} + y,z; \frac{1}{2},\frac{1}{2} + y,\frac{1}{2} - z)$; $y = 0.230$, $z = 0.900$
Ti in 8(f): $y = 0.126$, $z = 0.065$

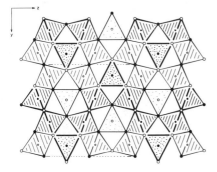

Figure 18. $CaTi_2O_4$, twinned ccp . . . , *4,4,* . . . projected on (100). All atoms are at $x = 0$ or $\frac{1}{2}$. CaO_6 trigonal prisms and TiO_6 octahedra are shown.

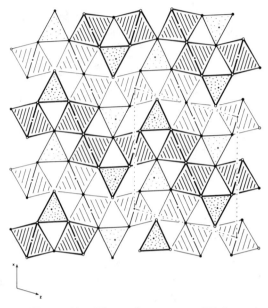

Figure 19. The real structure of Y_5S_7.

The structure of Y_5S_7 is of special interest, as it consists of twin blocks of two different sizes (5): . . . , 4,3,4,3, . . . (i.e., asymmetric twinning; cf. above). It can therefore be described as an intergrowth between the $CaIrO_3$ and $CaTi_2O_4$ types. Many binary and ternary sulfides, for example, MnY_4S_7, are isostructural. The structure of Y_5S_7 is shown in Figure 19, and the crystallographic data are given in Data Table 9.

Data Table 9 Y_5S_7 (12)

Monoclinic, space group $C2/m$, No. 12; $a = 12.768$, $b = 3.803$, $c = 11.545$ Å, $\beta = 104.82°$; $Z = 2$, $V = 541.9$ Å^3

Atomic Positions
Y(1) in 2(a): $0,0,0; \frac{1}{2},\frac{1}{2},0$
Y(2) in 4(i): $\pm(x,0,z; \frac{1}{2} + x,\frac{1}{2},z); x = 0.3021, z = 0.1903$
Y(3) in 4(i): $x = 0.1127, z = 0.4226$
S(1) in 4(i): $x = 0.6593, z = 0.051$
S(2) in 4(i): $x = 0.0401, z = 0.7846$
S(3) in 4(i): $x = 0.2590, z = 0.6476$
S(4) in 2(d): $0,\frac{1}{2},\frac{1}{2},; \frac{1}{2},0,\frac{1}{2},$

More complex asymmetrical twins also exist [e.g., . . . , *3,4,4*, . . . , . . . , *3,4,3,4,4*, . . .] in the systems $PbS + Bi_2S_3$ and $MnS + Y_2S_3$ (5, 13).

Otto and Strunz (14) were the first to describe the structure of lillianite, $Bi_2S_3 \cdot 3PbS$, as a twinned PbS (galena) structure. The structure was later refined by Takeuchi and Takagi. It consists of ccp S in twin blocks of . . . , 6,6,6, . . . (NaCl type) and is shown in Figure 20, with the structural data given in Data Table 10. Many sulfide structures are related, and this complex bit of structural mineralogy [up to twinned ccp . . . , 13,13,13 . . .] is described in recent reviews (5, 15).

Data Table 10   Lillianite, $Pb_3Bi_2S_6$ or $Bi_2S_3 \cdot 3PbS$ (16)

Orthorhombic, space group *Cmcm*, No. 63; $a = 4.104$, $b = 13.535$, $c = 20.451$ Å; $Z = 4$, $V = 1136._0$ Å^3

Atomic Positions

S(1) in 8(f): $\pm(0,y,z; 0,y,\frac{1}{2} - z; \frac{1}{2},\frac{1}{2} + y,z; \frac{1}{2},\frac{1}{2} + y,\frac{1}{2} - z)$; $y = 0.2386$,
                 $z = 0.0956$
S(2) in 4(b): $0,\frac{1}{2},0; 0,\frac{1}{2},\frac{1}{2}; \frac{1}{2},0,0; \frac{1}{2},0,\frac{1}{2}$
S(3) in 4(c): $\pm(0,y,\frac{1}{4}; \frac{1}{2},\frac{1}{2} + y,\frac{1}{4})$; $y = 0.6822$
S(4) in 8(f): $y = 0.9561$, $z = 0.1630$
M(1) in 8(f): $y = 0.5896$, $z = 0.1338$
M(2) in 8(f): $y = 0.8653$, $z = 0.0495$
M(3) in 4(c): $y = 0.3239$

It is argued that the Pb atoms, being larger, are more likely to occupy the trigonal prism sites and that therefore M(3) = Pb, M(1) = M(2) = (Pb,Bi), the latter being in octahedral coordination.

Figure 20.   The real structure of lillianite.

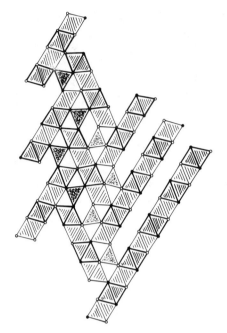

Figure 21. Twin operation on $CdCl_2$ to give warwickite.

There are also some related calcium iron oxides (not drawn): $CaFe_3O_5 = \ldots, 5,5,5, \ldots,$ $CaFe_4O_6 = \ldots, 6,6,6, \ldots$ (= lillianite) and $CaFe_5O_7 = \ldots, 7,7,7, \ldots$ (17).

Glide-Reflection Twinning

This operation involves a translation (parallel to the twin plane) in addition to the mirror reflection.

The warwickite structure is simply described if it is considered as twinned $CdCl_2$ [or (111) NaCl] layers. This is shown in Figure 21, and Figure 22 shows the real structure. The structural data are given in Data Table 11.

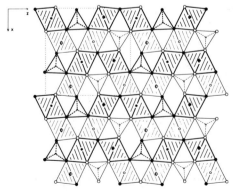

Figure 22. The real structure of warwickite-type $CoFeBO_4$.

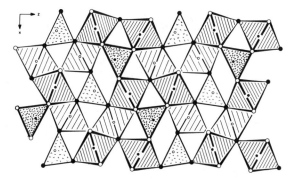

Figure 23. The real structure of Yb_3S_4 projected on (010).

Data Table 11   The Warwickite Structure Type of $CoFeBO_4$ (18)

Orthorhombic, space group *Pnma*, No. 62; $a = 9.234$, $b = 3.1252$, $c = 9.395$ Å; $Z = 4$, $V = 271.12$ Å^3

Atomic Positions
All atoms in $4(c)$: $\pm(x,\frac{1}{4},z; \frac{1}{2} + x,\frac{1}{4},\frac{1}{2} - z)$, with

	x	z
Fe	0.38322	0.56869
Co	0.39627	0.19406
B	0.33567	0.87732
O(1)	0.4822	0.8702
O(2)	0.2549	0.7517
O(3)	0.2662	0.0085
O(4)	0.4888	0.3824

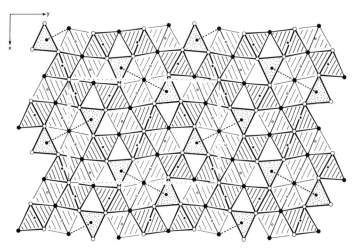

Figure 24. The real structure of $Ca_3In_2O_6$.

Figure 25. Three different ways in which trigonal prisms of anions can be occupied: MX_6 prism, MX_3 triangle, and M_2X_7 pair of corner-connected MX_4 tetrahedra.

This is a rather common A_2BX_4 structure, warwickite itself being $(Mg,Fe,Ti)_2BO_4$. The twin operation provides triangles of oxygen for the boron atoms (trigonal prisms for larger cations). Many compounds are isostructural or structurally related: Fe_2BO_4, $MgFeBO_4$, $NiFeBO_4$, and Y_2BeO_4 with Be in triangles. In Yb_3S_4 (Figure 23), $CaLn_2S_4$ with Ln = Ho, Er, Tm, Yb, or Lu and in $YbLn_2S_4$ with Ln = Ho, Er, or Tm, the larger cation takes the trigonal prism (instead of the triangular) site in the warwickite structure type. Many of these sulfides have the $CaTi_2O_4$ type structure at high temperature.

Larger NaCl-like twin blocks are similarly arranged in the structure of $Ca_3In_2O_6$ shown in Figure 24.

An alternative way of utilizing the trigonal prisms in the composition plane is to insert an Si–O–Si group, thus forming a pyrosilicate group, Si_2O_7, with a trigonal prism outline. Of course, only alternate prisms can be so occupied. We then get (from the warwickite or Yb_3S_4 types) the structure of cuspidine, $Ca_4Si_2O_7(F,OH)_2 = (Si_2O)_{1/2}Ca_2(O,OH,F)_4$.

The three different ways of occupying trigonal prisms are depicted in Figure 25.

Cyclic Twinning of CCP

In Figure 26 a trilling of ccp* is shown; two-thirds consists of tetrahedra, and one-third octahedra. It is obvious that this construction can be used to derive the previously described (Chapter V) structures of Si_3N_4 (and phenacite) and Nb_3Se_4. Similarly, the construction in Figure 27, derived by a "cyclic slip" as shown by the arrows in Figure 26, can be used to derive the structures of fluoborite (Figure 28*a*, *b*; also discussed in Chapter V), painite (Figure 28*b*; very similar to fluoborite, Figure 28*a*), and wightmanite (discussed in ref. 5) (Figure 29), and $BaBi_2S_4$ (19).

---

* Slightly distorted, since for perfect ccp the obtuse angle of each column is 109°28′ rather than 120°.

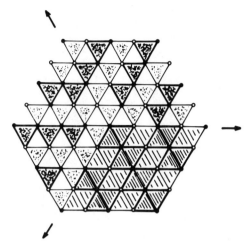

Figure 26. Trilling of ccp. In one trilling unit, only the octahedra are marked; in the other two, only tetrahedra.

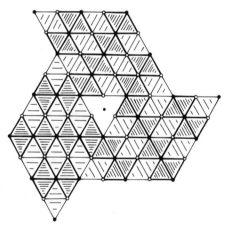

Figure 27. Cyclic translation of ccp, from which the wightmanite and the fluoborite structures can be derived.

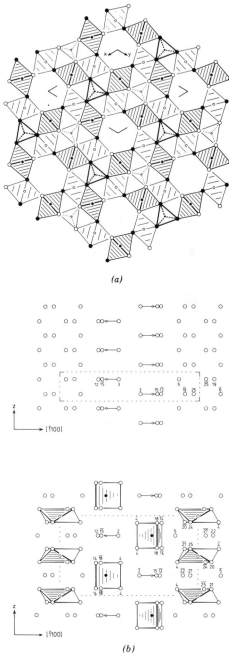

(a)

(b)

Figure 28. (*a*) Fluoborite, $Mg_3(F,OH)_3 \cdot BO_3$, projected on (0001) of its hexagonal unit cell. Large circles are anions; small circles are cations. BO_3 triangles and $Mg(OH,F)_6$ octahedra are shown. All atoms are at $z = \frac{1}{4}$ or $\frac{3}{4}$. (*b*) The filling of the tunnels in fluoborite (above) and in painite, $CaZrAl_9O_{15} \cdot BO_3$ (below); both structures are projected on (1120). BO_3 triangles are seen edge-on; CaO_6 octahedra and ZrO_6 trigonal prisms are also shown. (Heights are in units of $a[11\bar{2}0]/300$.)

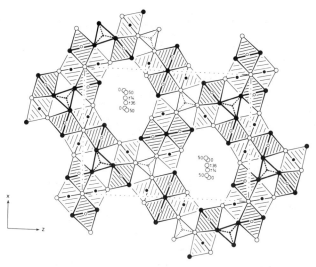

Figure 29. The structure of wightmanite, $Mg_5O(OH)_5 \cdot BO_3 \cdot nH_2O$, projected on (010) and showing BO_3 triangles, $Mg(O,OH)_6$ octahedra, and, in the large tunnels, the O atoms of the water molecules. Except for the last (whose heights are $\times b/100$), atoms are at $y = 0$ (open circles) or $y = \frac{1}{2}$ (filled circles).

Another Type of Glide-Reflection Twinning of ccp

The composition plane now consists of a distorted lamella of the $PbCl_2$ structure (Figure 30). One structure thus derived is the NH_4CdCl_3 type. Isostructural compounds include $RbCdCl_3$, $KCdCl_3$, $KMnBr_3$, $RbPbI_3$, $CsPbI_3$, $CeCrS_3$, and (Figure 31) $PbSnS_3$, with divalent Pb in trigonal prisms. Sn_2S_3, ottemanite, is also isostructural.

One can imagine other members of a family of structures $pAX_2 \cdot qBX$ based on this principle: $PbSnS_3 = (PbS_2)^{2-} \cdot (SnS)^{2+}$, i.e., $p/q = 1$. Another is

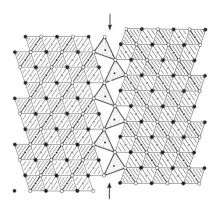

Figure 30. Glide-reflection twinning of B1 (= NaCl).

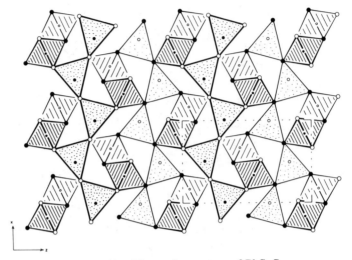

Figure 31. The real structure of PbSnS$_3$.

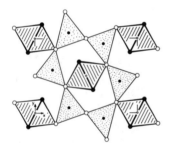

Figure 32. The real structure of La$_2$SnS$_5$.

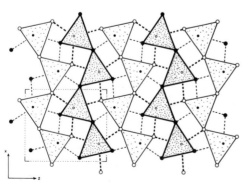

Figure 33. The real structure of PbCl$_2$.

La_2SnS_5 [$2(LaS_2)^- \cdot (SnS)^{2+}$, i.e., $p/q = 2$] shown in Figure 32, and, of course, there is the $PbCl_2$ type itself (which we will discuss later), in which $p/q = \infty$. The last is shown in Figure 33.

References

1. S. J. Geller, *J. Chem. Phys.* **33,** 676 (1960).

2. B. G. Hyde, T. J. White, M. O'Keeffe, and A. W. S. Johnson, *Z. Kristallogr.* **160,** 53 (1982).

3. C.-B. Ma and K. Sahl, *Acta Cryst. B* **31,** 2142 (1975).

4. A. E. Ringwood and A. Major, *Earth Planet. Sci. Lett.* **1,** 241 (1966).

5. B. G. Hyde, S. Andersson, M. Bakker, C. M. Plug, and M. O'Keeffe, *Prog. Solid State Chem.* **12,** 273 (1980).

6. S. Andersson and B. G. Hyde, *J. Solid State Chem.* **9,** 92 (1974).

7. B. Aronsson, M. Backman, and S. Rundqvist, *Acta Chem. Scand.* **14,** 1001 (1960).

8. W. H. Zachariasen, *Acta Cryst.* **1,** 265 (1948).

9. F. Rodi and D. Babel, *Z. Anorg. Allgem. Chem.* **336,** 17 (1965).

10. R. A. Howie, W. Moser, and I. C. Trevena, *Acta Cryst. B* **28,** 2965 (1972).

11. E. F. Bertaut and P. Blum, *Acta Cryst.* **9,** 121 (1956).

12. C. Adolphe, M. Guittard, and P. Laruelle, *C. R. Acad. Sci., Paris* **258,** 4773 (1964).

13. M. Bakker and B. G. Hyde, *Phil. Mag. A* **38,** 615 (1978).

14. H. H. Otto and H. Strunz, *Neues Jahrb. Miner. Abh.* **108,** 1 (1968).

15. E. Makovicky and S. Karup-Møller, *Neues Jahrb. Miner. Abh.* **130,** 264 (1977); **131,** 56 (1977); **131,** 187 (1977).

16. J. Takagi and Y. Takeuchi, *Acta Cryst. B* **28,** 649 (1972).

17. O. Evrard, B. Malaman, F. Jeannot, A. Courtois, H. Alebouyeh, and R. Gerardin, *J. Solid State Chem.* **35,** 112 (1980).

18. V. Venkatakrishnan and M. J. Buerger, *Z. Kristallogr.* **135,** 321 (1972).

19. B. Aurivillius, *Acta Chem. Scand. A* **37,** 399 (1983).

CHAPTER VII

Projection along the Shortest Axis of Mixed (h + c) Stackings of cp Atom Arrays—Mixed hcp + ccp

The following structures are described in this chapter: polytypes of SiC, ZnS, CdI_2; AlN (wurtzite type), and Al_4C_3, and their mixed compounds nAlN·Al_4C_3 such as Al_5C_3N ($n = 1$) and $Al_6C_3N_2$ ($n = 2$), and the related $YbFe_2O_4$; Th_3N_4; the "A-deficient perovskites" $CaTa_2O_6$, $LaNb_3O_9$, and $ThNb_4O_{12}$; $CaFe_2O_4$; rutile, and $IrSe_2$.

This projection corresponds to those used in the two previous chapters: HCP III (Chapter V) and CCP II (Chapter VI). The atomic array again consists of cp planes of atoms stacked one on top of the other, but now some have identical layers above and below—for example, $\cdots aba \cdots, \cdots bcb \cdots$, when the central layer is in an hcp environment and is therefore described as an **h** layer—and some have different layers above and below—for example $\cdots abc \cdots, \cdots bac \cdots$, when the central layer is in a ccp environment and is therefore described as a **c** layer. This is the notation of Jagodzinski.

Polytypism

There are, of course, an infinite number of possible combinations of **h** and **c** layers. Structures of a given compound that vary only in the nature of this combination are known as *polytypes*. Silicon carbide is a classical example based on the two structural types of ZnS—zinc blende (cubic = $\cdots \mathbf{ccc} \cdots$) and wurtzite (hexagonal = $\cdots \mathbf{hhh} \cdots$)—with which we have already dealt.

Silicon carbide (carborundum) is an important technical material, and it has been studied a great deal. It can be made in very many polytypic forms: Parthé (1) listed 77 different types in 1972, and many more have since been reported. For a detailed study of their nomenclature and structures, we refer to the special literature (1, 2) or, for example, Wyckoff. In commercial SiC, the so-called 6H form (Ramsdell notation; a six-layer structure of hexagonal symmetry) is the most abundant. All the polytypes are cubic, rhombohedral, or hexagonal, and the c_{hex} axes of their hexagonal unit cells (all a multiple of ~ 2.52 Å) range from 5.05 to ~ 1495 Å. (All have the same $a_{hex} \approx 3.08$ Å.) We shall describe only the 8H structure, which has the layer sequence $\cdots(abacbabc)\cdots = (\mathbf{chccchcc})\cdots = (\mathbf{hc}^3)^2$. The structure is shown in Figure 1, and the crystallographic data are given in Data Table 1.

Data Table 1   8H-SiC (3)

Hexagonal, space group $P6_3mc$, No. 186; $a = 3.079$, $c = 20.146$ Å; $Z = 8$, $V = 164.43$ Å^3

Atomic Positions

Si: $0,0,0$; $0,0,\frac{1}{2}$; $\frac{2}{3},\frac{1}{3},\frac{1}{8}$; $\frac{2}{3},\frac{1}{3},\frac{3}{8}$; $\frac{2}{3},\frac{1}{3},\frac{6}{8}$; $\frac{1}{3},\frac{2}{3},\frac{2}{8}$; $\frac{1}{3},\frac{2}{3},\frac{5}{8}$; $\frac{1}{3},\frac{2}{3},\frac{7}{8}$

C: $0,0,\frac{3}{32}$; $0,0,\frac{19}{32}$; $\frac{2}{3},\frac{1}{3},\frac{7}{32}$; $\frac{2}{3},\frac{1}{3},\frac{15}{32}$; $\frac{2}{3},\frac{1}{3},\frac{27}{32}$; $\frac{1}{3},\frac{2}{3},\frac{11}{32}$; $\frac{1}{3},\frac{2}{3},\frac{23}{32}$; $\frac{1}{3},\frac{2}{3},\frac{31}{32}$

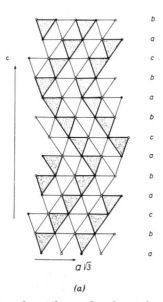

(a)

Figure 1. (*a*) The structure of one form of carborundum, 8H-SiC, projected on (11$\bar{2}$0) of the hexagonal unit cell. All atoms on 0 or $\frac{1}{2}$. For (*b*), a model of 8H-SiC, see color plates, *VII.1b*.

The figure reveals a . . . ,4,4, . . . zigzag sequence of cubic, zinc blende-type slabs joined (by corner-sharing between the tetrahedra) in such a way that a (hexagonal) wurtzite slab is formed at the boundary, its width being two layers. (Note that **h** layers are mirror planes for the atoms that define the tetrahedra—here and generally in all mixed stackings.)

The cause of polytypism is not yet known. Several proposals have been made, such as screw dislocations during the growth process and the presence of impurities, but clearly none of these is universally valid. The more recent ANNNI (axial next-nearest-neighbor interactions) model (4) is also more plausible.

Other isostructural compounds, including ZnS, display the same phenomenon, as well as other compounds of the CdI_2 type, such as CdI_2 itself and PbI_2, with octahedral layers. But we continue with some different types of "tetrahedral compounds."

Al_4C_3 and Yb_2FeO_4

The structure of Al_4C_3 is shown in Figure 2 (see color plates). It is a "tetrahedral structure" with a mixed cp array of C atoms: $\cdots(\mathbf{h}^2\mathbf{c})\cdots$. Between the **h** and **c** layers of C, half the tetrahedra are occupied by Al, so that a double layer of cubic ZnS type results—2AlC. Between adjacent **h** layers of C, *all* the tetrahedra are occupied; this is a {111} layer of antifluorite type—Al_2C. These types of layers alternate, giving $2AlC \cdot Al_2C = Al_4C_3$. Crystallographic data are given in Data Table 2. Across the antifluorite layers the orientation of the tetrahedra (parallel to $\mathbf{c}_{hex}$) is inverted.

Data Table 2 Al_4C_3 (5)

Rhombohedral, space group $R\bar{3}m$, No. 166; $a = 3.330$, $c = 24.89$ Å for the hexagonal unit cell; $Z = 6$, $V_{hex} = 239.0$ Å^3

Atomic Positions
All atoms are said to be in $6(c)$: $(0,0,0; \frac{1}{3},\frac{2}{3},\frac{2}{3}; \frac{2}{3},\frac{1}{3},\frac{1}{3}) \pm(0,0,z)$ except for C(1), $z = 0$, which is therefore in $3(a)$.

	Al(1)	Al(2)	Al(3)	Al(4)	C(2)	C(3)
z	0.705	0.129	0.869	0.296	0.781	0.217

Note that in the pairs Al(1,4), Al(2,3), and C(2,3) the z parameters are almost identical. Each such site is half-occupied. Clearly there is some disorder—probably occasional (0001) faults in the crystal that lead to twinning.

The determinations of the antitype structures of $YbFe_2O_4$ and $(Yb_{0.5}Eu_{0.5})Fe_2O_4$ are not complicated in this way. Their data are given in Data Table 3.

Data Table 3 **YbFe₂O₄ (6) and (Yb₀.₅Eu₀.₅) Fe₂O₄* (7)**

Rhombohedral, space group $R\bar{3}m$, No. 166; a = 3.455 (3.486), c = 25.054 (24.92) Å for the hexagonal unit cell; Z = 6, V_{hex} = 259.0 (262.3) Å³

Atomic Positions: $(0,0,0; \frac{1}{3},\frac{2}{3},\frac{2}{3}; \frac{2}{3},\frac{1}{3},\frac{1}{3})+$

Yb (Yb₀.₅Eu₀.₅) in 3(a):	0,0,0
Fe in 6(c):	$\pm(0,0,z)$; z = 0.2150 (0.2141)
O(1) in 6(c):	z = 0.2925 (0.2914)
O(2) in 6(c):	z = 0.1292 (0.1295)

Al₄C₃ · nAlN

AlN has the wurtzite structure. There is a homologous series of compounds formed by the reaction of AlN with Al₄C₃. They have the general formula Al₄C₃ · n AlN, and their structures (all hexagonal or rhombohedral) have n ranging from zero (Al₄C₃) through 1,2,3,4,5,6, to ∞ (AlN).* Their C layer + N layer stacking sequences are $\mathbf{h}^{2+n}\mathbf{c}$, and the structures are formed by simply inserting n layers of wurtzite-type AlN between the Al₂C antifluorite-type layers and the 2AlC zinc blende–type layers. (The last retain their central **c**-type layer of C atoms, and the single {111} antifluorite layers are bounded by C atoms.)

Details, including crystallographic data, are given by Jeffrey and Wu (5, 10). The structure of one of these intergrowth types, Al₆C₃N₂ = Al₄C₃ · 2AlN, is shown in Figure 3 (see color plates). As in Al₄C₃, the orientations of the tetrahedra are inverted across the antifluorite-type layer.

While many metals have the simpler close-packed structures—for example, Mg (hcp) and Cu (ccp)—and some have both polymorphs (e.g., Ca, Sr), it is well known that others have more complex cp stacking sequences. This is particularly true of some of the lanthanides and actinides, some of which have so-called double hexagonal close-packing of the atoms—the stacking sequence $\cdots(\mathbf{hchc})\cdots = \cdots(abcb)\cdots$ (e.g., La, Nd, Am). Similar and more complex stacking sequences can be observed among binary "interstitial" compounds.

Th₃N₄

In Chapter V (HCP III), we saw how the A-type rare earth sesquioxide structure of La₂O₃ is simply and accurately described as hcp La with all the

---

* There is a related but different series nYbFeO₃ · FeO (8, 9).

octahedral sites in alternate layers occupied by O and all the tetrahedral sites in the intervening layers also occupied by O atoms. A related example with a more complex sequence of cp cation layers is provided by the structure of Th_3N_4, shown in Figure 4. It is a nine-layer structure $\cdots(ababcbcac)\cdots =$ $(ch^2)^3$ of Th, with all the tetrahedral sites in every third layer and all the octahedral sites in the remaining (pairs of) layers occupied by N atoms. So, whereas A-La_2O_3 is an intergrowth of alternating single {111} layers of NaCl and CaF_2 types, Th_3N_4 consists of double layers of NaCl alternating with single layers of CaF_2 types.* The change in cation-layer stacking sequence from A-La_2O_3 to Th_3N_4 is a consequence of anion–anion repulsion. If it were not what it is (in both structures), then some of the anion-centered polyhedra would share faces rather than only edges across the cp cation layers. This avoidance rule leads, in these and similar cases (e.g., $Al_4C_3 \cdot nAlN$), to cp *anion*-layer sequences $\cdots abcabc \cdots$ (although it must be noted that the layer intervals do *not* correspond to close-packing) (11). Structural data for Th_3N_4 are given in Data Table 4. Note that Th_2N_3 is isostructural with A-La_2O_3.

Data Table 4 Th_3N_4 (12)

Trigonal, space group $R\bar{3}m$, No. 166; $a = 3.875$, $c = 27.39$ Å, for the hexagonal unit cell; $c/a = 0.962 \times (\sqrt{8/3} \times 9/2)$; $Z = 3$ (for hexagonal cell), $V = 356.2$ Å^3

Atomic Positions: $(0,0,0; \frac{1}{3},\frac{2}{3},\frac{2}{3}; \frac{2}{3},\frac{1}{3},\frac{1}{3})+$
Th(1) in 3(a):   0,0,0
Th(2) in 6(c): $\pm(0,0,z)$; $z = 0.2221$
N(1) in 6(c): $z = 0.1320$
N(2) in 6(c): $z = 0.3766$
Atomic Distances
N(1)–Th(2) = 2.31 Å (3×), 2.47 Å
N(2)–Th(1) = 2.53 Å (3×)
N(2)–Th(2) = 2.91 Å (3×)

Structures based on the same principle have been observed at high temperature in the systems $Ln_2O_3 + MO_2$ with Ln = La, Pr, Nd, etc., and M = Th, Ce (see Figure 5, color plates), the former having the A-La_2O_3 structure and the latter the fluorite-type structure. Related structures—intergrowths of {111} layers of NaCl-type with {111} layers of ccp cations (without anions)—occur in the subnitrides of Hf (N/Hf < 1). All these are described and discussed elsewhere (11).

---

* In comparison, Al_4C_3 and $Al_4C_3 \cdot nAlN$ (above) are anti-CaF_2 intergrown with cubic ZnS instead of cubic NaCl.

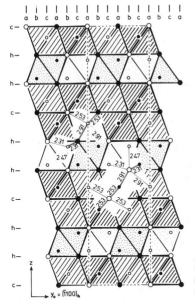

Figure 4. The structure of the Th_3N_4 projected on $(11\bar{2}0)$. All atoms are at 0 or $\frac{1}{2}$; N-centered Th_4 tetrahedra and Th_6 octahedra.

Structures Described by Parallel Reflection Operations

Using a different projection equivalent to that of HCP II in Chapter IV, it is simple to describe some structures related to the "orthorhombic perovskites" described in that chapter. These have a lower proportion of A (large) cations: $A(BO_3)_2$ and $A(BO_3)_3$ instead of ABO_3.

Figure 6 shows the structure of the orthorhombic (high-temperature, $T > 700°C$) form of $CaTa_2O_6$ as it is usually depicted. The Ca coordination is a slightly deformed O_6 trigonal prism, and the Ta atoms are in rather regular,

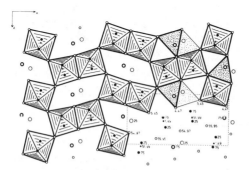

Figure 6. The orthorhombic $CaTa_2O_6$ structure projected on (010). Large circles are Ca; medium circles are O; small circles are Nb. Atom heights in units of $b/100$. NbO_6 octahedra and some CaO_6 trigonal prisms are shown.

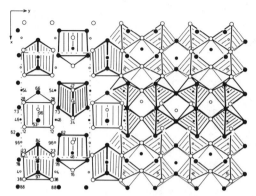

Figure 7. The orthorhombic $CaTa_2O_6$ structure projected on (001). Atoms as in Figure 6; heights in units of $c/100$. On the right, the corner-connected array of TaO_6 octahedra is shown; it is rather close to twinned **hc** anions. On the left, the CaO_6 trigonal prisms are depicted.

but tilted, O_6 octahedra. Figure 7 shows the same structure, but now in the projection referred to above.* It is very revealing; clearly it bears considerable similarity to that of $LuFeO_3$ (Figure 29 of Chapter IV). It is even closer to that of $PrFeO_3$ (cf. Figure 33*b* of Chapter XI). The array of edge- and corner-connected TaO_6 octahedra on the right of the figure reveals a (reflection) twinned (approx.) cp anion array with twin composition planes on (010) and (020), but the orientations of successive octahedra in the direction [100] shows that it is twinned **hc** (rather than the twinned hcp = **h** of $LuFeO_3$). The A sites, occupied by Ca, are like those in $LuFeO_3$—bicapped trigonal prisms of O atoms, CN(Ca) = 8—but only alternate ones are occupied. It is an *ordered* perovskite derivative, *not* a "nonstoichiometric perovskite" with A-site vacancies. (Along **c**, the "empty A-type sites" share *faces* with the occupied ones in one direction, and *edges* in the other. The former is due to the twinned **c** layer, the latter to the twinned **h** layer; cf. Figure 23 of Chapter IV and Figure 25 of Chapter IX.)

Isostructural compounds include $LaNbTiO_6$ and (slightly distorted to monoclinic symmetry) $SrNb_2O_6$. Crystallographic data for $CaTa_2O_6$ are given in Data Table 5.

---

* Incidentally, this is the projection down the *shortest* axis of the unit cell. Such a projection (down the shortest axis) is the almost universal rule for "seeing" a structure. Our appreciation of the structure of Fe_3C and $LuFeO_3$ (Chapter IV) and of $Ca_2Ta_2O_6$ would have emerged readily and rapidly if this rule had been followed!

Data Table 5   Orthorhombic CaTa₂O₆ (13)

Orthorhombic, space group *Pnma*, No. 62; $a = 11.068$, $b = 7.505$, $c = 5.378$ Å; $Z = 4$, $V = 446.72$ Å³

Atomic Positions

Ca in 4(c): $\pm(x,\frac{1}{4},z; \frac{1}{2} + x,\frac{1}{4},\frac{1}{2} - z)$; $x = 0.042$, $z = 0.540$

Ta in 8(d): $\pm(x,y,z; \frac{1}{2} + x,\frac{1}{2} - y,\frac{1}{2} - z; x,\frac{1}{2} - y,z; \frac{1}{2} + x,y,\frac{1}{2} - z)$; $x = 0.1412$, $y = 0.9944$, $z = 0.0376$

O(1) in 8(d): $x = 0.976$, $y = 0.035$, $z = 0.225$

O(2) in 8(d): $x = 0.213$, $y = 0.049$, $z = 0.383$

O(3) in 4(c): $x = 0.146$, $z = 0.967$

O(4) in 4(c): $x = 0.878$, $z = 0.838$

Atomic Distances

Ta–O = 1.85–2.11 Å (6×)

Ca–O = 2.43–2.57 Å (8×)

Figure 8 shows the structure of LaNb₃O₉, another "A-deficient perovskite derivative," this time A₁/₃BO₃. It is a logical sequel to LuFeO₃ and Ca₁/₂TaO₃: twinned **h²c** with one-third of the trigonal prisms in the composition planes occupied in a strictly ordered fashion (cf. Figure 25 of Chapter IX.) Crystallographic data are given in Data Table 6.

Note that a previous single-crystal X-ray analysis (15) gave an orthorhombic cell and a disordered structure: three oxygen positions, one Nb position, and one La position *two-thirds occupied,* clearly an average structure (from a multiply twinned crystal?). This is typical of many earlier determinations

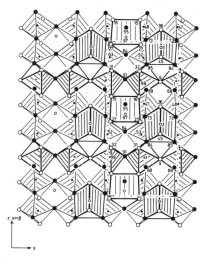

Figure 8. The monoclinic structure of LaNb₃O₉ projected along [100]. Atomic heights are in units of a/100 above (100). Compare Figure 7 (and Figure 29 of Chapter IV).

Data Table 6   LaNb₃O₉ (14)

Monoclinic, space group $P2_1/c$, No. 14; $a = 5.437$, $b = 7.664$, $c = 16.323$ Å, $\beta = 92.198°$; $Z = 4$, $V = 679.7$ Å³

Atomic Positions
All atoms in 4(e): $\pm(x,y,z; x,\frac{1}{2} - y,\frac{1}{2} + z)$, with

	x	y	z
La(1)	0.03146	0.75014	0.96824
Nb(1)	0.99228	0.03157	0.25136
Nb(2)	0.52100	0.99832	0.39937
Nb(3)	0.52735	0.48332	0.39963
O(1)	0.3166	0.0507	0.3105
O(2)	0.8462	0.9519	0.3555
O(3)	0.9091	0.2550	0.2575
O(4)	0.6871	0.9545	0.1885
O(5)	0.1582	0.0555	0.1447
O(6)	0.7312	0.9674	0.0164
O(7)	0.2689	0.0339	0.4834
O(8)	0.6391	0.2492	0.4176
O(9)	0.5524	0.2487	0.5950

of such "A-deficient perovskite derivatives." The crystal used to obtain the data in Data Table 6 was grown (clearly in a well-ordered state) by vapor-phase transport, not just by annealing. One's impression is simply that ordering in these perovskites is more difficult the smaller the stoichiometric ratio A/B. No ordered specimen of $ThNb_4O_{12}$ (= $Th_{1/4}NbO_3$) has yet been reported (16a, 16b).

The $CaFe_2O_4$ structure type (described in chapter VI) can also be described in terms of a twinned hc array of anions, as can be seen from Figure 9.

The structure of $IrSe_2$ provides, among other things, a rather simple example of parallel reflection (= rotation) twinning, which results in a mixed stacking of anion layers that are approximately close-packed. Figure 10 depicts the structure of rutile projected on (001) (as in Chapter V, Figure 3). In Figure 11 it has been idealized in the usual way to hcp anions, and this figure also shows the result of regularly twinning it on every fourth anion layer. (The twinning operation is a rotation of 180° about the vertical axis normal to these planes. Alternatively, it may be described as reflection in the plane containing this long axis and also the projection axis, which is vertical and normal to the paper.) This changes the idealized anion-layer stacking from **h** to **h³c**.

The actual structure of $IrSe_2$, shown in Figure 12, differs from the ideal (Figure 11) because of Se $\cdots$ Se bonding, very similar to the S $\cdots$ S bonding in

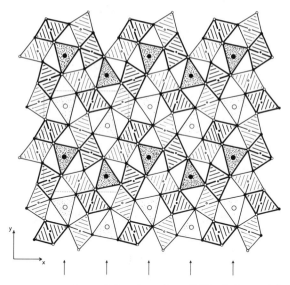

Figure 9. The structure of $CaFe_2O_4$ projected on (001). Anion octahedra are occupied by Fe, and trigonal prisms by Ca. The anion array is close to **hc**, twinned at the glide planes indicated by the arrows. Atoms at 0 or $\frac{1}{2}$.

marcasite, FeS_2, also discussed in Chapter V (Figure 4a). However, in $IrSe_2$, only one half of the Se atoms are bonded in this way. Its formula would appear to be best written as $Ir_2(Se_2)Se_2$, suggesting Ir_2^{3+} $(Se_2)^{2-}Se_2^{2-}$, or trivalent Ir. (This is consistent with its diamagnetic, semiconductor properties.) (Compare divalent Fe in marcasite.)

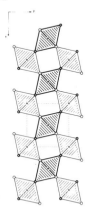

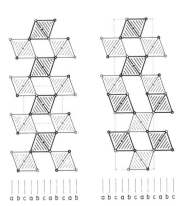

Figure 10. The (tetragonal) rutile structure projected on (001).

Figure 11. On the left, rutile projected on (001), but idealized to hcp anions (cf. Figure 10). On the right this idealized structure has been regularly twinned so that the anion array changes from **h** to **h³c**. The result is the idealized structure of $IrSe_2$ (cf. Figure 12).

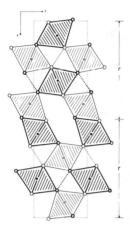

Figure 12. The structure of IrSe₂ projected on (001). Se–Se bonds are indicated by dotted lines.

Crystallographic data for IrSe₂ are given in Data Table 7.

Data Table 7   IrSe₂ (17)

Orthorhombic, space group *Pnam*,* No. 62; $a = 20.94$, $b = 5.93$, $c = 3.74$ Å; $Z = 8$, $V = 464.4$ Å³

Atomic Positions
All atoms in 4(c): $\pm(x, y, \frac{1}{4}; x + \frac{1}{2}, \frac{1}{2} - y, \frac{1}{4})$, with

	x	y
Ir(1)	0.924	0.573
Ir(2)	0.696	0.558
Se(1)	0.007	0.270
Se(2)	0.879	0.955
Se(3)	0.237	0.324
Se(4)	0.637	0.926

* Nonstandard setting (**acb**) of *Pnma* (**abc**).

References

1. E. Parthé, *Crystallochimie des Structures Tetraedriques.* Gordon & Breach, Paris, 1972.

2. A. R. Verma and P. Krishna, *Polymorphism and Polytypism in Crystals.* Wiley, New York: 1966.

3. L. S. Ramsdell and J. A. Kohn, *Acta Cryst.* **5**, 215 (1952).

4. J. Smith, J. Yoemans, and V. Heine, in *Modulated Structure Materials,* T. Tsakalakos, Ed. Martinus Nijhoff, Dordrecht, 1984, p. 95.

5. G. A. Jeffrey and V. Y. Wu, *Acta Cryst.* **20**, 538 (1966).

6. K. Kato, I. Kawada, N. Kimizuka, and T. Katsura, *Z. Kristallogr.* **141**, 314 (1975).

7. B. Malaman, O. Evrard, N. Tannières, J. Aubry, A. Courtois, and J. Protas, *Acta Cryst. B* **31**, 1310 (1975).

8. K. Kato, I. Kawada, N. Kimizuka, I. Shindo, and T. Katsura, *Z. Kristallogr.* **143**, 278 (1976).

9. N. Kimizuka, K. Kato, I. Shindo, I. Kawada, and T. Katsura, *Acta Cryst. B* **32**, 1620 (1976).

10. G. A. Jeffrey and V. Y. Wu, *Acta Cryst.* **16**, 559 (1963).

11. M. O'Keeffe and B. G. Hyde, *Struct. Bonding* **61**, 77 (1985).

12. A. L. Bowman and G. P. Arnold, *Acta Cryst. B* **27**, 243 (1971).

13. L. Jahnberg, *Acta Chem. Scand.* **17**, 2548 (1963).

14. J. Sturm, R. Gruehn, and R. Allmann, *Naturwiss.* **6**, 296 (1975).

15. P. N. Iyer and A. J. Smith, *Acta Cryst.* **23**, 740 (1967).

16a. M. A. Alario-Franco, I. E. Grey, J. C. Joubert, H. Vincent, and M. Labeau, *Acta Cryst. A* **38**, 177 (1982).

16b. M. Labeau, I. E. Grey, J. C. Joubert, H. Vincent, and M. A. Alario-Franco, *Acta Cryst. A* **38**, 753 (1982).

17. L. B. Barricelli, *Acta Cryst.* **11**, 75 (1958).

CHAPTER VIII

Primitive Cubic Arrays and Derived Structures

The following structures are considered in this chapter: α-Po, β-Po, Hg, cubic close-packing, As, Sb, Bi, Se, Te, bcc packing, CsCl (= β-brass), CaB_6, $NaZn_{13}$, Ti_2Ga_3, CaF_2 (fluorite), β-LnOF, γ-LaOF, $La_4Re_2O_{10}$, α-FeSi$_2$, β-FeSi$_2$, $PtHg_2$, $NiHg_4$, Ni_2Al_3, A-La$_2$O$_3$ (and La_2O_2S and Mg_3Sb_2), the τ structure of $(Cu,Ni)Al_x$, PtS, CuO, bixbyite and "anion-deficient fluorites" including stabilized zirconia, "anion-excess fluorites" $Ca_{14}Y_5F_{43}$ and KY_3F_{10}, $Cr_{23}C_6$, $BaHg_{11}$, UO_2F_2, α-UO$_3$, Li_2Si, $MoSi_2$, Th_2N_2Bi, Mo_3Al_8, Os_2Al_3, $MoNi_4$, $IrAl_3$, LaF_3, RuSi, Ru_4Si_3, Ru_5Si_3, Ru_2Si, $PbCl_2$.

There are relatively few structure types in which this sort of array occurs in a perfectly regular form, although often there are many compounds with a given structure type and many closely related structure types in which the cubes are deformed.

The paucity of structure types, as compared with, say, closest-packed anion arrangements, is in a sense inevitable. A primitive cubic (PC) array lacks variety; there is only one type of volume interstice, the cube, although we will add the square (cube face) and could, in principle, add the line (cube edge). Furthermore, we may recall that Pauling, in his paper "The principles determining the structure of complex ionic crystals," specifically excluded coordination polyhedra of large cations, CN > 6, from the field of application of his suggested principles (1) "on account of the ease with which these polyhedra are distorted"—which is consistent with our statement in the previous paragraph.

Primitive Cubic Arrays

α-Po

Polonium, in its α form (Figure 1), is the only element with the simple cubic structure, although some alloys such as Au + Te are said to retain a statistically primitive cubic structure if quenched rapidly from high temperature. Data are given in Data Table 1.

Data Table 1 α-Po (2)

Cubic, space group $Pm3m$, No. 221; $a = 3.345$ Å; $Z = 1$, $V = 37.43$ Å^3

Atomic Positions
Po in $1(a)$:   0,0,0

If the cube is regarded as a rhombohedron with angle $\alpha = 90°$, then the structures of many elements are seen to be related to that of α-Po. In β-Po the rhombohedron has been compressed along its body diagonal [111] until $\alpha = 98°$; in Hg it has been extended along [111] until $\alpha = 70°32'$. Greater extension, until $\alpha = 60°$, yields the fcc/ccp structure characteristic of many metals. Thus, the atomic arrays in the {111} planes of all these structures are 3^6 plane nets, as in close-packing, and they are stacked in the ccp sequence $\cdots abcabc\cdots$, but with the interplanar spacing much reduced (by $\sqrt{2}$ in α-Po).

The structures of As, Sb, Bi, Se, and Te (3) can be regarded as fairly minor distortions of that of α-Po. Some of the bonds are covalent, and some are of the van der Waals type and therefore longer. Infinite chains or layers of atoms can be recognized in these structures.

The common bcc structure type of many metals (the tungsten or A2 type) is, of course, derived from that of α-Po by centering all the cubes. (A body-centered cubic (bcc) packing is two interpenetrating primitive cubic arrays. Conversely, pc is two interpenetrating fcc arrays, etc.)

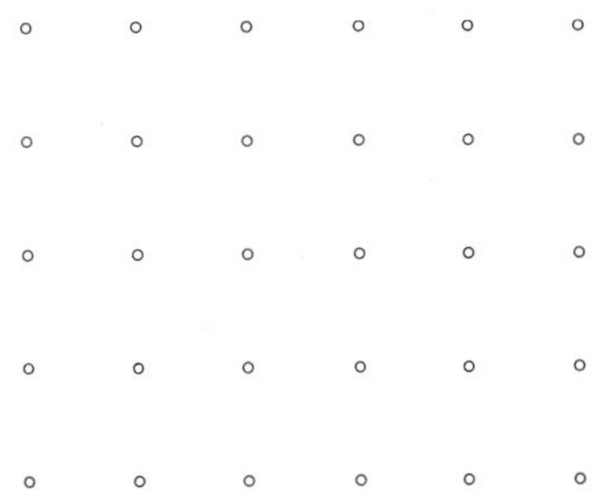

Figure 1. The primitive cubic structure of α-polonium projected on {100}.

CsCl

Data Table 2   CsCl

Cubic, space group $Pm3m$, No. 221; $a = 4.123$ Å; $Z = 1$, $V = 70.09$ Å^3

Atomic Positions
Cl in 1(a):   0,0,0
Cs in 1(b): $\frac{1}{2},\frac{1}{2},\frac{1}{2}$

Both the anions and the cations form primitive cubic arrays, with origins displaced by $[\frac{1}{2},\frac{1}{2},\frac{1}{2}]$ (see Figure 2 and Data Table 2). Many compounds crystallize with this structure [Schubert (4) gives a list of more than 150], such as the halides (except fluorides) of the larger univalent cations, including NH_4^+ and Tl^+, and many binary alloys of which β-brass, CuZn, is the prototype. CaB_6 can be regarded as CsCl type with [B_6] octahedra in place of the anion (and connected by B–B bonds in all three ⟨100⟩ directions). Similarly, in $NaZn_{13}$ the anions are replaced by centered icosahedra of zinc, [Zn_{13}]. However, both of these are very formal descriptions.

Under high pressure, many compounds with the NaCl structure transform to the CsCl type. This is the same transformation as that referred to in discussing α-Po deformations: the 60° rhombohedral primitive cell of NaCl is compressed along [111] to a 90° rhombohedron. Increasing the pressure increases the coordination number (CN) from 6 to 8 (5, 6).

The structure of Ti_2Ga_3 (Figure 3) may at first sight be regarded as a simple substitutional variant of CsCl, though tetragonally distorted. In a pc array of Ga, every fifth [001] row of interstices along [100] (and [010]) is occupied by Ga instead of Ti; Cs_5Cl_5 becomes $(Ti_4Ga)Ga_5 = 2Ti_2Ga_3$. The tetragonal distortion changes $c'/a' = c/(a/\sqrt{5})$ from 1.000 to 1.427. (Primed indices refer to the CsCl subcell; unprimed, to the tetragonal unit cell of

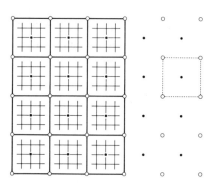

Figure 2. CsCl projected on {100}. Large open circles are anions at $z/c = 0$; small filled circles are cations at $z/c = \frac{1}{2}$ (or vice versa!). CsCl$_8$ cubes share all faces.

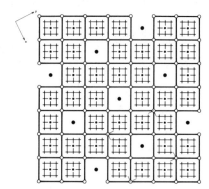

Figure 3. The structure of Ti_2Ga_3 projected on (001). Smaller circles are Ti, larger are Ga; open at $z/c = 0$, filled at $z/c = \frac{1}{2}$.

Ti$_2$Ga$_3$.) An intergrowth of columns of bcc Ga in a CsCl-like matrix of TiGa (an obvious sort of possibility that is quite common in distorted variants, e.g., U$_3$Si). However, the tetragonal distortion is so great that the "bcc" array is almost perfect ccp, for which the ideal $c'/a' = \sqrt{2}$ for the body-centered tetragonal (bct) cell (cf. Figure 5 of Chapter VI). This is the *Bain relation* between, for example, fcc and bcc Fe that is so important in the field of martensitic transformations in steels. Clearly, in the sort of projection used in Figure 3, care is needed to distinguish a mixed bcc array from a mixed ccp array. One needs the axial ratio of the bct unit cell: 1 for bcc, $\sqrt{2}$ for ccp. There are many such cases in which the projection looks like a pc derivative but is actually a ccp array of two atomic types [cf. Schubert (4), §2.36].

Fluorite

This is a very common structure type for compounds with anions (mainly) of elements in the first row of the periodic table and for metal alloys. Most M$_2$O oxides and other chalcogenides of the alkali metals take the antitype structure. Conventionally, this MX$_2$ structure is regarded as consisting of [MX$_8$] cubes sharing each edge with an adjacent cube (Figure 4). It can also be described as [XM$_4$] tetrahedra sharing each edge with an adjacent tetrahedron. The M array is perfect ccp; that is, the CaF$_2$ structure is ccp Ca with all tetrahedra occupied by F. [It is a rather striking fact that CaF$_2$ has a slightly *lower* molar volume than Ca (by 6.6%); the unit-cell volumes are 163.04 and 174.5 Å^3, respectively (i.e., they are very nearly equal); and the Ca array is identical in the two structures.] Crystallographic data for CaF$_2$ are given in Data Table 3.

Data Table 3   Fluorite, CaF$_2$

Cubic, space group $Fm3m$, No. 225; $a = 5.46295$ (25°C); $Z = 4$, $V = 163.035$ Å^3

Atomic Positions
Ca in 4(a):   0,0,0; fc
F in 8(c): $\pm(\frac{1}{4},\frac{1}{4},\frac{1}{4})$; fc

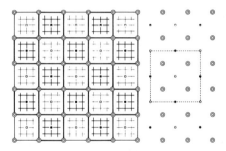

Figure 4. The fluorite, CaF$_2$, structure projected on {100}. Small circles are cations at $z/c = 0$ (open circles) and $z/c = \frac{1}{2}$ (filled circles). Double circles are anions at $z/c = \pm\frac{1}{4}$. CaF$_8$ cubes share all edges. Note that the Ca array is ccp (fcc unit cell of Ca).

All substances with the fluorite or antifluorite structure show a λ anomaly in their heat capacities below the melting point (7), at $T \approx 0.87 \, T_m$. This is the result of a diffuse transition associated with disordering of the anion array in fluorite (8) or the cation array in antifluorites. It is accompanied by a large increase in entropy ($\approx \Delta S$ for melting the anions) and in anion conductivity. They appear to belong to a class of compounds that are excellent solid electrolytes, in which only one of the ion arrays (in the case of CaF_2, the cations) is crystalline, the other being molten above the transition. (The classical examples of such electrolytes are the silver and cuprous sulfides and iodides. Many others are being explored, with the present emphasis on alternative methods for producing energy.)

There are some distorted variants of this structure type. EuOF (the β-LnOF type, Ln = lanthanoid) has a rhombohedral unit cell with $\alpha = 33°3'$ compared with $\alpha = 33°13'$ for the equivalent (primitive) cell in CaF_2. The [111] axis is very slightly elongated as a result of O and F segregating to alternate (111) anion planes. In γ-LaOF a similar segregation is to alternate (100) planes, resulting in a small tetragonal distortion; $c/(\sqrt{2}a) = 1.011$ instead of 1.000. A similar tetragonal distortion obtains in Pt_2Si (anti-ZrH_2 type), where $c/(\sqrt{2}a) = 1.066$.

The structure of $La_4Re_2O_{10}$ is an interesting derivative: it is a fluorite-type oxide in which four-fifths of the cation sites are (cubes) occupied by La and the remainder are occupied by Re_2 pairs. This again gives a tetragonal distortion, $c/a = 1.064$ for the equivalent fluorite cell.

α-FeSi₂

This represents an alternative way of filling half the interstices in a simple cubic array (Figure 5). The [FeSi₈] "cubes" are very nearly regular (within 0.5%); it is the empty cubes that are squashed along **c** (due to Si–Si bonds). According to ref. 11, α-FeSi₂ is slightly iron-deficient.

Data for iron disilicide, α-FeSi₂, are given in Data Table 4. The Si array would be exactly primitive cubic if z were 0.250 and $c/a = 2.000$ instead of 1.911.

Data Table 4   Iron Disilicide, α-FeSi₂ (9, 10)

Tetragonal, space group $P4/mmm$, No. 123; $a = 2.684$, $c = 5.128$ Å, $c/a = 1.911$; $Z = 1$, $V = 36.94$ Å³

Atomic Positions
Fe in 1(a):   0,0,0
Si in 2(h): $\pm(\frac{1}{2},\frac{1}{2},z)$; $z = 0.270$

There is a low-temperature form β-FeSi₂ in which the [FeSi₈] "cubes" are joined by edge-sharing, as in the fluorite type, rather than face-sharing, as in

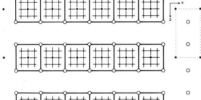

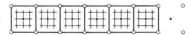

Figure 5. α-FeSi$_2$ projected on (010). Small filled circles are Fe at $y/b = 0$; large open circles are Si at $y/b = \frac{1}{2}$.

α-FeSi$_2$. The cubes are very severely distorted, although the fluorite-type structure is still recognizable (12).

PtHg$_2$

There may be some doubt about this structure (the ξ phase of the Pt + Hg system), particularly the extent of ordering of Pt and Hg; but it is an interesting further possibility for ordered filling of half the interstices in the pc array (Figure 6). Here the "cubes" share faces to form columns, which are then joined by edge-sharing. (The "cubes" are, in fact, $3.31_5 \times 3.31_5 \times 2.91_3$ Å^3— a tetragonal compression of ~12%.) Structure data are given in Data Table 5.

Data Table 5 PtHg$_2$ (13)

Tetragonal, space group $P4/mmm$, No. 123; $a = 4.68$, $c = 2.91$ Å, $c/a = 0.621$; $Z = 1$, $V = 63.99$ Å^3

Atomic Positions
Pt in 1(a): 0,0,0
Hg in 2(e): $0,\frac{1}{2},\frac{1}{2}; \frac{1}{2},0,\frac{1}{2}$

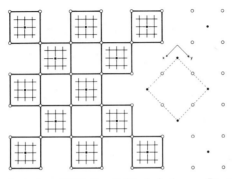

Figure 6. PtHg$_2$ projected on (001). Small filled circles are Pt at $z/c = 0$; large open circles are Hg at $z/c = \frac{1}{2}$.

NiHg₄

Several compounds have this structure, including $MnHg_4$, $Cr_{1+}Ga_{4-}$, and $PtHg_4$, for which data are given in Data Table 6. This is the β phase in the system Pt + Hg. A primitive cubic array of Hg is occupied by a bcc array of Pt. $[PtHg_8]$ cubes share each vertex with a similar cube (Figure 7).

Data Table 6　PtHg₄ (13)

Cubic, space group $Im3m$ (not $I432$), No. 229; $a = 6.18_7$ Å; $Z = 2$, $V = 236.8$ Å³

Atomic Positions
Pt in 2(a):　　$0,0,0; \frac{1}{2},\frac{1}{2},\frac{1}{2}$
Hg in 8(c):　　$\pm(\frac{1}{4},\frac{1}{4},\frac{1}{4}; \frac{1}{4},\frac{3}{4},\frac{3}{4}; \frac{3}{4},\frac{1}{4},\frac{3}{4}; \frac{3}{4},\frac{3}{4},\frac{1}{4})$

Ni₂Al₃, La₂O₃, and Some Related Structures

In this structure (Data Table 7) two-thirds of the interstices in a pc array of aluminum atoms are occupied by nickel atoms. Each $[NiAl_8]$ cube shares three faces with adjacent cubes, so that layers parallel to the cube faces are of the (ideal) form shown in Figure 8a. Similar layers are stacked with appropriate offset to give a pseudocubic unit cell of $3 \times 3 \times 3$ CsCl subcells. *Empty* cubes share only edges and corners, not faces. Crystallographic data are given in Data Table 7.

Data Table 7　Ni₂Al₃

Hexagonal (trigonal), space group $P\overline{3}m1$, No. 164; $a = 4.036$, $c = 4.901$ Å, $c/a = 1.214$; $Z = 1$, $V = 69.14$ Å³

Atomic Positions
Ni in 2(d):　　$\pm(\frac{1}{3},\frac{2}{3},z)$; $z = -0.149$
Al in 1(a):　　$0,0,0$
Al in 2(d):　　$\pm(\frac{1}{3},\frac{2}{3},z)$; $z = 0.352$

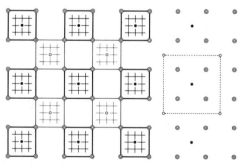

Figure 7. PtHg₄ projected on {100}. Small circles are Pt at height 0 (open) or $\frac{1}{2}$ (filled). Double circles are Hg at $\pm\frac{1}{4}$.

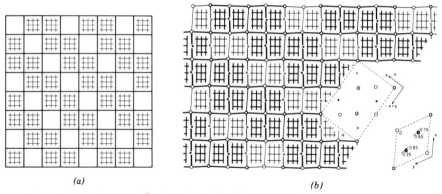

Figure 8. (*a*) Idealized (1$\bar{1}$01) *layer* of Ni_2Al_3. $NiAl_8$ cubes share faces and edges within the layer. (*b*) The real Ni_2Al_3 structure (complete) projected on (11$\bar{2}$0). Larger circles are Al; smaller circles are Ni open at height 0, filled at height $\frac{1}{2}$ along ($a/3$) × [11$\bar{2}$0]. Lower right shows the unit cell and contents projected on (0001), with heights in units of **c**.

The structure is hexagonal and is shared by Ni_2Ga_3, Ni_2In_3, Pd_2Al_3, Pd_2In_3, Pt_2Ga_3, Pt_2In_3, and In_2Au_3. The [$NiAl_8$] polyhedra would be perfect cubes if the parameters were $c/a = \sqrt{3/2} = 1.225$, $z(Ni) = -1/6 = -0.167$, and $z(Al) = 1/3 = 0.333$. In fact, $c/a = 1.214$ is slightly reduced, and the z parameters are slightly high, but the differences are quite small. The real structure is displayed in Figure 8*b*, projected onto (11$\bar{2}$0) of the hexagonal cell. (Heights are in fractions of $a[11\bar{2}0]/3 = $ the orthohexagonal **a** axis.) Note that it is composed of 3^6 layers of cations or of anions parallel to the basal plane (0001). The sequence is $a \quad \gamma \quad b \quad \alpha \quad c \qquad a \quad \gamma \quad b \quad \alpha \quad c$.

This sort of distortion of a hexagonal cell is of general interest: for ideal hcp, $c/a = 2\sqrt{2/3} = 1.6330$; for ccp, the equivalent hexagonal cell has $c/a = 3\sqrt{2/3}) = 2.4495$. Cubic close-packing is transformed to pc by reducing c/a by a factor of exactly 2, to $c/a = (3/2)\sqrt{2/3} = \sqrt{3/2} = 1.2247$. A further compression by a factor of 2, to $c/a = \sqrt{3/8} = 0.6124$, converts pc to bcc. (But recall that ccp is also transformed to bcc by the much smaller compression of $\sqrt{2}$ in the **c** direction of the fcc or bct unit cells.)

The (approx.) pc anion array in the A-type rare earth sesquioxide structure is distorted in the opposite sense: the space group and atomic positions are the same, but, for La_2O_3, $c/a = 1.557$ (so that the cation array is not far from ideal hcp; cf. the sequence $a \quad \gamma \quad b \quad \alpha \quad c$). This elongation is largely responsible for the CN of La being reduced from 8 to 7. (Note how the La atoms are slightly off-center in their anion "cubes"; see Figure 9.) This last structure particularly may be described in several approximate ways: as [LaO_8] "cubes," as monocapped octahedra [LaO_7], as distorted [LaO_6] trigonal prisms (see Figure 9), or as [OLa_4] tetrahedra plus [OLa_6] octahedra (see Figure 7 of Chapter V).

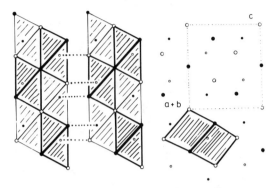

Figure 9. The A-type (La_2O_3) structure projected on ($11\bar{2}0$). Smaller circles are La; larger circles are O. Heights are 0 or $\frac{1}{2}$. LaO_{6+1} monocapped octahedra, and LaO_8 "cubes" are drawn. Note that the cation packing is hcp.

Many other compounds crystallize in this form: lanthanoid and actinoid sesquioxides, oxysulfides and oxyselenides (M_2O_2X), Th_2N_3, $CeLi_2P_2$, and, in the antiform, $Na(LiZn)O_2$, Li_2CeN_2, Mg_3Bi_2, and Mg_3Sb_2. In the last, the extension is approximately the same as in La_2O_3: $c/a = 1.582$. Not surprisingly, the M_2O_2X compounds have higher c/a ratios: about 1.71–1.75 (M–X bonds being longer than M–O bonds).

There is an interesting group of alloys $(Cu,Ni)Al_x$ with structures closely related to that of Ni_2Al_3. They all have hexagonal unit cells (some of rhombohedral symmetry) based on ordered sequences of primitive cubic subcells (CsCl type) and empty cubes—very slightly distorted in a rhombohedral fashion, with $2.894 \leq a_{subcell} \leq 2.908$ Å, $90.3472° \geq \alpha_{subcell} \geq 90.1011°$—stacked along the triad axes. In Ni_2Al_3 (the δ structure of this system), all the empty cubes lie on one of the three triads $0,0,z$; $\frac{1}{3},\frac{2}{3},z$ and $\frac{2}{3},\frac{1}{3},z$. Together they comprise the "τ structure," with Al : Cu : Ni lying in the domain 61 : 27 : 12 to 55 : 38 : 7, i.e., $(Cu,Ni)Al_x$ with $1.564 \geq x \geq 1.222$. Eight different ordered structures have been characterized: τ_5, τ_{11}, τ_6, τ_{13}, τ_7, τ_{15}, τ_8, and τ_{17}, with $a = a_0$ in all cases, and $c = 5, 11, 6, \ldots$, etc. times c_0, respectively. (The lower numbers are halved because these cells have the higher, rhombohedral symmetry.)

The structures appear to be determined by the free electron concentration. They are a very early, very thoroughly documented example of a "continuous sequence of ordered structures" or a so-called infinitely adaptive structure (14). The authors concluded (15) that the structures occupy a single-phase field in the Cu + Ni + Al phase diagram but warned of the danger inherent in defining a phase in terms of a unit cell of structure.

An additional paper, giving details of these interesting structures and their determination, was promised (15) but appears not to have been published. We have not been able to trace it.

Cooperite, PtS

In this structure (Figure 10 and Data Table 8), the sulfide anions are close to a simple cubic arrangement. The cations lie in cube faces, displaced from the Cs sites in CsCl by $\frac{1}{2}$[100] or $\frac{1}{2}$[010], which gives them their normal (almost) square-planar coordination. No cube faces parallel to (001) are occupied. The c axis is approximately doubled, compared with CsCl; $a = b = 1.138 \times (c/2)$.

Data Table 8   Cooperite, PtS (4)

Tetragonal, space group $P4_2/mmc$, No. 131; $a = 3.48$, $c = 6.11$ Å, $c/a = 1.756$; $Z = 2$, $V = 74.0$ Å^3

Atomic Positions
Pt in 2(c): $0,\frac{1}{2},0; \frac{1}{2},0,\frac{1}{2}$
S in 2(f): $\frac{1}{2},\frac{1}{2},\frac{1}{4}; \frac{1}{2},\frac{1}{2},\frac{3}{4}$

The [PtS$_4$] squares are alternately parallel to and normal to the plane of the drawing, forming mutually perpendicular PtS$_2$ strips connected through common S atoms. The S atoms are in [Pt$_4$] tetrahedra, and it is clear that the geometry is a compromise between regular tetrahedral coordination of S (in a ccp array of Pt) and regular square-planar coordination of Pt.

PtO and PdO have the same structure, which can also be described as tetragonally distorted ccp cations with half the tetrahedral holes occupied by anions. The c/a ratio, which is $\sqrt{2} = 1.414$ for perfect ccp and 2.000 for perfect MX$_4$ squares, is 1.754 for PdO, 1.75$_7$ for PtO, and 1.75$_6$ for PtS—about halfway between the two ideals. Figure 11a shows PdO in the same projection; Figure 11b shows it in a different projection (along $\mathbf{c}$). Both are in terms of [OPd$_4$], anion-centered tetrahedra, rather than [PdO$_4$] "squares."

In tenorite (CuO), there is an additional monoclinic distortion: $a = 4.653$ Å ($\equiv \sqrt{2}a_{PdS}$); $b = 3.410$ Å ($\equiv \sqrt{2}a_{PdS}$); $c = 5.108$ Å ($\equiv c_{PdS}$); $\beta = 99°29'$ (90° in PdS).

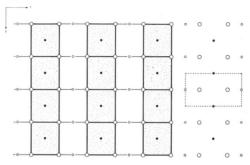

Figure 10.   The structure of cooperite, PtS, projected on (100). Small circles are Pt at $x/a = 0$ (open) or $\frac{1}{2}$ (filled). Large circles are S at $x/a = \frac{1}{2}$.

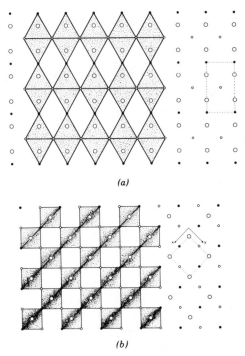

(a)

(b)

Figure 11. (a) The structure of PdO projected on (010). Small circles are Pd at $y/b = 0$ (open) or $y/b = \frac{1}{2}$ (filled); large open circles are oxygen atoms at $y/b = 0$. OPd$_4$ tetrahedra are shown. (b)The same, projected on (001).

Some Fluorite-Derived Structures with Modified Anion Arrays

Anion-Deficient Fluorites

There are a considerable number of derivatives known as "anion-deficient fluorite-type structures" (16, 17). They have compositions in the range M_2O_3 to MO_2, conforming to the general formula M_nO_{2n-2}, with integral $n = 4, 7, 9, 10, 11, 12, \infty$ (18). The lowest member is the C-type rare earth sesquioxide, or bixbyite structure commonly described as fluorite type with 25% of the anion sites unoccupied. The anion arrangement is strictly ordered, and the bcc unit cell has $a = 2a_{\text{fluorite}}$ (19, 20).* The other members are restricted to oxides of some of the lanthanoids and actinoids (Ce, Pr, Tb, Cm) and related ternary systems such as $Zr_3Sc_4O_{12}$ (22). They are most easily appreciated as ordered arrangements of anions in some of the tetrahedral sites of a ccp

---

* Pure α-Mn$_2$O$_3$ is very slightly distorted to orthorhombic with $a:b:c = 1.0012:1.0020:1$. Bixbyite is (Fe,Mn)$_2$O$_3$; it is strictly cubic—less than 1 at % of Fe transforms orthorhombic α-Mn$_2$O$_3$ to the cubic structure (21).

array of cations. (Recall that *all* these tetrahedral sites are occupied in the fluorite type.) In M_7O_{12} the empty anion sites form rows parallel to [111] so that the structure is rhombohedral; there is a slight elongation of the cubic [111] axis. In the C-type sesquioxide, similar rows occur in all four $\langle 111 \rangle$ cubic directions. Of the other structures, some are yet to be determined. All are difficult to depict by drawings.

Some structures appear to be restricted to ternary systems. The M_7O_{13} type of $Zr_5Sc_2O_{13}$ (22) is derived from the M_7O_{12} type by filling half the empty $[M_4]$ tetrahedra; along each such [111] row, alternate pairs of tetrahedra are filled and empty.

Additional ordered structures are observed in the system $CaO + HfO_2$ (23a, 23b). These are $Ca_2Hf_7O_{16}$ (corresponding to $n = 9$), $CaHf_4O_9$ ($n = 10$), and $Ca_6Hf_{19}O_{44}$—that is, $MO_{1.778}$, $MO_{1.800}$, and $MO_{1.760}$, respectively. Structure determinations have been carried out by a combination of single-crystal electron diffraction (to determine unit cells and space groups) and X-ray powder diffraction (24). Again they are ordered arrangements of empty $[M_4]$ and anion-centered $[OM_4]$ tetrahedra in a ccp M array, with a pronounced, and only very slightly distorted, fluorite subcell. In the first and third, the $\frac{1}{2}\langle 111 \rangle_{fl}$ pairs of vacant sites appear so that, in these structures as in M_7O_{13}, there is six-, seven-, and eight-fold coordination of M. (In M_7O_{12} there is only six- and seven-fold coordination.)

In all these cases, "disordered anion-deficient fluorite types" also exist, at least at high temperature. (These are clearly analogous to that produced by the order/disorder transformation characteristic of the fluorite type itself, the disorder temperature being composition-dependent.) Their exact structural nature is still a puzzle. All these materials are excellent, refractory, anion electrolytes. Best known, perhaps, are the "stabilized zirconias" and thorias: divalent or trivalent metal oxides (commonly CaO or Y_2O_3) are "dissolved" in ZrO_2 or ThO_2 to produce cubic or pseudocubic solid solutions. Their exact constitution is one of the important, outstanding problems in solid state chemistry (all the more pressing at present because of the great technological interest in solid state cells). There is growing evidence for some degree of ordering in these electrolytes, at least at low temperature and after sufficient annealing. This has recently culminated in the unequivocal identification, by electron diffraction, of short-range order in both "cubic stabilized zirconia" (CSZ) and the hafnia analogue. Furthermore, the type of order is obviously closely related to the long-range order in the $(Ca, Hf)O_x$ structures discussed above. Shorter annealing times lead to a domain texture, that is, to the phase separation of ordered structures, or a single multiply-twinned structure, on a fine scale (25a, 25b). This can give rise to favorable mechanical properties as in zirconia "partly stabilized" by MgO. In all cases, including the highly conducting fluorite types, the cation array is undisturbed, being virtually identical to that in fluorite itself.

The MgO partly stabilized zirconia (PSZ) is being developed as a strong, light ceramic for such applications as car engine blocks that can run at higher temperatures than conventional blocks so that water-cooling is unnecessary.

Anion-Excess Fluorites

There are also various anion-excess "fluorites." Their traditional description has been "fluorite type plus interstitial anions." But, in many cases at least, this description is now known to be entirely inadequate: the increased anion density is accommodated by transforming the pc anion array in a systematic way. Two basic methods are now recognized and substantiated.

In the first, *alternate* (100) anion layers (square arrays, i.e., 4^4 nets) are changed to denser triangular (3^6) nets. For perfectly regular nets, the density ratio $3^6/4^4$ would be $2/\sqrt{3} = 1.155$, so that the fluorite stoichiometry MX_2 would be changed to $MX_{2.155}$. In reality, such structures occupy the composition range $2.13 \leq X/M \leq 2.22$ when, for example, $M = Y$, $X = (O,F)$, as in the system YOF + YF$_3$. These structures (26a, 26b, 27) will be considered in detail in Chapter XII.

In the second method, some empty anion cubes are transformed, by rotating all square faces by 45° as in Figure 12, to cuboctahedra. The number of vertices or anions is then, in each case, increased from 8 to 12 and, by the addition of another anion at the center of the cuboctahedron (which thus becomes a small volume of ccp anions), to 13. Four or five additional anions are thus added at each "defect." Simultaneously, the six [MX$_8$] cubes sharing faces with the (original) empty cube become six [MX$_8$] square antiprisms. Beyond these the fluorite structure is unaffected, so that coherent intergrowth of fluorite with such "defects" is possible. Depending on the density of defects (and their arrangement, for the structures are ordered), a homologous series M_nX_{2n+p}, $p = 4$ or 5, may be generated (28). Several examples with $13 \leq n \leq 19$ ($2.385 \geq X/M \geq 2.263$) are known, and their structures have been determined. These include the mineral tveitite, $Ca_{14}Y_5F_{43}$ (29) and

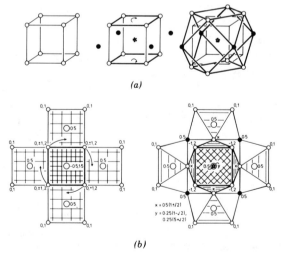

(a)

(b)

Figure 12.   Generation of a cuboctahedron from a cube by rotating all square faces by 45°: (*a*) perspective view, (*b*) plan = (100) projection.

$Eu_{14}Cl_{32}O$ (H. Bärnighausen, private communication). Another example is KY_3F_{10}, in which structure the cuboctahedra are not centered ($p = 4$, $n = 8$). There are several other compounds with related structures. Very recently the long-standing problem of the "U_4O_9" structure has been solved in the same terms (30), and a radically new light shed on the classical ideas of "Willis clusters" and interstitial anion sites, OI, OII, and OIII in UO_{2+x} (31).

Note that all these structures are somewhat anion-rich compared with those in the first group ($MX_{\sim2.3}$ compared with $MX_{\sim2.2}$).

Note added in proof: A definitive analysis of the question of occupancy of the anion site at the center of the cuboctahedral cluster—i.e. whether $p = 4$ or 5, or some (statistical) intermediate value—has recently become available (49*a*, 49*b*). It is based on a series of virtuoso single-crystal X-ray structure determinations carried out over the last ten years, mainly by Professor H. Bärnighausen and his students (49*a–c*).

$Cr_{23}C_6$

This is another example of such a fluorite-related structure with cuboctahedra. $Cr_{23}C_6$ is an important carbide occurring in certain steels. As shown in Figure 13*a*, it is composed of cuboctahedra sharing their square faces with square antiprisms, with C in the $[Cr_8]$ square antiprisms and additional Cr in distorted truncated tetrahedra (see Figure 13*a*). Crystallographic data are given in Data Table 9.

The transformation process described above (a cube + six surrounding cubes → cuboctahedron + six surrounding square antiprisms) provides an elegant method for transforming Cr metal (which is bcc) to $Cr_{23}C_6$ when carbon atoms diffuse into the metal (see Figure 13*b*).

Data Table 9 $Cr_{23}C_6$ (32)

Cubic, space group $Fm3m$, No. 225; $a = 10.65$ Å; $Z = 4$, $V = 1208$ Å^3

Atomic Positions: $(0,0,0; 0,\frac{1}{2},\frac{1}{2}; \frac{1}{2},0,\frac{1}{2}; \frac{1}{2},\frac{1}{2},0)+$
Cr in 4(*a*): 0,0,0
Cr in 8(*c*): $\pm(\frac{1}{4},\frac{1}{4},\frac{1}{4})$
Cr in 32(*f*): $\pm(x,x,x; \bar{x},x,x; x,\bar{x},x; x,x,\bar{x}); x = 0.3809$
Cr in 48(*h*): $\pm(0,x,x; x,0,x; x,x,0; 0,x,\bar{x}; \bar{x},0,x; x,\bar{x},0); x = 0.1699$
C in 24(*e*): $\pm(x,0,0; 0,x,0; 0,0,x); x = 0.2765$
Atomic Distances
Cr–Cr: 2.41 Å, 2.56 Å, 2.64 Å
Cr–C: 2.12 Å

$BaHg_{11}$

In $Cr_{23}C_6$ the building blocks (one cuboctahedron + six square antiprisms) share edges. (By analogy with tveitite, the formula can be written as

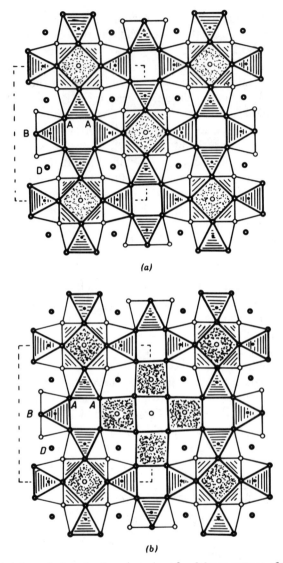

(a)

(b)

Figure 13. (*a*) A bounded projection, $\frac{1}{4} \leq z/c \leq \frac{3}{4}$, of the structure of $Cr_{23}C_6$. Double circles A ($z = \frac{1}{2} \pm 0.119$), B ($z \approx \pm\frac{1}{3}$), and D ($z = \pm\frac{1}{4}$) represent two superimposed Cr atoms. Single circles indicate Cr at $z = \frac{1}{2}$ (including one in the center of each cubocta-hedron). D is surrounded by a truncated tetrahedron. Small filled circles are C atoms at $z = \frac{1}{2}$ (in square antiprisms). The mode of transformation in Figure 12 is clear from a comparison of the centers of Figures 13*a* and *b*. (*b*) $Cr_{23}C_6$ partly transformed to bcc.

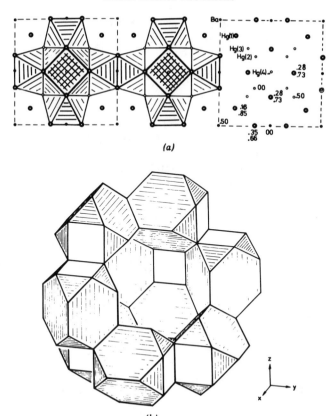

Figure 14. (*a*) The real structure of BaHg$_{11}$: the framework of square antiprisms is shown. (*b*) The framework in (*a*) is interpenetrated by the framework of elongated truncated dodecahedra shown here in clinographic projection.

Cr$_2$C$_6$Cr$_{21}$.) In the structure of BaHg$_{11}$ they share faces (see Figure 14*a*). The remaining part of the BaHg$_{11}$ structure (shown in Figure 14*b*) interpenetrates this array. It is built of elongated truncated dodecahedra and creates a large polyhedron (a truncated octahedron) around the Ba atoms. Data for BaHg$_{11}$ are given in Data Table 10.

Data Table 10   BaHg$_{11}$ (33)

Cubic, space group *Pm3m*, No. 221; *a* = 9.62 Å; *Z* = 3, *V* = 890.28 Å^3

Atomic Positions
Hg in 1(*b*): $\frac{1}{2},\frac{1}{2},\frac{1}{2}$
Ba in 3(*d*): $\frac{1}{2},0,0$; $0,\frac{1}{2},0$; $0,0,\frac{1}{2}$
Hg in 8(*g*): $\pm(x,x,x; \bar{x},x,x; x,\bar{x},x; x,x,\bar{x})$; $x = 0.155$
Hg in 12(*i*): $\pm(0,x,x; x,0,x; x,x,0; 0,x,\bar{x}; \bar{x},0,x; x,\bar{x},0)$; $x = 0.345$
Hg in 12(*j*): $\pm(\frac{1}{2},x,x; x,\frac{1}{2},x; x,x,\frac{1}{2}; \frac{1}{2},x,\bar{x}; \bar{x},\frac{1}{2},x; x,\bar{x},\frac{1}{2})$; $x = 0.275$

Parallel Translation Derivatives

UO_2F_2

In UO_2F_2 the uranium atoms are each coordinated by eight anions—two oxygens and six fluorines—at the corners of a rhombohedrally distorted "cube" compressed along the triad axis = c_h = $[111]_{rh}$ by the strong U–O bonds characteristic of the uranyl group $(O–U–O)^{2+}$. Each oxygen atom is bonded to only one uranium atom, each fluorine to three metal atoms. Only F–F edges are shared between "cubes," and all such edges are common to two cubes. (Crystallographic data are given in Data Table 11.)

Ignoring the rhombohedral distortion, the structure consists of isolated, infinite (111) layers of fluorite type stacked along c_h (Figure 15). The primitive cubic array is disturbed by each UO_2F_2 layer being slipped by $\frac{1}{3}[01\bar{1}0]_h$ with respect to its neighbor. The stacking sequence of the 3^6 anion nets along the c_h axis is therefore $\cdots/abca,cabc,bcab/\cdots$ instead of $\cdots/abcabcabcabc/\cdots$ (/ = slip plane), i.e., the sequence is h^2c^2 instead of c. (Note that the anions are not close-packed; the interplanar spacing is far too small.) Appropriate and obvious CS will convert this MX_4 structure type to Zachariasen's α-UO_3 type (= anti-Li_3N), thus extending the formal CS family NaCl/CsCl, La_2O_3, CaF_2, "α-UO_3" previously discussed (35) to UO_2F_2, i.e., MX, $MX_{1.5}$, MX_2, MX_3 to MX_4. Other, more complex, stackings and stoichiometries have also been reported to occur in La_2O_3 + ThO_2 (36a, 36b) and related systems (37); see Chapter VII and ref. 38.

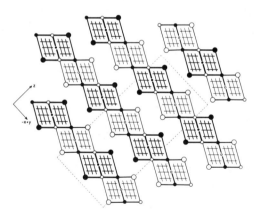

Figure 15. The structure of UO_2F_2 projected on $(11\bar{2}0)_{hex}$. Small circles are U; medium circles are F; large circles are O; open at height 0, filled at height $\frac{1}{2}$, the unit projection vector being $[11\bar{2}0]_{hex}/3$. (Unit-cell origin has been translated by $\sim -0.172c_{hex}$ in the plane of the drawing.) UO_2F_2 "cubes" are shown.

Data Table 11   UO$_2$F$_2$ (34)

Rhombohedral, space group $R\bar{3}m$, No. 166; in the equivalent hexagonal cell a_h = 4.206 Å, c_h = 15.692 Å, c/a = 3.731; Z = 3, V = 240.4 Å^3

Atomic Positions

U in 3(a):   0,0,0; $\frac{1}{3},\frac{2}{3},\frac{2}{3}$; $\frac{2}{3},\frac{1}{3},\frac{1}{3}$

O in 6(c): $\pm$(0,0,z; $\frac{1}{3},\frac{2}{3},\frac{2}{3} + z$; $\frac{2}{3},\frac{1}{3},\frac{1}{3} + z$); z = 0.122

F in 6(c): z = 0.294

Li$_2$Si

In this structure, shown in Figure 16 (with data in Data Table 12), the Li array is similar to the anion array in UO$_2$F$_2$ (except that the rhombohedral distortion, due to bonding in the uranyl group, is now almost absent). However, in this case the cubes are unoccupied. Instead, trigonal prisms, empty in UO$_2$F$_2$ (see Figure 15), are now occupied (by the Si atoms). So these two structures are complementary—that is, the interstices occupied in one are empty in the other—although they are also antitypes.

Li$_2$Si can also be described as twinned primitive cubic Li . . . ,*1,3,1,3*, ... The wider twin bands are easily identified. [Compare NbAs$_2$ (Chapter IX) and CrB = twinned (*1*), and IrAl$_3$ = twinned (*3*).]

The short Si–Si spacing of 2.37 Å suggests a Si–Si bond. It is almost the same as the bond length in silicon itself, l(Si–Si) = 2.35 Å.

Finally, we may also note that the Li + Si array in the twin plane region is approximately ccp (see Figure 17).

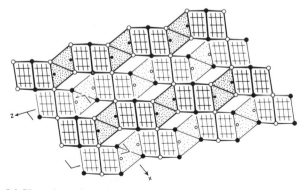

Figure 16.   Li$_2$Si projected on (010), showing Li$_8$ cubes and SiLi$_6$ trigonal prisms.

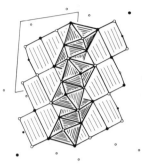

Figure 17. Li$_2$Si projected on (010), showing ccp Li + Si in twin-plane region.

Data Table 12   Li$_2$Si (39)

Monoclinic, space group $C2/m$, No. 12; $a = 7.70$, $b = 4.41$, $c = 6.56$ Å, $\beta = 113.4°$; $Z = 4$, $V = 204.4$ Å^3

Atomic Positions
All atoms in $4(i)$: $(0,0,0; \frac{1}{2},\frac{1}{2},0) \pm (x,0,z)$, with

	x	z
Li(1):	0.375	0.120
Li(2):	0.788	0.365
Si:	0.067	0.197

Atomic Distances
$d(\text{Si}\cdots\text{Si}) = 2.37$ Å, 4.41 Å (2×)
$d(\text{Si}\cdots\text{Li}) = 2.59\text{–}2.77$ Å (seven)

In a number of other structures, "cubically" coordinated cations are connected to form layers, and these layers are unconnected. In a manner analogous to the slip that disturbs the pc anion array in UO$_2$F$_2$, the same thing occurs in these structures. They are metal alloys, and we consider them next.

MoSi$_2$

This is the type compound; in Data Table 13 we give the data for the isostructural OsAl$_2$ (40). This structure, shown in Figure 18, is related to that of α-FeSi$_2$ (Figure 5). It also contains isolated CsCl-like layers stacked along the c axis, but now adjacent layers are offset, and the empty space between them is reduced by collapse of pc to lamellae of ccp. Compared with α-FeSi$_2$, this interlayer space is collapsed by displacing adjacent layers toward each other by the vector $\mathbf{R} = \frac{1}{2}\langle 111 \rangle_{\text{CsCl}}$. As a result, the coordination of each Os atom is increased from 8 to 10. The [OsAl$_8$] "cubes" are tetragonally distorted by compression in the **c** direction, their edges being 3.16 × 3.16 × 2.62 Å^3. Note that OsAl is the CsCl type ($a = 3.001$ Å).

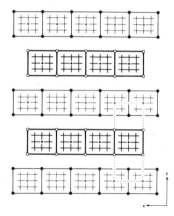

Figure 18. The structure of OsAl$_2$ projected on (010). Large circles are Al, small are Os; open at $y/b = 0$, filled at $y/b = \frac{1}{2}$. Layers of CsCl—face-sharing cubes—are stacked along **c**.

Data Table 13   OsAl$_2$ (40)

Tetragonal, space group $I4/mmm$, No. 139; $a = 3.162$, $c = 8.302$ Å, $c/a = 2.626$; $Z = 2$, $V = 83.01$ Å^3

Atomic Positions
Os in 2(a): $0,0,0; \frac{1}{2},\frac{1}{2},\frac{1}{2}$
Al in 4(e): $\pm(0,0,z; \frac{1}{2},\frac{1}{2},\frac{1}{2} + z)$; $z = 0.34_2$
Atomic Distances
Os–Al: 2.59 Å (8×), 2.84 Å (2×)

Several disilicides take this structure type, often with reduced c/a—for example, 2.45 for MoSi$_2$ compared with 2.63 for OsAl$_2$. (For regular cubes and ideal **R**, $c/a = 3.000$.) MoPt$_2$ and a number of isostructural compounds are orthorhombic variants with $a' \approx c' \equiv a$ and $b' \equiv c$. (Primes refer to the MoPt$_2$ cell, unprimed indices to MoSi$_2$.) The ratio c/a is still ~2.5, but now the ratio $c'/a' = 1.42$ is very close to $\sqrt{2}$; so these are really superstructures of ccp rather than pc. Other distortions are discussed by Pearson in his book (41); OsAl$_2$ and related structures are discussed by Edshammar (40).

Th$_2$N$_2$Bi and U$_2$N$_2$Sb are stuffed versions of the MoSi$_2$ type in which, for example, Th occupies the Si sites, Bi occupies the Mo sites, and N atoms are in the Th$_4$ tetrahedra. (Their structures should be compared with that of A-La$_2$O$_3$.) La$_2$O$_2$Se is isostructural, and TlCu$_2$S$_2$, TlFe$_2$S$_2$, etc. are antitypes. It may be noted that La$_2$O$_2$S and, of course, La$_2$O$_2$O have the A-La$_2$O$_3$-type structure. Clearly the two structures are closely related: the latter are intergrowths of (anti-) CaF$_2$ plus NaCl types, and the former an intergrowth of (anti-) CaF$_2$ plus CsCl types.*

---

* Another, Sc$_2$O$_2$S, is a similar intergrowth of (anti-) CaF$_2$ plus WC (= NiAs) type layers, the latter with trigonal prismatic coordination of the anions. Recall (Chapter VII) that Al$_4$C$_3$, etc., are intergrowths of (anti-) CaF$_2$ + cubic ZnS.

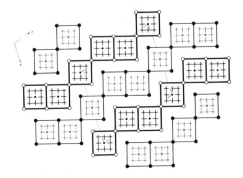

Figure 19. The structure of Mo_3Al_8 slightly idealized and projected on (010). Large circles are Al, small are Mo; open at $y/b = 0$, filled at $y/b = \frac{1}{2}$.

The monoclinic structure of Mo_3Al_8 is related to the $MoSi_2$ type; in it, some face-sharing between $[MoAl_8]$ "cubes" has been replaced by edge-sharing (Figure 19). This suggests the possibility of further structures and compositions being produced by varying the size of the face-shared strips, and, indeed, there are a number of related alloy structures. The $OsAl_2$ structure is a very slightly distorted, mixed bcc array (of Os and Al atoms). For ideal bcc, the c/a ratio for the CsCl-like subcell is 1.00; in $OsAl_2$ it is 1.202. In Mo_3Al_8 it is ~1.25—about halfway from bcc toward ccp (the Bain transformation again), when $c/a = \sqrt{2} = 1.414$. Other examples of this group are (a hypothetical?) MX_3 (Figure 20) and several other structures that we shall consider next. They all form mixed arrays lying between ideal bcc and ideal ccp.

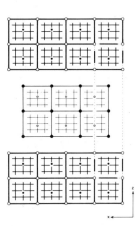

Figure 20. A hypothetical MX_3 structure related to Mo_3Al_8, etc., consisting of columns of face-shared MX_8 cubes joined into layers by edge-sharing, with close-packing between the layers.

Figure 21. The structure of Os_2Al_3 projected on (010). Compare especially Figure 18.

Os₂Al₃

The building principle here is exactly analogous to that in $OsAl_2$, but now the CsCl-like (001) layers are two cubes thick (Figure 21). The relative displacement of adjacent layers increases the Os coordination from 8 to 9 (+ 5 more Os atoms). The $[OsAl_8]$ "cubes" are $3.10 \times 3.10 \times 2.68$ Å³, i.e., $c/a = 0.86$ for the CsCl subcell. Crystallographic data are in Data Table 14.

Data Table 14   Os₂Al₃ (40)

Tetragonal, space group $I4/mmm$, No. 139; $a = 3.106$, $c = 14.184$ Å, $c/a = 4.567$, $Z = 2$, $V = 136.8_4$ Å³

Atomic Positions
Os in 4(e): $\pm(0,0,z; \frac{1}{2},\frac{1}{2},\frac{1}{2} + z)$; $z = 0.391_2$
Al(1) in 4(e): $z = 0.18_9$
Al(2) in 2(a): $0,0,0; \frac{1}{2},\frac{1}{2},\frac{1}{2}$
Atomic Distances
Os–Al: 2.68 Å (4×), 2.47 Å (4×), 2.87 Å

The sequence OsAl, Os_2Al_3, $OsAl_2$ suggests the possibility of additional structures intermediate between the first and the last.

MoNi₄

This structure, shown in Figure 22, resembles unconnected columns of CsCl type, $1 \times 1 \times \infty$ subcells in size, collapsed together like the slabs of CsCl in $MoSi_2$ but now in *two* directions. (Data are given in Data Table 15.) CS will give $MoSi_2$ or $PtHg_2$, and then CsCl. Taken together, however, the Mo and Ni atoms form a mixed array that is again almost perfect ccp (c/a for the "CsCl" subcell = 1.393, compared with $\sqrt{2} = 1.414$ for perfect ccp). The "cubes" are elongated along **c**, being $2.56 \times 2.56 \times 3.56$ Å³ ($1 \times 1 \times 1.39 \approx \sqrt{2}$), almost exactly the elongation that transforms bcc to bct ($\equiv$ fcc)—the Bain relation again. This emphasizes the need to consider dimensions in the projection direction rather carefully!

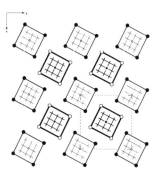

Figure 22. The structure of $MoNi_4$ projected on (001) drawn as columns of $MoNi_8$ face-sharing cubes (but see text).

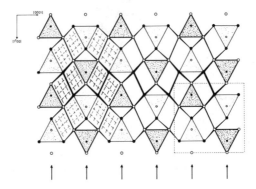

Figure 23. The real structure of IrAl$_3$, projected along **a**. Smaller atoms are Ir; all atoms on 0 or $\frac{1}{2}$. Edshammar's 11-polyhedron is obvious in the figure, in which the structure is depicted as twinned pc.

Data Table 15   MoNi$_4$ (4)

Tetragonal, space group $I4/m$, No. 87; $a = 5.720$, $c = 3.564$ Å

Atomic Positions
Mo in 2(a): $0,0,0; \frac{1}{2},\frac{1}{2},\frac{1}{2}$
Ni in 8(h): $\pm(x,y,0; \frac{1}{2} + x,\frac{1}{2} + y,\frac{1}{2}; \bar{y},x,0; \frac{1}{2} - y,\frac{1}{2} + x,\frac{1}{2})$; $x = 0.200$, $y = 0.400$

Twinned Primitive Cubic Arrays (Reflection Derivatives)

IrAl$_3$

The hexagonal structure of IrAl$_3$ is shown in Figures 23 and 24. It is clearly a unit-cell-twinned structure related to that of Re$_3$B (Chapter VI), but the

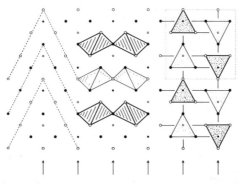

Figure 24. IrAl$_3$ drawn to show relation with Re$_3$B. The deformed cubes in Figure 23 are now depicted as deformed octahedra.

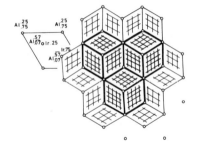

Figure 25. The real IrAl₃ structure projected along **c**. Ir is always on $z = \frac{1}{4}$ or $\frac{3}{4}$, i.e., in the mirror planes. The empty Al₈ cubes are between the mirror planes.

empty polyhedra in the twin bands are closer to cubes than to octahedra—the array that is twinned is closer to pc than to ccp. Ir atoms are in Edshammar's (40) 11-coordination polyhedron, which we have already observed for Ti(2) in the ω-phase structure and for In in Ni₂In. Crystal data are given in Data Table 16.

The structure type of IrAl₃ is Na₃As, but a number of other compounds, for example, LaF₃ and Cu₃P, have closely related structures.

Figure 25 shows the twin bands of IrAl₃ projected along the **c** axis of the hexagonal cell.

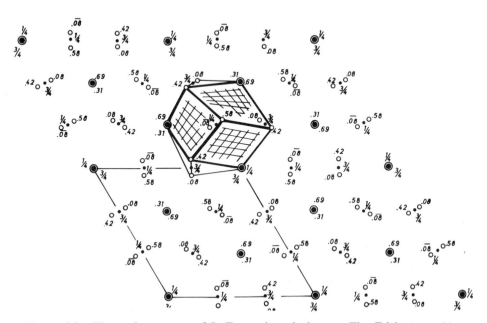

Figure 26. The real structure of LaF₃, projected along **c**. The Edshammar 11-polyhedron is drawn. Small circles at $\frac{1}{4}$ and $\frac{3}{4}$ are La; larger circles are fluorines.

Data Table 16   **IrAl$_3$ (42)**

Hexagonal, space group $P6_3/mmc$, No. 194; $a = 4.246$, $c = 7.756$ Å*; $Z = 2$, $V = 139.83$ Å^3

Atomic Positions
Ir in 2(c): $\pm(\frac{1}{3},\frac{2}{3},\frac{1}{4})$
Al(1) in 2(b): $\pm(0,0,\frac{1}{4})$
Al(2) in 4(f): $\pm(\frac{1}{3},\frac{2}{3},z; \frac{1}{3},\frac{2}{3},\frac{1}{2} - z)$; $z = 0.575$

* The corresponding orthohexagonal unit cell has $a = 4.264$, $b = 7.354$, $c = 7.756$ Å.

LaF$_3$

In the structure of LaF$_3$ the **c** axis is the same as in IrAl$_3$ but the a axis is $\sqrt{3}$ times as long, the result of a small distortion of the anion array. In Figure 26 the complete structure of LaF$_3$ is shown projected along **c** (with the Edshammar polyhedron picked out); while Figure 27 shows the projection of only one twin band in LaF$_3$. It is to be compared with the analogous projection for IrAl$_3$ shown in Figure 25. Structural data are given in Data Table 17.

Data Table 17   **LaF$_3$ (43a, 43b, 43c)**

Hexagonal, space group $P\bar{3}c1$, No. 165; $a = 7.185$, $c = 7.351$ Å; $Z = 6$, $V = 379.49$ Å^3

Atomic Positions
La in 6(f): $\pm(x,0,\frac{1}{4}; 0,x,\frac{1}{4}; \bar{x},\bar{x},\frac{1}{4})$, with $x = 0.3401$
F in 2(a): $\pm(0,0,\frac{1}{4})$
F in 4(d): $\pm(\frac{1}{3},\frac{2}{3},z; \frac{1}{3},\frac{2}{3},\frac{1}{2} + z)$, with $z = 0.313$
F in 12(g): $\pm(x,y,z; \bar{y},x - y,z; y - x,\bar{x},z; \bar{y},\bar{x},\frac{1}{2} + z; x,x - y,\frac{1}{2} + z;$
               $y - x,y,\frac{1}{2} + z)$; $x = 0.312$, $y = -0.05$, $z = 0.581$

Atomic Distances
La–F = 2.42–2.49 Å (7×), 2.64 Å (4×), 3.01 Å (4 times)

Figure 28 shows another way of deriving the IrAl$_3$ structure. Slabs of the structure occur in the ordinary hcp array. If these slabs are collapsed to-

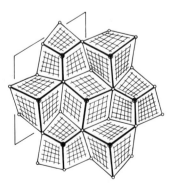

Figure 27. Part of the real structure of LaF$_3$, which shows empty "cubes" forming a layer of a pc structure of fluorine atoms. Such layers are joined by a twin operation at $z = \frac{1}{4}$ and $z = \frac{3}{4}$, so that Edshammar's 11-polyhedra are formed, containing La atoms.

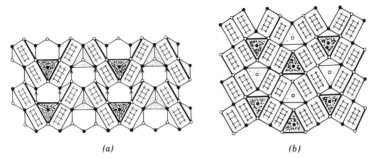

Figure 28. (*a*) HCP I. Arrows indicate shear operation to give (*b*). (*b*) Slightly idealized IrAl$_3$, or LaF$_3$, derived by shear operation on HCP I (*a*).

gether by the translation vectors shown in Figure 28*a*, we get the structure shown in Figure 28*b*. The generated cubes are distorted. (The two diagonals on the faces parallel to the projection axis are, respectively, $\sqrt{8/3} = 1.633$ and $2/\sqrt{3} = 1.155$ times the close-packed atom separation, a ratio of $\sqrt{2}$. For the ideal structure, the calculated $c/a = 3\sqrt{3}/(2\sqrt{2}) = 1.837$; for perfect cubes, it is $3\sqrt{3}/2 = 2.598$. These are to be compared with the observed values of $c/a = 1.827$ for IrAl$_3$ and $\sqrt{3}c/a = 1.772$ for LaF$_3$; the collapsed hcp description is perhaps more accurate than that of twinned pc. However, it must be remembered that the twin bands are between pc and ccp, which reduces c and increases a. Both these changes reduce the c/a ratio below the ideal value for twinned pc . . . ,3,3,3, ... [In similar terms the ω-phase structure is approximately twinned pc . . . ,2,2,2, ... This is the smallest possible twinned pc structure with Edshammar polyhedra in the twinned band.]

Glide Reflection Twinning

In Chapter VI we considered the NH$_4$CdCl$_3$ and La$_2$SnS$_5$ structures as deriving from the NaCl type by the operation of glide-reflection twinning. In this chapter the analogue of NaCl (B1) is CsCl (B2) (Figure 2). RuSi is isostructural and, applying this operation (Figure 29; cf. Figure 30 of Chapter VI), one easily obtains the structures of Ru$_4$Si$_3$, Ru$_5$Si$_3$, and Ru$_2$Si.

The first is shown in Figure 30. Figure 31 shows Rh$_5$Ge$_3$, which is isostructural with Ru$_5$Si$_3$ and Rh$_5$Si$_3$ and the antitype of U$_3$S$_5$ and U$_3$Se$_5$. (It is analogous to anti-Ln$_2$SnS$_5$, Figure 32 in Chapter VI.) The last, Ru$_2$Si, is isostructural with Co$_2$Si (Figure 32). This is very close to the PbCl$_2$ type (Figure 33), the difference being a slight topological distortion so that the tricapped prisms in the latter become tetracapped in the former. [Details have been discussed elsewhere (44).]

Many silicides, phosphides, and so on are isostructural with Co$_2$Si, and many dihalides and disulfides such as BaCl$_2$, ThS$_2$, and US$_2$ are isostructural with PbCl$_2$. Although they are given different Strukturbericht symbols (C37 and C23, respectively), they are scarcely distinguishable. Hence the

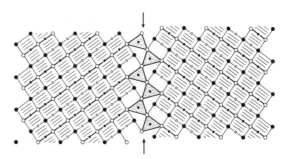

Figure 29. Glide reflection twinning of CsCl.

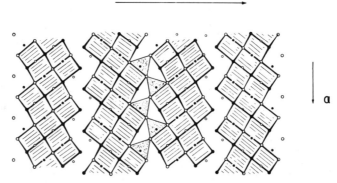

Figure 30. The real structure of Ru₄Si₃. The large RuSi slabs of CsCl structure are obvious, and one C37-like layer of Ru₂Si is also drawn. The cubes are projected along a face diagonal.

α-PbO₂/PbCl₂ relation discussed in Chapter III is paralleled by a Co₂N (= anti-α-PbO₂)/Co₂Si relationship.

The data for Ru₄Si₃, Rh₅Ge₃, and Co₂Si, are given in Data Tables 18–20.

Data Table 18 Ru₄Si₃ (45)

Orthorhombic, space group *Pnma*, No. 62; $a = 5.1936$, $b = 4.0216$, $c = 17.1343$ Å; $Z = 4$, $V = 357.88$ Å³

Atomic Positions
All atoms in 4(*c*): $\pm(x,\frac{1}{4},z; \frac{1}{2} + x,\frac{1}{4},\frac{1}{2} - z)$, with

	x	z
Ru(1)	0.3475	0.23561
Ru(2)	0.0075	0.10659
Ru(3)	0.1739	0.52958
Ru(4)	0.3418	0.83548
Si(1)	0.0044	0.39479
Si(2)	0.1478	0.96601
Si(3)	0.3776	0.68869

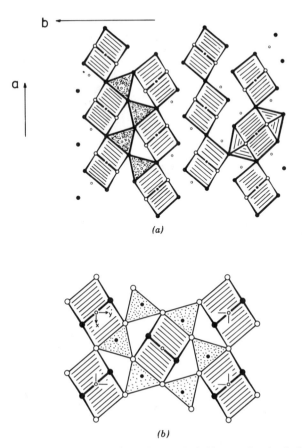

(a)

(b)

Figure 31. The real structure of Rh$_5$Ge$_3$. (a) A bicapped cube is indicated, and a slab of Rh$_2$Ge. The CsCl-like slabs are much smaller than in Figure 30. (b) The trigonal prism walls are joined to form hexagonal columns.

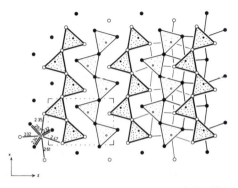

Figure 32. The real structure of Co$_2$Si.

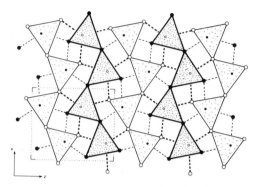

Figure 33. The PbCl₂ structure. See also Chapter III.

Data Table 19   Rh₅Ge₃ (46)

Orthorhombic, space group *Pbam*, No. 55; $a = 5.42$, $b = 10.32$, $c = 3.96$ Å;
$Z = 2$, $V = 221.5$ Å³

Atomic Positions
Ge(1) in 2(*a*) $0,0,0; \frac{1}{2},\frac{1}{2},0$
Ge(2) in 4(*h*): $\pm(x,y,\frac{1}{2}; \frac{1}{2} + x,\frac{1}{2} - y,\frac{1}{2})$; $x = 0.388$, $y = 0.152$
Rh(1) in 2(*c*): $0,\frac{1}{2},0; \frac{1}{2},0,0$
Rh(2) in 4(*g*): $\pm(x,y,0; \frac{1}{2} + x,\frac{1}{2} - y,0)$; $x = 0.152$, $y = 0.220$
Rh(3) in 4(*h*): $x = 0.330$, $y = 0.393$

Data Table 20   Co₂Si (47)

Orthorhombic, space group *Pnma*, No. 62; $a = 4.918$, $b = 3.738$, $c = 7.109$ Å;
$Z = 4$, $V = 130.69$ Å³

Atomic Positions
Co(1) in 4(*c*): $\pm(x,\frac{1}{4},z; \frac{1}{2} + x,\frac{1}{4},\frac{1}{2} - z)$; $x = 0.038$, $z = 0.218$
Co(2) in 4(*c*): $x = 0.174$, $z = 0.562$
Si(1) in 4(*c*): $x = 0.702$, $z = 0.611$
Atomic Distances
Co–Si = 2.32 − 2.62 Å (10×)
Co–Co = 2.50 − 2.68 Å (8×)
Si–Si = 3.15 Å (2×) (shortest)

Also related to C37 and C23 by a slight topological distortion of the sheets of edge-shared columns of trigonal prisms are the structures of MoP₂ (a consequence of P–P bonding) (Figure 34) and Ni₂In. In these the trigonal prisms are respectively monocapped and pentacapped. The increasing number of caps in the sequence MoP₂ (CN = 6 + 1), PbCl₂ (6 + 3), Co₂Si (6 + 4),

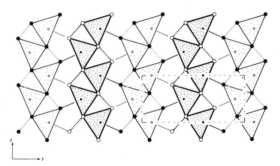

Figure 34.   The MoP$_2$ structure, with P–P bonds indicated by dotted lines.

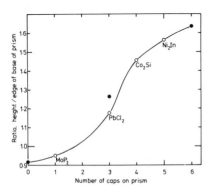

Figure 35.   Height/base ratio of trigonal prisms related to the number of capping atoms. (The filled circles are the results of calculations using an electrostatic model.)

Ni$_2$In (6 + 5) (and 6 + 6 in hcp!) correlates strongly with elongation of the trigonal prism parallel to its pseudo-three-fold axis (cf. Figure 35) (44, 48). In this series the size of the coordinated atom is steadily increasing until, in hcp as in ω-Ti, it is the same as that of the surrounding, coordinating atoms.

References

1. L. Pauling, *J. Am. Chem. Soc.* **51,** 1017 (1929).

2. W. H. Beamer and R. L. Maxwell, *J. Chem. Phys.* **17,** 1293 (1949).

3. J. Donohue, *The Structures of the Elements.* Wiley, New York, 1974.

4. K. Schubert, *Kristallstrukturen zweikomponentiger Phasen.* Springer-Verlag, Berlin, 1964.

5. E. Zintl and G. Brauer, *Z. Elektrochem.* **41,** 102 (1935).

6. B. G. Hyde, L. A. Bursill, M. O'Keeffe, and S. Andersson, *Nature Phys. Sci.* **237,** 35 (1972).

7. A. S. Dworkin and M. A. Bredig, *J. Chem. Phys.* **72,** 1277 (1968).

8. L. E. Nagel and M. O'Keeffe, in *Fast Ion Transport in Solids,* W. van Gool, Ed. North-Holland, Amsterdam, 1973, p. 165.

9. G. Phragmen, *J. Iron Steel Inst.* **1926,** 397.

10. B. Aronsson, *Acta Chem. Scand.* **14,** 1414 (1960).

11. F. A. Sidorenko, P. V. Gel'd, and M. A. Shumilov, *Fiz. Metal. Metalloved.* **9,** 861 (1960).

12. Y. Dusausoy, J. Protas, R. Wandji, and B. Roques, *Acta Cryst.* B **27,** 1209 (1971).

13. E. Bauer, H. Nowotny, and A. Stempfl, *Monatsh. Chem.* **84,** 692 (1953).

14. J. S. Anderson, *J. Chem. Soc. Dalton* **1973,** 1107.

15. S. S. Lu and T. Chang, *Sci. Sinica* **6,** 431 (1957).

16. G. Brauer, in *Progress in the Science and Technology of the Rare Earths,* L. Eyring, Ed. Pergamon, New York. Vol. 1, 1964, p. 152; Vol. 2, 1966, p. 312; Vol. 3, 1968, p. 434.

17. D. J. M. Bevan, in *Comprehensive Inorganic Chemistry,* J. C. Bailar *et al., Eds.* Pergamon, Oxford, 1973, p. 522.

18. B. G. Hyde, D. J. M. Bevan, and L. Eyring, *Phil. Trans. Roy. Soc. Lond.* A **259,** 583 (1966).

19. R. Norrestam, *Arkiv Kemi* **29,** 343 (1968).

20. B. H. O'Connor and T. M. Valentine, *Acta Cryst.* **25,** 2140 (1969).

21. S. Geller, *Acta Cryst.* B **27,** 821 (1971).

22. M. R. Thornber, D. J. M. Bevan, and J. Graham, *Acta Cryst.* B **24,** 1183 (1968).

23a. J. G. Allpress, H. J. Rossell, and H. G. Scott, *Mater. Res. Bull.* **9,** 455 (1974).

23b. H. J. Rossell and H. G. Scott, in *Diffraction Studies of Real Atoms and Real Crystals* (International Crystallographic Conference, Melbourne, 19–23 August, 1974). Australian Academy of Science, Canberra, 1974, p. 225.

24. H. J. Rossell and H. G. Scott, *J. Phys. Colloque C7,* **38,** C7 (1977).

25a. H. J. Rossell and H. G. Scott, in *Diffraction Studies of Real Atoms and Real Crystals* (International Crystallography Conference, Melbourne, 19–23 August, 1974). Australian Academy of Science, Canberra, 1974, p. 249.

25b. H. J. Rossell, J. R. Sellar, and I. J. Wilson, *J. Electron Microsc.* **35** (*Suppl.*), 1667 (1986).

26a. B. G. Hyde, A. N. Bagshaw, S. Andersson, and M. O'Keeffe, *Ann. Rev. Mater. Sci.* **4,** 43 (1974).

26b. E. Makovicky and B. G. Hyde, *Struct. Bonding* **46,** 101 (1981).

27. D. J. M. Bevan and A. W. Mann, *Acta Cryst.* B **31,** 1406 (1975).

28. D. J. M. Bevan, O. Greis, and J. Strähle, *Acta Cryst.* A **36,** 889 (1980).

29. D. J. M. Bevan, J. Strähle, and O. Greis, *J. Solid State Chem.* **44,** 75 (1982).

30. D. J. M. Bevan, I. E. Grey, and B. T. M. Willis, *J. Solid State Chem.* **61,** 1 (1986).

31. B. T. M. Willis, *J. Phys.* **25,** 431 (1964).

32. A. L. Bowman, G. P. Arnold, E. K. Storms, and N. G. Nereson, *Acta Cryst.* B **28,** 3102 (1972).

33. G. Peyrond, *Struct. Rep.* **16,** 24 (1952).

34. W. H. Zachariasen, *Acta Cryst.* **1,** 277 (1948).

35. B. G. Hyde, *Acta Cryst. A* **27,** 617 (1971).

36a. F. Sibieude, *J. Solid State Chem.* **7,** 7 (1973).

36b. F. Sibieude, G. Schiffmacher, and P. Caro, *J. Solid State Chem.* **23,** 361 (1978).

37. C. M. van der Walt and H. N. J. Louw, *Acta Metall.* **16,** 777 (1968).

38. M. O'Keeffe and B. G. Hyde, *Struct. Bonding* **61,** 77 (1985).

39. H. Axel, H. Schäfer, and A. Weiss, *Angew. Chem. Int. Ed.* **4,** 358 (1965).

40. L. E. Edshammar, X-ray studies on binary alloys of aluminium with platinum metals, thesis, University of Stockholm, 1969.

41. W. B. Pearson, *The Crystal Chemistry and Physics of Metals and Alloys.* Wiley, New York, 1972.

42. L. E. Edshammar, *Acta Chem. Scand.* **21,** 1104 (1967).

43a. A. Zalkin, D. H. Templeton, and T. E. Hopkins, *Inorg. Chem.* **5,** 1466 (1966).

43b. B. Maximov and H. Schulz, *Acta Cryst. B* **41,** 88 (1985).

43c. A. Zalkin and D. H. Templeton, *Acta Cryst. B* **41,** 91 (1985).

44. B. G. Hyde, S. Andersson, M. Bakker, C. M. Plug, and M. O'Keeffe, *Prog. Solid State Chem.* **12,** 273 (1980).

45. I. Engström and T. Johnsson, *Arkiv Kemi* **30,** 141 (1969).

46. S. Geller, *Acta Cryst.* **8,** 15 (1955).

47. S. Geller, *Acta Cryst.* **8,** 83 (1955).

48. M. O'Keeffe and B. G. Hyde, *Acta Cryst. B* **33,** 3802 (1977).

49a. H. Bärnighausen, private communication.

49b. A. Lumpp, Die röntgenographische Bestimmung einer komplexen Überstruktur des Fluorit-Typs für eine neue Neodym(II,III)-Chlor-Verbindung, thesis, Universität Karlsruhe, 1988.

49c. M. Eitel, Die röntgenographische Bestimmung einer komplexen Überstruktur des Fluorit-Typs für eine Neodym(II,III)-Chlor-Verbindung an Hand eines pseudomeroedrischen Drillingskristalls, thesis, Universität Karlsruhe, 1985.

CHAPTER IX

Primitive Hexagonal Arrays and Some Derived and Related Structures

The following structures are considered in this chapter: AlB_2 (= β-$ThSi_2$), $Cd(OH)_2$, NiAs, WC, CrB (= β-MoB), TlI, SnS, KOH, W_2B_5, ReB_3, $CaSi_2$, Sc_2O_2S, β-MoS_2, α-MoS_2, TiP, $TlInS_2$, Ta_3B_4, V_2B_3, V_5B_6, W_2CoB_2, W_3CoB_3, $NbAs_2$, Dy_3Ni_2, Li_2Si, Re_3B, $Y(OH)_3$, pyrargyrite $SbAg_3S_3$, proustite $AsAg_3S_3$, α-MoB, α-$ThSi_2$, α-$CdSi_2$, monoclinic $YO(OH)$ = $HoO(OH)$, $SrBr_2 \cdot H_2O$, PbFCl, $PbCl_2$, tetragonal $YbO(OH)$, NbAs, FeB, high-temperature TbNi, $Gd_{0.75}Y_{0.25}Ni$, β-CaCu, $Gd_{0.4}Tb_{0.6}Ni$, SrAg, $Gd_{0.3}Tb_{0.7}Ni$, α-CaCu, $Gd_{0.2}Tb_{0.8}Ni$, TbNi, $Ce_6Ni_2Si_3$, "Ce_2NiSi," $Ce_{15}(Ni,Si)_{17}$, Fe_2P = β_1-K_2ThF_6 = Ni_6Si_2B, Th_7S_{12}, $Ru_{20}Si_{13}$, MoAlB, Mn_2AlB_2, Cr_3AlB_4, Mo_2BC, $ZrSi_2$, UBC, CoGe, Ni_3Sn_4, α-$PdBi_2$, chalcostibite/wolfsbergite $CuSbS_2$, emplectite $CuBiS_2$, Sr_2PbO_4, B-Sm_2O_3, o-Ni_4B_3, bismuthinite Bi_2S_3, Hf_3P_2, the bismuthinite–aikinite series, meneghinite $CuPb_{13}Sb_7S_{24}$, and the homologous series $Sb_2S_3 \cdot n$PbS, α-Gd_2S_3, Cr_3C_2, K_2AgI_3

The simplest stacking of planar, triangular nets (3^6 = close-packed) is $\cdots a\,a\,a\cdots$, a single-layer repeat. [Note, by way of comparison, that hcp is a double-layer repeat, ccp a triple-layer repeat, double hexagonal packing ($\cdots a\,b\,a\,c\cdots$) a quadruple-layer repeat, and so on.] The only polyhedral interstice it contains is the trigonal prism (although there are also square, triangular, and linearly coordinated interstices). A few simple alloy structures are obtained by inserting a second type of atom into some or all of these prisms, of which there are two per stacked atom.*

---

* Inserting *identical* atoms into the appropriate half of the prisms and increasing c/a from 1.000 to a multiple of $n\sqrt{8/3}$ = $1.633n$ produces close-packed arrays: hcp for $n = 1$, hc for $n = 2$, etc.

Aluminum Boride, Strukturbericht Type C32

Data Table 1   AlB₂ (1)

Hexagonal, space group $P6/mmm$, No. 191; $a = 3.009$, $c = 3.262$ Å, $c/a = 1.084$; $Z = 1$, $V = 25.58$ Å^3

Atomic Positions
Al in 1(a):   0,0,0
B in 2(d): $\pm(\frac{1}{3},\frac{2}{3},\frac{1}{2})$

If every trigonal prism in a primitive hexagonal (ph) array of aluminum atoms is occupied by a boron atom, we get the AlB₂ structure shown in Figure 1. (Data are given in Data Table 1.) The stacking sequence of (0001) layers is $\cdots \alpha$ (bc) α (bc) $\alpha \cdots$, where, as usual, Greek letters represent metal layers and Roman letters nonmetal layers, the pairs in parentheses representing two close-packed (3^6) layers of B at the same height (which gives a 6^3, "honeycomb," layer of smaller edge length). The BAl₆ trigonal prisms are almost regular [edge lengths are 3.009 Å (six) and 3.262 Å (three)], and each shares all five faces with neighboring trigonal prisms. The AlB₁₂ hexagonal prisms [edge lengths 1.737 Å (twelve) and 3.262 Å (six)] share all eight faces with adjacent hexagonal prisms. The B atoms are bonded together in the 6^3 nets: d(B–B) = 1.737 Å compared with an average of ~1.8 Å in boron.

Very many metal borides and silicides and intermetallic compounds are isostructural. As mentioned earlier (Chapter V), this structure is to be distinguished from that of the ω phase by its c/a ratio (~1 for AlB₂, ~0.6 for ω).

It may be noted that the Cd(OH)₂ structure can also be described as a ph Cd array with OH groups in each trigonal prism. The orientations of the latter pucker their 6^3 layers, splitting them into two 3^6 layers, i.e., $\cdots \alpha$ bc α bc $\alpha \cdots$. The c/a ratio for Cd(OH)₂ lies between the ideal values for AlB₂ and that for the usual description (based on hcp hydroxyl ions, Chapter III): 1.366 compared with 1.000 and 1.633, respectively.

MX Structures Topologically Related to the AlB₂ Type

There are many possible geometrical arrangements in which half the trigonal prism sites in a ph array are occupied. We can distinguish those in which the

Figure 1.   AlB₂ projected on (0001): large circles are Al (at $z = 0$), small circles are B (at $z = \frac{1}{2}$).

arrangement (topology) precludes collapse of the ph array—which may be imagined as a denser (cp) array expanded ("propped" up; it is a low-density packing) by the occupancy of interstitial sites by atoms too large for, say, octahedral coordination—and those that allow collapse of the former. The two simplest structures are those of NiAs and WC.

NiAs = Anti-RhB, Strukturbericht type B8, B8$_1$, or B8$_a$

In Chapter III this was described as hcp As with Ni atoms in all the octahedral holes, i.e., $\cdots$ $a \gamma b \gamma a \gamma b \gamma a$ $\cdots$. This would suggest that it could equally well be described as a ph array of Ni with As atoms in one-half of the trigonal prismatic interstices. However, the c/a ratio for NiAs is 1.39*; for "half-filled AlB$_2$," one would ideally expect a value of $\sim$2.000, and for hcp As a value of 1.633. Clearly, then, the As$_6$ octahedra are compressed along **c**, and the Ni$_6$ trigonal prisms even more compressed (edge lengths being in the ratio 0.7/1), so that it is *not* a compromise between regular trigonal prisms and regular octahedra.

Data for NiAs are given in Chapter III, Data Table 5. The structure is shown schematically in Figure 2. Each AsNi$_6$ prism shares each of its edges with adjacent prisms, alternate prisms being empty. There are an enormous number of isostructural compounds (3).†

Before describing other structures in this group, it is worth reconsidering the topology of the trigonal prisms in a ph array, i.e., the ways in which they are connected. It has already been pointed out that each prism shares each of its five faces (three square, or at least rectangular, and two triangular) with adjacent prisms. It is convenient, for the system set out in Chapter I, to consider larger elements of the array. Kripyakevich (4) identified four semi-infinite units:

(i) Rectilinear chains of triangular-face-sharing prisms (parallel to **c**), which he termed *columns*.

(ii) Rectilinear chains of prisms sharing square faces, and parallel to $\langle 11\bar{2}0 \rangle$, termed *strips*.

(iii) *Layers* parallel to (0001) and formed of strips sharing square faces.

(iv) *Walls* parallel to $\{1\bar{1}00\}$ and formed of strips sharing triangular faces, or columns sharing two sets of square faces.

[In terms of Chapter I, both (i) and (ii) are columns or rods, and (iii) and (iv) are slabs, sheets, layers or lamellae.]

---

* The mean value is 1.345 for a list of B8$_a$ structures given by Matthias (2).

† Very recently, since this was written, it has been discovered that the real structure of NiAs is a small distortion of this ideal structure (45a–d). [Larger distortions, as in the MnP-type structure, have long been known (46).]

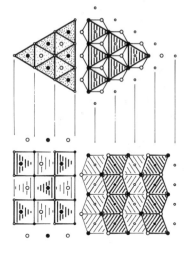

Figure 2. The NiAs structure idealized in two ways: on the left as regular AsNi$_6$ trigonal prisms, and on the right as regular NiAs$_6$ octahedra. [Note that a is the same for both, but c is different, the (ideal) c/a values being 2.000 and 1.633, respectively, compared with the actual value of 1.391.] In the top part of the figure the projection is on (0001); in the bottom part it is on (11$\bar{2}$0). Larger circles are As, smaller are Ni: top, As at $z = \pm\frac{1}{4}$, Ni at $z = 0, \frac{1}{2}$; bottom, open at height 0, filled at height $\frac{1}{2}$ (along $\frac{1}{3}$ [11$\bar{2}$0]).

WC

In this structure (Figure 3), the prisms in alternate columns of a ph array of tungsten atoms are occupied by carbon. The remaining columns are empty. CW$_6$ prisms are joined by common triangular faces and by "vertical" edges. The stacking sequence is $\cdots\ \alpha\,b\,\alpha\,b\,\alpha\ \cdots$ (or $\cdots\ \alpha\,c\,\alpha\,c\,\alpha\ \cdots$, etc.), so that the C atoms are also in a ph array, and the WC structure is its own antitype. In NiAs and WC the occupied prisms form identical (0001) unit layers that are differently stacked in the two structures. Structural data for WC are given in Data Table 2.

WC is important technically, being extremely hard and refractory. Many compounds, including nitrides, phosphides, sulfides, selenides, and tellurides, are isostructural. Not all are stoichiometric. (And note that many transition metal carbides have the NaCl-type structure.)

Data Table 2   WC (5)

Hexagonal, space group $P\bar{6}m2$, No. 187; $a = 2.906$, $c = 2.837$ Å, $c/a = 0.976$; $Z = 1$, $V = 20.75$ Å^3

Atomic Positions
W in 1(a):   0,0,0
C in 1(d): $\frac{1}{3},\frac{2}{3},\frac{1}{2}$

The AlB$_2$ type is a parent for a number of derived structures. Although it does not happen in the structures of WC and NiAs, it is possible to arrange the unoccupied prisms in "B-deficient AlB$_2$" in such a way—complete

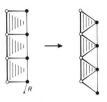

Figure 3. WC projected on (0001). Large circles are W (at $z = 0$); small circles are C (at $z = \frac{1}{2}$).

Figure 4. Relative slip by the displacement vector **R** transforms a column or wall of trigonal prisms (left) into half-octahedra and tetrahedra (right); i.e., ph packing to ccp.

strips, layers, or walls—that they can collapse without distorting the occupied prisms (Figure 4). If this occurs, the volume is reduced, and elements of a denser, close-packed array are formed. This is an easy, alternative way of looking at structures such as those of CrB, FeB, and related types, which we consider next.

Structures Produced by Parallel Translation of AlB₂ Elements

CrB

The structure of CrB is shown in Figure 5. Previously described (in Chapter VI) as twinned ccp Cr $\cdots$, *1,1,1,* $\cdots$, with B in the composition planes, it is now also seen to be unit walls of AlB₂ type repeated in parallel fashion by the translation vector $\frac{1}{2}[110]_{CrB}$. Structural data are given in Data Table 3.

Data Table 3   CrB (6)

Orthorhombic, space group *Cmcm*, No. 63; $a = 2.969$, $b = 7.858$, $c = 2.932$ Å; $Z = 4$, $V = 68.40$ Å³

Atomic Positions

All atoms in 4(*c*): $\pm(0,y,\frac{1}{4}; \frac{1}{2},\frac{1}{2} + y,\frac{1}{4})$

Cr: $y = 0.146$

B: $y = 0.440$

It can also be regarded as AlB₂ in which half the prisms are empty and the empty walls are collapsed. Between the AlB₂ type walls the Cr atoms are approximately ccp (actually slightly denser, by ~22%.) Each B atom is 7-coordinated, the trigonal prisms being monocapped.

The low-temperature structure of TlI (also that of InI) is an antitype of CrB, but with the Tl atom pushed toward one square face of the prism,

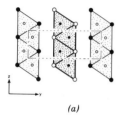

(a)

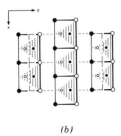

(b)

Figure 5. The structure of CrB projected on (a) (100) and (b) (001). Large circles are Cr and small circles are B: (a) open at $x = \frac{1}{2}$, filled at $x = 0$; (b) open at $z = \frac{1}{4}$, filled at $z = \frac{3}{4}$.

almost into the prism cap, as a consequence of the stereochemically active lone pair of electrons on Tl^+. (Cf. Chapter X, on lone pair structures.) The structures of GeS (Figure 6) and α-SnS are very similar to that of CrB (7), although more often (and with equal accuracy) described as "deformed NaCl" types; cf. Chapter XI. They also differ from CrB because of the lone pairs of electrons on the cations.

[Note also that the high-temperature form, β-SnS, is TlI type, and that the high-temperature form of TlI is B2, CsCl type. Simple martensitic mechanisms for the low $\rightleftarrows$ high transformations are readily visualized: they involve slipping the adjacent AlB_2-like walls, by $a/2$ for TlI $\rightarrow$ CsCl. A mechanism for NaCl $\rightarrow$ α-SnS $\rightarrow$ β-SnS (= TlI) is depicted in Figure 6c.]

The KOH and α-NaOH structures are even more accurately antitypes of TlI than is CrB. The prisms are defined by cations (Na, K) in place of I; the oxygen atoms are in the Tl sites, and the H atoms in the positions of the lone pair of electrons in TlI. β-NaOH is a deformed version of α-NaOH sheared parallel to the walls of trigonal prisms so that now it is the *anions* that define the more regular prisms. The two structures are depicted in Figures 7a–c.

Many borides of binary alloys are isostructural with CrB, including β-MoB. (Cf. α-MoB below.)

Tungsten Boride, Rhenium Boride and Related Structures

The other simple geometrical possibility for emptying half the trigonal prisms in AlB_2 is to eliminate a proportion of the (0001) sheets of B atoms (6^3 nets)—to empty, say, *alternate* layers of prisms—which could then collapse

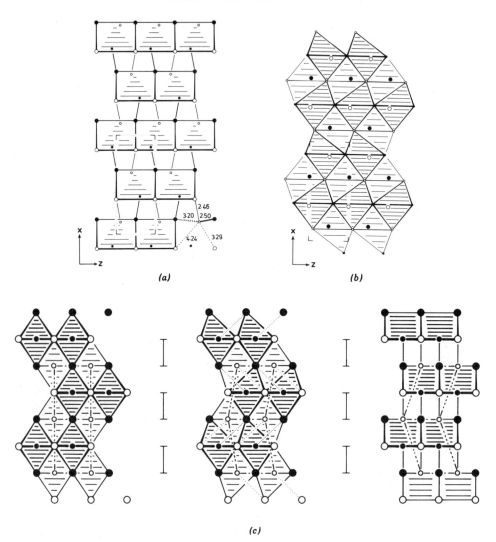

Figure 6. The structure of GeS (α-SnS type) projected on (010) of *Pnma*. Small circles are Ge; large circles are S: open at $y = \frac{1}{4}$, filled at $y = \frac{3}{4}$. (*a*) Cation-"centered" polyhedra, GeS$_6$ trigonal prisms (bond lengths in Å); (*b*) anion-"centered" polyhedra, SGe$_6$ octahedra. The former emphasizes the relation to the TlI ($= \beta$-SnS, CrB) type; the latter, to the NaCl type. (*c*) A mechanism for NaCl (left) $\rightarrow$ α-SnS (center) $\rightarrow$ β-SnS (TlI) is depicted schematically. (Note that the layers marked by "bars" remain undistorted throughout, but they are sheared as in Figure 4.)

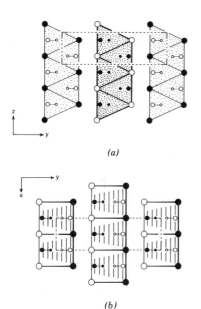

(a)

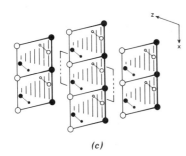

(b)

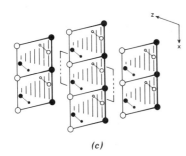

(c)

Figure 7. (a) α-NaOH projected on (100); open circles at $x = 0$, filled at $x = \frac{1}{2}$. (b) α-NaOH projected on (001); open circles at $z = \frac{1}{4}$, filled at $z = \frac{3}{4}$. (c) β-NaOH projected on (010); open circles at $y = \frac{1}{4}$, filled at $y = \frac{3}{4}$. In all cases the large circles are Na, medium are O, and small are H. O–H bonds are indicated by dotted lines. [Compare (a) and (b) with Figure 5.]

to (0001) lamellae of cp. Stacking sequences would be

$$\cdots \quad \alpha(bc)\alpha \quad \beta(ca)\beta \quad \alpha(bc)\alpha \quad \beta(ca)\beta \quad \cdots,$$
$$\cdots \quad \alpha(bc)\alpha \quad \beta(ca)\beta \quad \gamma(ab)\gamma \quad \alpha(bc)\alpha \quad \cdots,$$

or other polytypes.

We know of no such structures, but there are filled derivatives with additional atoms in the interstices of the cp lamellae. For example, if *all* these interstices (octahedral *and* tetrahedral) are occupied, one gets, for the first of the above stacking sequences, the W_2B_5 structure (Figure 8), and for the second, that of Mo_2B_5. In both cases the cp layers are lamellae of the ReB_3 type (*not* Re_3B). ReB_3 itself (Figure 9) has hcp Re with all octahedral and tetrahedral interstices occupied by B:

$$\cdots \quad \alpha bca\beta acb\alpha bca\beta \quad \cdots$$

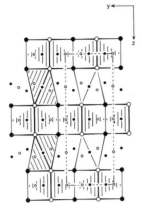

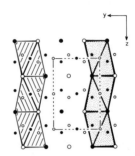

Figure 8. The "W_2B_5" structure projected on $(11\bar{2}0)$ of the hexagonal unit cell = (100) of the corresponding ortho-hexagonal cell (outlined). Large circles are W, small circles are B: open at $x_0 = 0$, filled at $x_0 = \frac{1}{2}$. The structure consists of alternate $(0001)_h = (001)_o$ layers of AlB_2 type (trigonal prisms) and ReB_3 type (of which some octahedra are shown on the left and some tetrahedra on the right, both being very elongated in the **c** direction). Cf. Figure 9.

Figure 9. The structure of ReB_3 on $(11\bar{2}0)_h = (100)_o$. Large circles are Re, small are B; open at $x_0 = 0$, filled at $x_0 = \frac{1}{2}$. Some BRe_6 octahedra are shown on the left, and some BRe_4 tetrahedra on the right; all (elongated) octahedral and tetrahedral sites between the cp Re layers are occupied by B atoms.

If, in the cp lamellae of the second stacking sequence above, only the tetrahedral holes are filled, the $CaSi_2$ structure results. But both Si layers are slightly puckered (as in Si itself) so that the sequence is now

$$\cdots \ \alpha \ (bc) \ \alpha b \ \alpha \beta \ (ca) \ \beta c \ b\gamma \ (ab) \ \gamma a \ c\alpha \ (bc) \ \alpha \ \cdots \ .$$

The structure of Sc_2O_2S (8), already referred to in Chapter VII, also fits in here. It has a stacking sequence related to the first one given above, namely,

$$\cdots \ \alpha b \ \alpha \beta \ c \ \beta a \ b\alpha \ c \ \alpha b \ \alpha \beta \ c \ \beta a \ b\alpha \ \cdots \ ,$$

and differs from La_2O_3 or La_2O_2S in having an anion S in trigonal prismatic instead of octahedral coordination. That is, it is an intergrowth of WC or NiAs with fluorite-type lamellae (instead of NaCl-type plus fluorite-type). It differs from W_2B_5 in having half its trigonal prisms and all its octahedra empty: $WB_2 \cdot WB_3 \rightarrow ScS \cdot ScO_2$.

There are a number of related structures in which the trigonal prism layers are only half-occupied (NiAs- or WC-like), and the interstices in the intervening cp layer pairs may or may not be occupied. These have been considered in some detail, and large numbers of examples listed, by Hulliger (9), and so we will mention only a few examples here.

β-MoS$_2$ (natural molybdenite) has layers of MoS$_6$ prisms of the NiAs- or WC-type stacked in a two-layer sequence with no additional cations between the cp pairs of S layers. That is, the cp layer sequence is

$$\cdots \quad b \; \alpha \; b \quad a \; \beta \; a \quad b \; \alpha \; b \quad \cdots .$$

The structure is shown in Figure 10. A synthetic α-MoS$_2$ has the obvious alternative, rhombohedral structure, with the layer-stacking sequence

$$\cdots \quad a \; \beta \; a \quad b \; \gamma \; b \quad c \; \alpha \; c \quad a \; \beta \; a \quad \cdots .$$

An interesting related structure is that of InSe. It differs from α-MoS$_2$ only in having In–In pairs in each trigonal prism (β_2, γ_2, α_2, oriented parallel to **c**).

The structure of TiP, shown in Figure 11, also belongs here; it is a filled antiform of β-MoS$_2$, consisting of alternate layers of WC-type, with PTi$_6$ trigonal prisms, and anti-NaCl type, with PTi$_6$ octahedra. It is drawn this way in the upper part of the figure. In the lower part we show the alternative description with cation-centered polyhedra: all octahedra, with the P atom

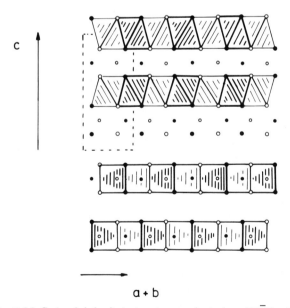

Figure 10. The β-MoS$_2$ (molybdenite) structure projected on ($11\bar{2}0$); all atoms are at 0 or $\frac{1}{2}$. The lower part shows the MoS$_6$ trigonal prisms, the upper part the empty S$_6$ octahedra in the intervening layers, from which it is obvious that the arrangement of cp S is slightly distorted.

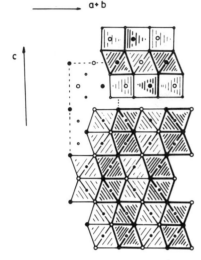

Figure 11. The structure of TiP projected on $(11\bar{2}0)$; all atoms are at 0 or $\frac{1}{2}$. Top, PTi_6 trigonal prisms and octahedra; bottom, TiP_6 octahedra.

layers in double hexagonal stacking $\cdots$ *abacab* $\cdots$ = **hc**. The full stacking sequence in this structure is

$$\cdots \ a\ \beta(c\ \beta\ a\ \gamma\ b\ \gamma\ a\ \beta)c\ \beta\ \cdots\ .$$

$TlInS_2$ (10) is the antistructure, with TlS_6 trigonal prisms and InS_6 octahedra.

Ta_3B_4

The Ta_3B_4 structure is based on the same principle as the CrB type, but the translation repeat unit is now a double wall of AlB_2 type (Figure 12). Again it can be regarded as AlB_2 with walls of empty prisms collapsed to cp (in this case one-third of the walls, compared with one-half in CrB). Structural data are given in Data Table 4.

Data Table 4 Ta_3B_4, $D7_b$ (11)

Orthorhombic, space group *Immm*, No. 71; $a = 3.29$, $b = 14.0$, $c = 3.13$ Å; $Z = 2$, $V = 144.0$ Å^3

Atomic Positions:
Ta(1) in 2(c): $0,0,\frac{1}{2}; \frac{1}{2},\frac{1}{2},0$
Ta(2) in 4(g): $\pm(0,y,0; \frac{1}{2},\frac{1}{2} + y,\frac{1}{2}); y = 0.180$
B(1) in 4(g): $y = 0.375$
B(2) in 4(h): $\pm(0,y,\frac{1}{2}; \frac{1}{2},\frac{1}{2} + y,0); y = 0.444$

In principle, a whole family of structures, each with repeat units n AlB_2-type walls thick, would appear to be possible: $M_{n+1}B_{2n}$. The only members

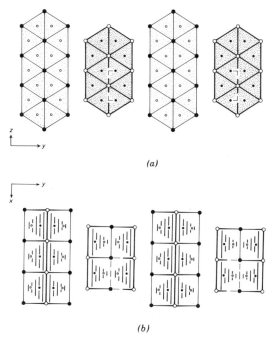

(a)

(b)

Figure 12. The structure of Ta_3B_4 projected on (a) (100) and (b) (001). Large circles are Ta; small circles are B: open at 0, filled at $\frac{1}{2}$.

known so far are the AlB_2, CrB, Ta_3B_4, and V_2B_3 types, all of which occur in the systems M + B, M = V, Nb, Ta, and Cr. They are related by a simple CS operation.

Intergrowths at intermediate compositions are also possible in principle. Few have been reported—one is $V_5B_6 = 2VB + V_3B_4$ (Figure 13)—although nonstoichiometry is not uncommon and is presumably due to *disordered* intergrowths.

W_2CoB_2 and W_3CoB_3

The structure of W_2CoB_2 (Figure 14) can be imagined as being derived from the AlB_2 type by regular omission of two thirds of the B atoms. [The "Al"

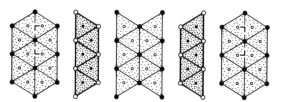

Figure 13. The structure of V_5B_6 (idealized). Compare with Figures 5a and 12a.

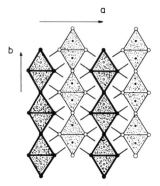

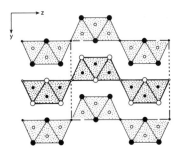

Figure 14. W₂CoB₂ on (001). W atoms are in the shared faces between B-centered trigonal prisms; Co atoms are in their shared edges. All atoms are at 0 or ½. This structure is readily derived from that of NbAs₂ (Figure 16) by CS.

Figure 15. The structure of W₃CoB₃ projected on (100). Large circles are W, medium are B, and small are Co; open at $x = 0$, filled at $x = \frac{1}{2}$. Cf. Figure 5a (CrB) and Figure 14 (W₂CoB₂).

array is, in fact, $(W_{2/3}Co_{1/3})$.] The emptied trigonal prisms lie in zigzag strips, which then collapse to cp layers. Several other ternary borides are isostructural (13).

Structural data are given in Data Table 5.

Data Table 5   W₂CoB₂ (12)

Orthorhombic, space group *Immm*, No. 71; $a = 7.07_5$, $b = 4.56_1$, $c = 3.71_7$ Å; $Z = 2$, $V = 102.5$ Å³

Atomic Positions: $(0,0,0; \frac{1}{2},\frac{1}{2},\frac{1}{2})+$
Co in 2(a):   0,0,0
W in 4(f): $\pm(x,\frac{1}{2},0)$; $x = 0.205$
B in 4(h): $\pm(0,y,\frac{1}{2})$; $y = 0.30$

The structure of W₃CoB₃, shown in Figure 15, is similarly derivable from the AlB₂ type. Data are given in Data Table 6.

Data Table 6   W₃CoB₃ (14)

Orthorhombic, space group *Cmcm*, No. 63; $a = 3.173$, $b = 8.422$, $c = 10.728$ Å; $Z = 4$, $V = 286.7$ Å³

Atomic Positions: $(0,0,0; \frac{1}{2},\frac{1}{2},0)+$
Co in 4(b): $0,\frac{1}{2},0$; $0,\frac{1}{2},\frac{1}{2}$
W(1) in 4(c): $\pm(0,y,\frac{1}{4})$; $y = 0.500$
W(2) in 8(f): $\pm(0,y,z; 0,y,\frac{1}{2} - z)$; $y = 0.210$, $z = 0.106$
B(1) in 4(c): $y = 0.795$
B(2) in 8(f): $y = 0.907$, $z = 0.114$

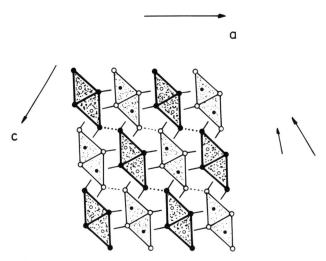

Figure 16. The structure of NbAs$_2$ projected on (010); all atoms are at 0 or $\frac{1}{2}$. Obvious As–As bonds are indicated by dotted lines. Other As–As bonds may deform the prisms by shortening one edge.

Column Derivatives

Having considered parallel translation derivatives of walls of AlB$_2$ (and a few very similar structures) we now turn to analogous *column* derivatives. These include NbAs$_2$ (Figure 16), in which the building element is a double column. CrP$_2$, OsGe$_2$, etc. are isostructural. Dy$_3$Ni$_2$ has columns twice as large as those in NbAs$_2$ namely, four trigonal prisms (in a linear strip).

NbAs$_2$ is readily transformed to CrB by CS in two obvious ways (the alternative CS vectors being indicated by arrows in Figure 16). A different, but obvious, CS vector transforms NbAs$_2$ to the W$_2$CoB$_2$ structure, shown in Figure 14. The latter consists of NbAs$_2$ double columns joined into sheets by edge-sharing between adjacent prisms. This structural unit is also, as we have already seen, readily excised from the AlB$_2$ structure.*

Note that both of these structures have been described earlier as twinned ccp (although the trigonal prisms are somewhat distorted by As–As bonding in NbAs$_2$): NbAs$_2$ as twinned ccp . . . , *1,3,1,3*, . . . and W$_2$CoB$_2$ as . . . , *1,2,1,2*, ... The (pseudo) mirror planes are (001) and (110), respectively. W$_3$CoB$_3$ is twinned ccp . . . , *1,1,2,1,1,2*, . . . = (*1²,2*).[2]

Compare Figure 16 with Figure 16 of Chapter VIII, the structure of Li$_2$Si. The latter is readily transformed into the former by the As $\cdots$ As bonding.

The structures of Re$_3$B (anti-PuBr$_3$), Y(OH)$_3$ = UCl$_3$, ZrSe$_3$, and TaSe$_3$ can also be produced by parallel translation of single columns of AlB$_2$ type.

---

* Clearly, NbAs$_2$ can also be readily converted to Dy$_3$Ni$_2$ by CS.

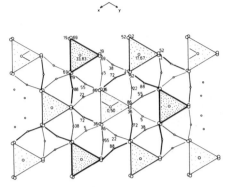

Figure 17. Pyrargyrite (SbAg$_3$S$_3$) projected on (111) of its rhombohedral cell = (0001)$_h$. Large circles are S, medium circles are Sb, and small circles are Ag; heights are given in units of c_h/100. Ag–S bonds and SbS$_6$ trigonal prisms are drawn. Contrast this with Figures 15 and 16 of Chapter VI.

The first two have been dealt with earlier, although from a different point of view (see Chapters VI and III). The second can also be produced by *cyclic* translation (see below).

Another interesting arrangement of single columns occurs in the isostructural pyrargyrite, SbAg$_3$S$_3$ (Figure 17), and proustite, AsAg$_3$S$_3$. Compare these with the structures wittichenite, BiCu$_3$S$_3$, and skinnerite, SbCu$_3$S$_3$.

Other Derivatives of AlB$_2$

α-MoB

The structure of (low-temperature) α-MoB (Figure 18), is closely related to that of CrB (= β-MoB; Figure 5), being also composed of unit walls of AlB$_2$ type. However, adjacent walls are now rotated by 90° with respect to each other. Whereas CrB is generated by a translation operation, α-MoB is generated by a screw operation: translation + rotation. It is a twinned CrB type. Mechanistically, the twinning is plausibly attained by a slip of (010) $\frac{1}{2}$[101] on alternate unit cells of CrB, doubling the c axis. Crystallographic data are given in Data Table 7.

Data Table 7 α-MoB (15)

Tetragonal, space group $I4_1/amd$, No. 141; $a = 3.105$, $c = 16.97$ Å; $Z = 8$, $V = 163.6$ Å^3

Atomic Positions
Atoms in 8(e): $\pm(0,0,z; 0,\frac{1}{2},\frac{1}{4} + z; \frac{1}{2},\frac{1}{2},\frac{1}{2} + z; \frac{1}{2},0,\frac{3}{4} + z)$
Mo: $z = 0.197$
B: $z = 0.352$

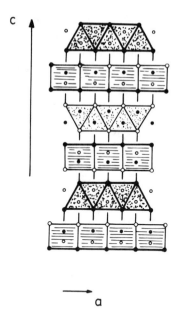

c

a

Figure 18. The α-MoB structure projected on (010). Larger circles are Mo; smaller are B: open at $y = \frac{1}{2}$, filled at $y = 0$. Compare with Figures 5 a, b.

One might expect a family related to α-MoB as Ta_3B_4, V_2B_3, and AlB_2 are to CrB. The only other known member is the analogue of AlB_2, the α-ThSi₂ structure shown in Figure 19. (β-ThSi₂ is AlB₂-type.) Pearson reports (though Hansen contradicts) that α-ThSi₂ is often Si-deficient (up to 30%) which could indicate α-MoB + α-ThSi₂ intergrowths. There are many isostructural compounds. (α-GdSi₂ is very slightly deformed α-ThSi₂; orthorhombic, with $a/b = 1.02$, instead of tetragonal.)

In the same way as WC and NiAs may be regarded as AlB₂ with half the trigonal prisms unoccupied, so there are half-occupied analogues of the derivative structures, such as CrB. A few of these have already been discussed earlier in this chapter—TiP, etc. We now discuss some others.

Monoclinic YO(OH)

A number of lanthanoid oxyhydroxides, for example, those of Tb, Ho, Er, Yb, crystallize with this structure type. Neutron powder diffraction data

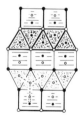

Figure 19. The α-ThSi₂ structure projected on (010); all atoms are at 0 or $\frac{1}{2}$. Compare with Figure 18, from which structure this one is readily derived by CS, and also Figure 1. (The broken lines indicate part of the three-dimensional, 3-connected silicon net.)

allowed the deuterium positions in YO(OD) to be determined, showing that there is no hydrogen bonding in the structure. More accurate data (except for H) were determined by single crystal X-ray diffraction analysis of HoO(OH). Both sets of data are included in Data Table 8, and the structures are shown in Figure 20. (Compare CrB in Figure 5.)

Data Table 8   (a) YO(OD) (16) and (b) HoO(OH) (17)

Monoclinic, space group $P2_1/m$, No. 11; (a) $a = 5.95$, $b = 3.65$, $c = 4.30$ Å, $\alpha = 109.1°$; (b) $a = 5.96$, $b = 3.64$, $c = 4.31$ Å, $\beta = 109.1°$; $Z = 2$, $V = 88.2$ Å^3 (a), 88.4 Å^3 (b)

Atomic Positions
All atoms in 2(e): $\pm(x,\frac{1}{4},z)$

For YO(OD):

	x	z
Y	0.8102	0.6686
O(1)	0.935	0.230
O(2)	0.434	0.245
D	0.357	0.011

For HoO(OH):

	x	z
Ho	0.8101	0.6633
O(1)	0.943	0.235
O(2)	0.437	0.235
H	—	—

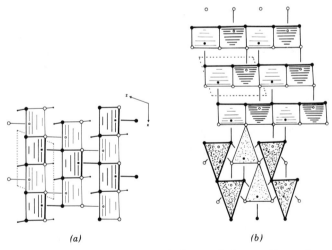

(a)   (b)

Figure 20. (*a*) YO(OD) projected on (010). Large circles are O, medium circles are Y, and smallest circles are D; open at $y = \frac{1}{4}$, filled at $y = \frac{3}{4}$. YO$_6$ trigonal prisms and O–D bonds are drawn. (*b*) HoO(OH). The relationship to PbFCl type is indicated at the bottom.

The relation to the CrB structure is clear. The equivalent pseudo-orthorhombic cell would have an axis $4c \sin \beta \equiv 2\mathbf{b}(CrB)$.

In Figure 20b a topological resemblance to the PbFCl structure is shown.

SrBr₂·H₂O

Some years ago the structure shown in Figure 21 (Strukturbericht type C53) was reported as that of SrBr₂ (18). It later became clear that the substance investigated was in fact the monohydrate, SrBr₂·H₂O (19). The relevant data are given in Data Table 9 (the earlier set in parentheses); and the structure in Figure 21 is shown without H₂O.

Data Table 9   SrBr₂·H₂O (18, 19)*

Orthorhombic, space group *Pnma*, No. 62; $a = 11.39$ (11.42), $b = 4.28$ (4.3), $c = 9.19$ (9.20) Å; $Z = 4$, $V = 448$ (452) Å³

Atomic Positions
All atoms in 4(*c*): $\pm(x,\frac{1}{4},z; \frac{1}{2} + x,\frac{1}{4},\frac{1}{2} - z)$

	x	z
Sr	0.3094 (0.311)	0.3845 (0.389)
Br(1)	0.1019 (0.103)	0.1143 (0.119)
Br(2)	0.6150 (0.614)	0.8456 (0.842)
O	0.325 (−)	0.879 (−)

* Data in parentheses are from ref. 18.

Comparison of Figure 21 with Figure 20 [for YO(OD) and HoO(OH)] shows that C53 is an internally twinned form of HoO(OH). This arises from a

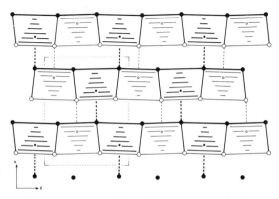

Figure 21.   SrBr₂·H₂O projected on (010) of *Pnma*. Small circles are Sr; large circles are Br; open at $y = \frac{1}{4}$, filled at $y = \frac{3}{4}$. (H₂O molecules are not shown.) Dotted lines are bonds from Sr to the Br atoms capping the SrBr₆ trigonal prisms.

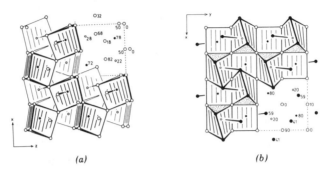

Figure 22. Tetragonal YbO(OH) projected on (*a*) (010) and (*b*) (001). Smaller circles are Yb, larger circles are O (H not shown); heights in units of *b*/100 or *c*/100. (Compare with Figure 27 of Chapter IV.)

difference in the sequence of filled and empty prisms, and makes $\mathbf{a} \equiv \mathbf{b}(CrB)$. Consequently, the relation analogous to that between HoO(OH) and PbFCl is now between C53 and C23 (PbCl$_2$). The last is therefore twinned PbFCl!

[The structure of anhydrous SrBr$_2$ is rather complex; it is tetragonal and is related to that of α-US$_2$ (20, 21). Its "molecular" volume is 97 Å^3, compared with 112 Å^3 for the monohydrate.]

It should be noted that although there is a tetragonal lanthanoid oxyhydroxide structure, it is quite different. The example of YbOOH, shown in Figure 22, is a slightly distorted twinned hcp anion array, . . . , 2,2,2, ... It is rather similar to ZrO$_2$ (Chapter IV).

NbAs

This is the analogous derivative of α-ThSi$_2$, with half the trigonal prisms empty. As α-ThSi$_2$ is an internally twinned AlB$_2$, so NbAs is an internally twinned derivative of WC or NiAs. Its structure is shown in Figure 23.

FeB and Other Twinned Derivatives of CrB

The structure of FeB is shown in Figure 24, and data are given in Data Table 10. It has previously been described (Chapter IV, Figure 28) as twinned hcp Fe . . . , *1,1,1,* . . . . Now it is seen to consist of *strips* of AlB$_2$ connected by edge-sharing of the trigonal prisms. A unit strip of AlB$_2$ is repeated by *a*-glide reflection on (001) (or, equivalently, by exactly the same translation repeat as the one that generates CrB from an AlB$_2$ *wall*) and *n*-glide reflection on (100), setting *Pnma*. (*Caution!* The setting *Pbnm* is commonly used for this space group.) It is readily derived from the AlB$_2$ structure by emptying the B atoms from alternate strips of trigonal prisms in the latter and allowing the empty prisms (no longer "propped up") to collapse to close-packing.

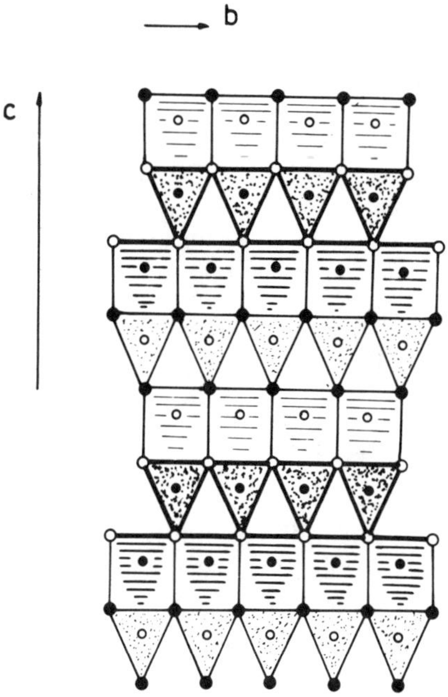

b

c

Figure 23. NbAs projected on (100). Large circles are Nb; small are As; open at $x = \frac{1}{2}$, filled at $x = 0$. (Compare with Figure 19.)

Data Table 10 FeB, B27 (22)

Orthorhombic, space group *Pnma*, No. 62; $a = 5.495$, $b = 2.946$, $c = 4.053$ Å; $Z = 4$, $V = 65.61$ Å³

Atomic Positions

All atoms in 4(c):	$\pm(x,\frac{1}{4},z; \frac{1}{2} + x,\frac{1}{4},\frac{1}{2} - z)$
Fe:	$x = 0.180$, $z = 0.125$
B:	$x = 0.036$, $z = 0.61$

Recall that CrB is described as twinned ccp Cr . . . , *1,1,1*, . . . or as a repeated parallel translation of *walls* of AlB₂. In fact, the (100) layers of FeB (Figure 24) are identical to the (110) layers of CrB (Figure 5). In the latter, however, they are repeated by translation parallel to **a**, and in the former by

Figure 24. FeB projected on (010). Large circles are Fe; small circles are B; open at $y = \frac{1}{4}$, filled at $y = \frac{3}{4}$. (Compare with Figure 28 of Chapter IV.)

TABLE 1 Some Twinned CrB Structures

		nh/nc
GdNi (23) = CrB	c	0
TbNi (high-temperature) (24)	hc^2	0.50
$Gd_{0.75}Y_{0.25}Ni$ (23)	h^2c^3	0.67
β-CaCu (25)	hc^2hc	0.67
$Gd_{0.4}Tb_{0.6}Ni$ (26)	$h^2c^3h^2c^2$	0.80
SrAg (25)	hc	1.00
$Gd_{0.3}Tb_{0.7}Ni$ (26)	h^2c^2	1.00
α-CaCu (25)	h^2chc	1.50
$Gd_{0.2}Tb_{0.8}Ni$ (26)	h^2c	2.00
TbNi (low-temperature) (24)	h^4c^2	2.00
BaAg (25), DyNi (23), FeB	h	∞

an n-glide on (100). FeB can therefore also be described as (another type of) twinned CrB. In fact, there is a whole family of such (n-glide) twinned CrB structures. Since they are intergrowths of CrB and FeB, they can also be described as twinned mixed (**h** and **c**) cp stackings (see Table 1).

Some of these are shown schematically in Figure 25. (Compare Figures 23b and d of Chapter IV.)

Cyclic Translation Derivatives of AlB₂ and Related Column Elements

Cyclic translation (about a $\bar{3}$ axis) of a unit column of AlB_2 generates the structure of $Y(OH)_3$ (or UCl_3) previously considered at the end of Chapter III and depicted there in Figure 39. Figure 26 shows how a simple CS operation relates this structure to that of $NbAs_2$ (which, in turn, is related to other structures by CS as described earlier, p. 229).

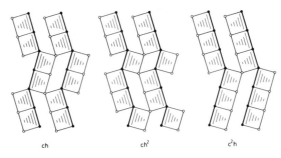

ch   ch²   c²h

Figure 25. Examples of the topologies of the trigonal prism layers in the composition planes of some twinned mixed (**h** + **c**) stackings. Compare Figure 24 (twinned **h**) and Figure 5b (twinned **c**).

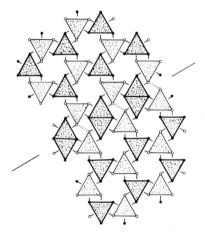

Figure 26. CS operation on the $Y(OH)_3$ structure to produced a lamella of $NbAs_2$ type. Note how the short "As–As" distances (i.e., bonds), indicated by double dotted lines, result automatically.

The same cyclic translation using a larger element of AlB₂ generates the structure of $Ce_6Ni_2Si_3$ shown in Figure 27. The AlB₂ elements are $Ce_6(NiSi_3)$ (triangular columns of four trigonal prisms) with Ni or Si in the trigonal prisms; and the octahedra at the origin ($\bar{3}$ axis), which form columns by sharing faces, are all occupied by Ni. A still larger AlB₂ element (triangular columns of nine trigonal prisms) yields the structure of "Ce_2NiSi" shown in

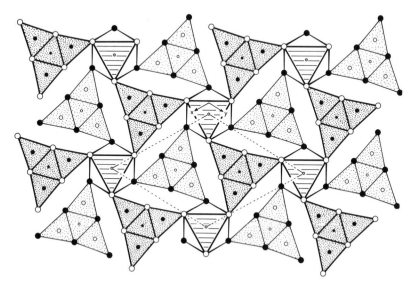

Figure 27. The structure of $Ce_6Ni_2Si_3$ projected on (0001): large circles are Ce, medium circles are (Ni,Si), small circles are Si; open at $z = \frac{3}{4}$, filled at $z = \frac{1}{4}$, dotted (Ni at $000z$) at 0 and $\frac{1}{2}$. AlB₂-like columns, four trigonal prisms in cross section, are cyclically twinned about the 6_3 axis at $000z$.

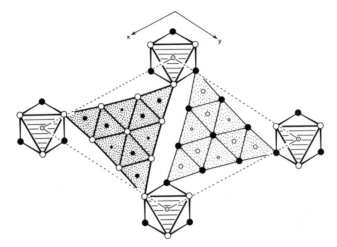

Figure 28. The structure of Ce_2NiSi projected on (0001). Large circles are Ce, medium circles are (Ni,Si), small circles are Si; open at $z = \frac{3}{4}$, filled at $z = \frac{1}{4}$, dotted (Ni at $000z$) at $z = 0$ and $\frac{1}{2}$. Compare Figure 27; the AlB_2-like columns are now nine trigonal prisms in cross section.

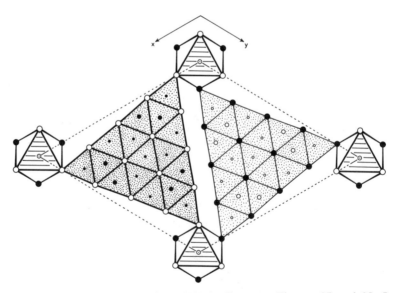

Figure 29. $Ce_{15}Ni_4Si_{13}$ projected on (0001). Compare Figures 27 and 28. Large circles are Ce, medium circles are $(Ni_{0.5}Si_{0.5})$ at $\pm\frac{1}{4}$, Ni at 0 and $\frac{1}{2}$, small circles are Si; open at $z = \frac{3}{4}$, filled at $z = \frac{1}{4}$, dotted at $z = 0$ and $\frac{1}{2}$. The AlB_2-like columns are now 16 trigonal prisms in cross section.

Figure II.39. A model of the spinel structure.

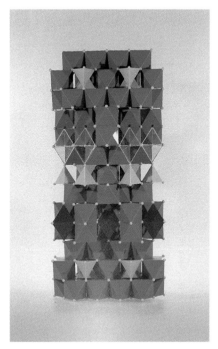

Figure II.40. A model of the structure of $BaFe_{12}O_{19}$.

Figure II.41. The structure of $NaAl_{11}O_{17}$ (or β-alumina).

Figure III.47. The structure of Ti_5O_9, which is $n = 5(121)_r$. Alternate rutile slabs are colored red and blue.

Figure IV.15. Octahedral model of Al_2O_3. Blue and yellow parts are rutile.

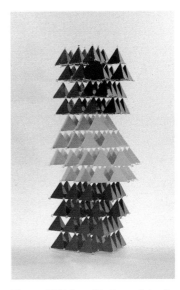

Figure VII.1. (*b*) A model of 8H-SiC.

Figure VII.2. A model of the Al_4C_3 structure: Al atoms in C_4 tetrahedra.

Figure VII.3. A model of the $Al_6C_3N_2$ structure: Al atoms in the tetrahedra.

Figure VII.5. Models of structures in the system $La_2O_3 + ThO_2$. Left to right (with cp cation stacking sequence in parentheses): NaCl type (c); La_2O_3 (h); $M_{17}X_{24}$ (h^{4}c); M_3X_5 (h^{2}c); CaF_2 type (c). In terms of the c_h axis of their hexagonal unit cells the model heights are respectively c_h; $5c_h/2$; $16c_h/15$; $7c_h/9$; c_h.

Figure XIV.2. (*b*)
A polyhedral model
of the γ-brass structure.

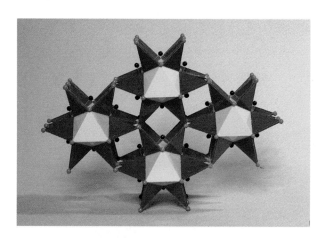

Figure XIV.3. (*b*)
A polyhedral model
of part of the
Mn_5Si_3 structure
viewed approxi-
mately along **c**.

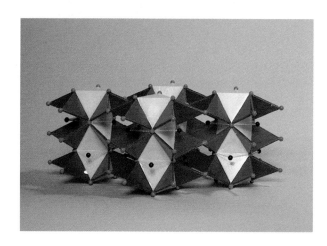

Figure XIV.3. (*c*)
A polyhedral model
of part of the
Mn_5Si_3 structure
viewed at right
angles to **c**, approxi-
mately along $[11\bar{2}0]$.

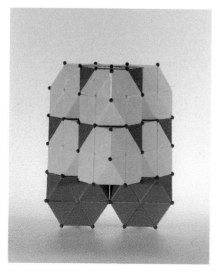

(a) MgCu$_2$.

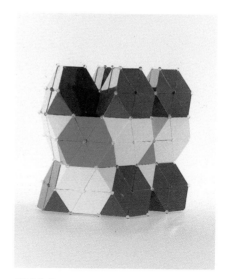

(b) MgZn$_2$.

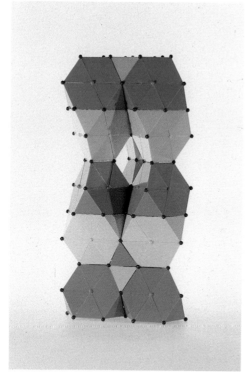

(c) MgNi$_2$.

Figure XIV.30. Polyhedral models of the three simplest Friauf–Laves phases.

Figure XV.3. Polyhedral model of garnet. Note the rods of alternating octahedra and trigonal prisms sharing faces. Such rods pack as in Figure 2 of Chapter XV.

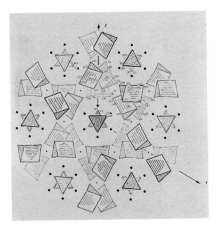

Figure XV.4. The structure of $Y_3Fe_5O_{12}$ projected on (111). The empty trigonal prisms are outlined. Prisms of different colors lie on rods parallel to different $\langle 111 \rangle$ directions.

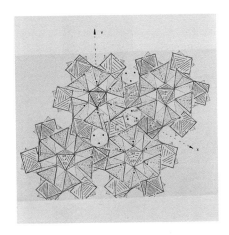

Figure XV.5. The network of octahedra and tetrahedra in the structure of $Y_3Fe_5O_{12}$ drawn in the same projection as in Figure 4 of Chapter XV.

Figure XV.6. The structure of benitoite, $BaTiSi_3O_9$: Ti in red octahedra, Si in yellow tetrahedra.

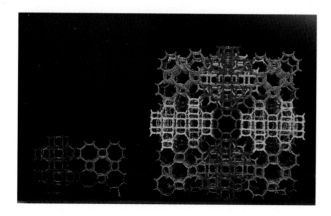

Figure XV.17. The blue unit on the left can be traced in the center of Figure 16 of Chapter XV. It is very closely related to merlinoite and builds the structure of paulingite on the right—a formidable sixling. The red, green, and yellow units are all identical and equal to the blue unit.

Felspar

Figure XV.24. Model of the feldspar structure with regular tetrahedra viewed along **b.**

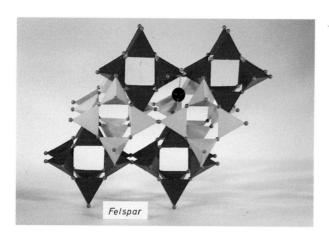

Figure XV.28. Cuboctahedron unit of Linde A.

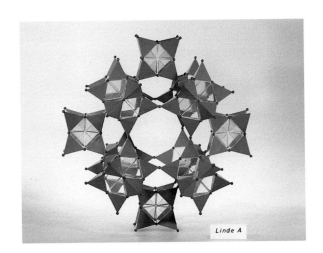

Figure XV.29. Polyhedral model of Linde A.

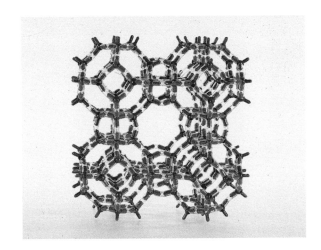

Figure XV.30. Rod model of Linde A. Oxygens are in the connectors.

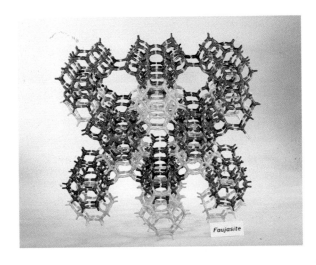

Figure XV.31. Rod model of faujasite.

Figure 28 and built on exactly the same principle. An even larger AlB$_2$ element (triangular columns of 16 trigonal prisms) yields the Ce$_{15}$(Ni,Si)$_{17}$ structure shown in Figure 29. These are all members of a family Ce$_{n^2+3n+2}$ (Ni,Si)$_{2n^2+2}$ with n = 2, 3, 4 being the number of Ce$_6$ trigonal prisms in the edge of the triangular column of AlB$_2$ type. (In each case there is partial ordering of Si and Ni in the trigonal prisms.) Unit cell data for these structures are given in Data Tables 11–13.

Data Table 11   Ce$_6$Ni$_2$Si$_3$ (27)

Hexagonal, space group $P6_3/m$, No. 176; a = 12.112, c = 4.323 Å; Z = 2, V = 549.2 Å^3

Atomic Positions
See ref. 27.

Data Table 12   Ce$_2$NiSi (= Ce$_{10}$Ni$_5$Si$_5$) (28)

Hexagonal, space group $P6_3/m$, No. 176; a = 16.12, c = 4.309 Å; Z = 10, V = 969.7 Å^3

Atomic Positions
See ref. 28.

Data Table 13   Ce$_{15}$Ni$_4$Si$_{13}$ (29)

Hexagonal, space group $P6_3/m$, No. 176; a = 20.27, c = 4.306 Å; Z = 2, V = 1532 Å^3

Atomic Positions
See ref. 29.

It may be noted that CeSi$_2$ has the α-ThSi$_2$-type structure (Figure 19)— not AlB$_2$ type, but closely related to it (by internal twinning, see above).

The exact n = 1 member of this series A$_6$B$_4$, is unknown. But the Fe$_2$P type (discussed in Chapter III) is a close analogue whose structure differs only in having atoms along $00z$ (at the $\bar{3}$ axes) in tricapped trigonal prisms instead of octahedra. The relationship is emphasized in Figure 30. [Note the topological relation between columns of face-sharing octahedra and (triangular) face-sharing, tricapped trigonal prisms, extensively discussed toward the end of Chapter III (especially Figures 27–30) and in Chapter XI (Figure 15).] There are therefore only half as many atoms at these positions, and the stoichiometry is changed (from M$_6$X$_4$ to M$_6$X$_3$ = 3M$_2$X).*

---

* Compare the structures of Fe$_2$P and Y(OH)$_3$/UCl$_3$.

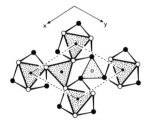

Figure 30. The structure of Fe_2P projected on (0001) and drawn to emphasize its relationship with the $Ce(Ni,Si)_x$ structures shown in Figures 27–29 and Th_7S_{12} (Figure 31), etc. (Origin shifted by $[\frac{1}{3}, 0, -\frac{1}{3}, -\frac{1}{4}]$.) Large circle is Fe, small circle is P; open at $z = \frac{3}{4}$, filled at $z = \frac{1}{4}$. The P coordination at $00z$ is a tricapped trigonal prism, not an octahedron (cf. Figures 27–29). Compare with Figure 33 of Chapter III.

Of course, all these structures too can be regarded as deriving from AlB_2 by emptying walls of prisms and collapsing the empty walls. In generating CrB, Ta_3B_4, etc., the empty walls run in only one direction, but in these column structures they run in all three equivalent directions. They are (CS) "column structures." (Compare the one- and two-dimensional CS structures in Chapter II.)

By emptying alternate columns of trigonal prisms in the triangular columns of the $Ce(Ni,Si)_x$ structures, we obtain an analogous series based on rotational translation of WC (instead of AlB_2). The general formula then changes from $M_{n^2+3n+2}X_{2n}2Y_2$ to $M_{n^2+3n+2}X_{n^2+n}Y_2$, where the M atoms form the columns of trigonal prisms, the X atoms are in the trigonal prisms, and the Y atoms are in the octahedra that lie in columns parallel to **c** at the origin of the cell. If these octahedra are only 50% occupied (or if the Y atoms are in tricapped trigonal prisms at these positions), the latter formula is $M_{n^2+3n+2}X_{n^2+n}Y$. The known examples of the last are

$n = 1*$: $M_6X_2Y = \beta_1$-$F_6K_2Th \equiv Ni_6Si_2B$; or $M_6(X,Y)_3 = 3M_2(X,Y) \equiv$ Fe_2P

$n = 2$: $M_{12}X_6Y$ or $M_{12}(X,Y)_7 = Th_7S_{12}$, $S_2Br_{10}Pb_7$, $\sim Cr_{12}P_7$

$n = 3$: $M_{20}X_{12}Y$ or $M_{20}(X,Y)_{13} = Rh_{20}Si_{13}$

$n = 4$: $M_{30}X_{20}Y$ or $M_{30}(X,Y)_{21} = 3M_{10}(X,Y)_7 -$ unknown

$n = 1, 2$, and 3 structures are shown in Figures 30, 31, and 32, respectively. Fe_2P ($n = 1$) has the P atoms at $00z$ in trigonal prisms. In $Rh_{20}Si_{13}$, the corresponding Si atoms occupy half the octahedra, apparently in a disordered way. Discussions of $Cr_{12}P_7$, $V_{12}P_7$, etc. all follow Zachariasen's original discussion of Th_7S_{12} (in 1949) in putting the corresponding atoms (at $00z$) into half the possible tricapped trigonal prisms (centered at $z = \frac{1}{4}$ or $\frac{3}{4}$) with corresponding minor adjustments in the positions of the coordinating atoms (see Figure 31). This would appear to be only an approximate solution to the "disorder" problem; and it may be relevant that the real structure of $Cr_{12}P_7$ appears to have a sevenfold supercell, rather than the subcell shown in Figure 31 (in which the structure was solved; see ref. 30).

$Bi_{19}S_{27}Br_3$ (= $M_{19}X_{30}$ or $M_{12\ 2/3}X_{20}$) (31) has the $n = 3$ structure, but with

---

* This structure is common to cyclically twinned AlB_2 and WC types.

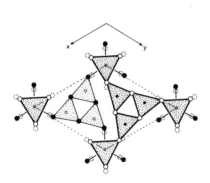

Figure 31. The Th_7S_{12} type structure of $Cr_{12}P_7$ projected on (0001). Larger circles are Cr, smaller are P; open at $z = \frac{3}{4}$, filled at $z = \frac{1}{4}$, dotted at $z = \pm \frac{1}{4}$, but these last P sites are only half-occupied, as a consequence of which two (alternative) sets of adjacent Cr positions are said to be also half-occupied—these are the paired full and broken circles (at $z = \frac{1}{4}$ or $\frac{3}{4}$). One set of these has been selected and emphasized: P at $0,0,\frac{1}{4}$ and the adjacent full circles representing the neighboring Cr atoms. (Compare Figure 27.) Note the trigonal prisms along [000z].

only one-third (instead of one-half) of the octahedra at $00z$ occupied. [Minor cation displacements are also apparent, as expected for Bi(III); cf. Chapter X.]

We complete this chapter with some miscellaneous structures related to those already considered but not conveniently falling into any of the above groups.

Intergrowths of AlB₂ with Other Types of Layers

Some of these have been discussed previously (p. 220), where the intervening layers were cp. Structures such as CrB itself consist of cp layers alternating with AlB₂-like walls. Here we consider thicker intervening layers. Examples are MoAlB (Figure 33), which consists of single CrB layers interleaved with additional bcc layers, each two atoms thick (making four layers of bcc in all). In Mn_2AlB_2 (Figure 34), the latter are three atoms thick, i.e., single

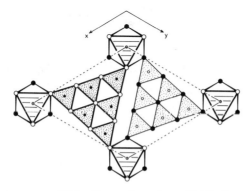

Figure 32. The structure of $Rh_{20}Si_{13}$ projected on (0001). Larger circles are Rh, smaller are Si; open at $z = \frac{3}{4}$, filled at $z = \frac{1}{4}$, dotted at $z = 0$ and $\frac{1}{2}$ (but these last are only half-occupied). Compare Figure 28. Note the octahedra along [000z].

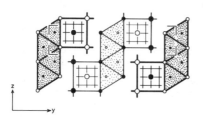
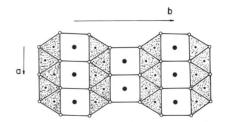

Figure 33. MoAlB projected on (100). Large, medium, and small circles are Al, Mo, and B, respectively; open at $x = 0$, filled at $x = \frac{1}{2}$. In the 4-layer "bcc" lamellae, only some of the Al(Al$_4$Mo$_4$) cubes can be shown. (The intermediate lamellae are, of course, CrB type.)

Figure 34. The structure of Mn$_2$AlB$_2$ projected on (001). Large, medium, and small circles are Al, Mn, and B, respectively; filled at $z = 0$, open at $z = \frac{1}{2}$. It may be described as an intergrowth of CrB- and CsCl-like lamellae. Compare Figures 5 and 18 of Chapter VIII and Figures 5a, 33, and 35.

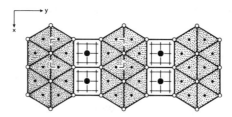

Figure 35. The structure of Cr$_3$AlB$_4$ projected on (001). Large, medium, and small circles are, respectively, Al, Cr, and B: open at $z = 0$, filled at $z = \frac{1}{2}$. It is an intergrowth of Ta$_3$B$_4$-type (Figure 12a) and CsCl-type lamellae.

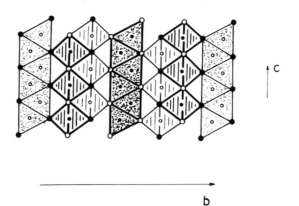

Figure 36. The structure of Mo$_2$BC projected on (100), showing the BMo$_6$ trigonal prisms and CMo$_6$ octahedra. It is an intergrowth of NaCl and CrB types. (Cf. Figure 33.)

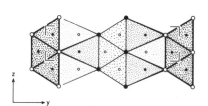

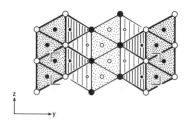

Figure 37. $ZrSi_2$ projected on (100). Large circles are Zr; small circles are Si, open at $x = 0$, filled at $x = \frac{1}{2}$.

Figure 38. UBC projected on (100). Large, medium, and small circles are U, B, and C, respectively; open at $x = 0$, filled at $x = \frac{1}{2}$. Compare with Figures 33 and 37.

layers of bcc or CsCl type. Cr_3AlB_4 (32) (Figure 35) is similar to Mn_2AlB_2, except that the CsCl layer is intergrown with Ta_3B_4 type: a double wall of AlB_2 instead of a single wall (CrB type).

Clearly the opportunities for disorder, nonstoichiometry, and polymorphism are legion.

The structure of Mo_2BC has two layers of ccp (metal) atoms between the CrB-like walls. The octahedra thus generated are occupied by C atoms (Figure 36); Mo_2BC is an intergrowth of CrB and NaCl types. (Also note that β-MoB is CrB type and that α-MoC, a high-pressure form, is NaCl type. So $Mo_2BC = MoB \cdot MoC$.)

The structure of $ZrSi_2$ (Figure 37) is a filled ("stuffed") version of CrB, with additional Si in tetrahedra between the AlB_2-like walls. (Note that one form of ZrSi is CrB type.) UBC (Figure 38) is another type of filled CrB, with additional (C) atoms now in half-octahedra/square pyramids in the cp layer. (Topologically, it is very similar to MoAlB, Figure 33.)

Some Miscellaneous Trigonal Prism Structures

CoGe, Ni_3Sn_4, and $PdBi_2$

These structures are based on (100) layers of FeB or, equivalently (110) layers of CrB type (one AlB_2 strip thick). These are intergrown with unit layers of another type.

The structure of CoGe, shown in Figure 39, is an intergrowth of $(100)_{FeB} \equiv (110)_{CrB}$ layers with $(11\bar{2}0)$ layers of the NiAs type. The structure of Ni_3Sn_4 is very similar (Figure 40), but now alternate octahedra in the NiAs-type layer are unoccupied; that is, it is an intergrowth of FeB (or CrB) with $(11\bar{2}0)$ layers of the $Cd(OH)_2$ type. α-$PdBi_2$ is similar, but now *all* the octahedra are empty (Figure 41). [β-$PdBi_2$, stable in the temperature range 125–150°C, is $MoSi_2$ type (Chapter VIII, Figure 18); clearly related to the α form by a small topological distortion.]

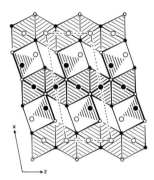

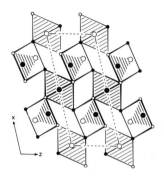

Figure 39. CoGe projected on (010). Large circles are Co; small circles are Ge; open at $y = 0$, filled at $y = \frac{1}{2}$. Alternate (100) layers are FeB/CrB (trigonal prisms) and NiAs (octahedra) types. Compare Figures 40–42.

Figure 40. Ni$_3$Sn$_4$ projected on (010). Large circles are Ni; small circles are Sn, open at $y = 0$, filled at $y = \frac{1}{2}$. Alternate (100) layers are now FeB/CrB (trigonal prisms, as in CoGe) and Cd(OH)$_2$ (octahedra) types. Compare Figures 39, 41, and 42.

It is easy to see that these three structures are column derivatives (in the sense defined in Chapter I) of AlB$_2$, produced by emptying alternate walls and alternate layers of trigonal prisms in the latter.

Chalcostibite or Wolfsbergite, CuSbS$_2$, and Emplectite, CuBiS$_2$

These are isostructural and are structurally similar to CoGe, etc. The hcp layers alternating with FeB/CrB are now of the wurtzite type; the octahedra are empty, but half the tetrahedra are occupied (Figure 42). In addition, the FeB/CrB layers of trigonal prisms alternate in their orientation; each is rotated 180° with respect to its adjacent layers (a twin operation). This changes the symmetry from monoclinic to orthorhombic.

The Sb/Bi atoms are, as usual, displaced from the centers of the trigonal prisms by their lone pairs of electrons. (See Chapter X.)

Sr$_2$PbO$_4$ and B-Sm$_2$O$_3$

These two structures bear more than a passing resemblance to the CoGe and wolfsbergite groups, and they are very closely related to each other. They both contain elements of anti-FeB-like structure (but with only *half* the trigonal prisms—of anions now—occupied in each strip) plus elements of NaCl type.

The Sr$_2$PbO$_4$-type structure of δ-Mn$_2$GeO$_4$ is shown in Figure 43. It is a high-pressure polymorph of Mn$_2$GeO$_4$ (olivine is the stable form at 1 atm), and the structure type appears to be important in geophysics as a very-high-pressure polymorph of olivines; either as a post-spinel phase or as an alter-

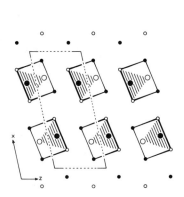

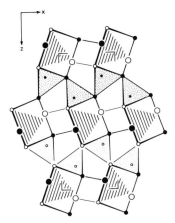

Figure 41. α-Bi$_2$Pd on (010). Large circles are Pd; small circles are Bi; open at $y = 0$, filled at $y = \frac{1}{2}$. Alternate (100) layers are now FeB/CrB (prisms) and hcp (empty octahedra). Compare Figures 39, 40, and 42.

Figure 42. The structure of emplectite, CuBiS$_2$, projected on (010). Large, medium, and small circles are Bi, S, and Cu, respectively; open at $y = \frac{1}{4}$, filled at $y = \frac{3}{4}$. Alternate (001) layers are FeB/CrB (trigonal prisms; better SnS) and ZnO (tetrahedra); but now the first occurs in two twin orientations (instead of just one as in Figures 39–41).

native to spinel (34). There are several isostructural compounds such as Na$_2$MnCl$_4$ and Ca$_2$SnO$_4$ (as well as Sr$_2$PbO$_4$).

Figure 43 shows the structure to consist of zigzag FeB-like sheets of MnO$_6$ trigonal prisms parallel to **a**. The prisms in these sheets are joined to each other and to similar adjacent sheets by edge-sharing. The columns of cp oxygens thus formed contain octahedra that are occupied by Ge atoms. (The structure can, in fact, also be described as a glide-reflection twinned NaCl type. It can also be related to a twinned TiP; cf. above and Figure 11.) Structural data are given in Data Table 14.

Data Table 14 δ-Mn$_2$GeO$_4$ (33)

Orthorhombic, space group *Pbam*, No. 55; $a = 5.257$, $b = 9.270$, $c = 2.951$ Å; $Z = 2$, $V = 143.8$ Å^3

Atomic Positions
Ge in 2(a):   0,0,0; $\frac{1}{2},\frac{1}{2},0$
Mn in 4(h): $\pm(x,y,\frac{1}{2}; \frac{1}{2} + x,\frac{1}{2} - y,\frac{1}{2})$; $x = 0.069$, $y = 0.318$
O(1) in 4(h): $x = 0.245$, $y = 0.041$
O(2) in 4(g): $\pm(x,y,0; \frac{1}{2} + x,\frac{1}{2} - y,0)$; $x = 0.364$, $y = 0.314$

Several rare earth sesquioxides are isostructural with B-type Sm$_2$O$_3$. The structure of Eu$_2$O$_3$ is shown in Figure 44. NaCl-like octahedral strips are

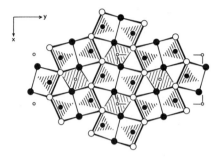

Figure 43. The δ-Mn$_2$GeO$_4$ structure projected on (001): small, medium and large circles are Ge, Mn, and O, respectively; open at $z = 0$, filled at $z = \frac{1}{2}$. MnO$_{6+1}$ monocapped trigonal prisms are edge-connected to each other and to GeO$_6$ octahedra.

joined into pairs by edge-sharing and are slightly distorted to A-La$_2$O$_3$-like elements (monocapped octahedra, LnO$_{6+1}$, cf. Chapter VIII, Figure 9). Note also the large elements/columns of Sr$_2$PbO$_4$-type structure. As in Sr$_2$PbO$_4$, the trigonal prisms are monocapped; i.e., these cations too have CN = 7. Structural data for Eu$_2$O$_3$ are given in Data Table 15.

As with A-La$_2$O$_3$, the B-Sm$_2$O$_3$ structure can also be described in anti-structure terms, as an hcp array of Sm, with anions inserted into the interstices. But such a description is neither as simple nor as elegant in this case. There are five crystallographically distinct anions, instead of the two in A-La$_2$O$_3$; and there are four types of anion coordination: CN = 6, octahedral; CN = 4, tetrahedral; and CN = 5, square pyramidal and trigonal bipyramidal. Figure 45 illustrates this description and also indicates the very small anion shuffles that will convert B-type to A-type Ln$_2$O$_3$.

The δ-Ho$_2$S$_3$ rare earth sesquisulphide structure type is rather similar.

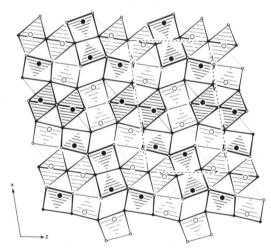

Figure 44. B-type Eu$_2$O$_3$ projected on (010) and described in terms of cation-centered polyhedra: monocapped trigonal prisms and octahedra. Large circles are Eu; small circles are O; open at $y = 0$, filled at $y = \frac{1}{2}$. The dotted lines are weak bonds (valence ≈ 0.07) to extra oxygens capping the EuO$_6$ octahedra; cf. A-type La$_2$O$_3$ (Figure 9 of Chapter VIII). In addition, there are elements of FeB type (Figure 24) and Sr$_2$PbO$_4$ type (Figure 43).

Data Table 15   B-Type Eu$_2$O$_3$ (35)

Monoclinic, space group $C2/m$, No. 12; a = 14.1105, b = 3.6021, c = 8.8080 Å, β = 100.037°; Z = 6, V = 440.84 Å^3

Atomic Positions: $(0,0,0; \frac{1}{2},\frac{1}{2},0)+$
Eu(1) in 4(i): $\pm(x,0,z)$; x = 0.63740, z = 0.4897
Eu(2) in 4(i): x = 0.68972, z = 0.13760
Eu(3) in 4(i): x = 0.96635, z = 0.18763
O(1) in 4(i): x = 0.1291, z = 0.2855
O(2) in 4(i): x = 0.8248, z = 0.0267
O(3) in 4(i): x = 0.7961, z = 0.3732
O(4) in 4(i): x = 0.4734, z = 0.3431
O(5) in 2(b): $0,\frac{1}{2},0$

Atomic Distances
Eu(1)—O = 2.28 − 2.62 Å (7×) (average = 2.43 Å)
Eu(2)—O = 2.29 − 2.74 Å (7×) (average = 2.41 Å)
Eu(3)—O = 2.24 − 2.54 Å (6×), [3.13 Å] (average = 2.36, [2.47] Å)

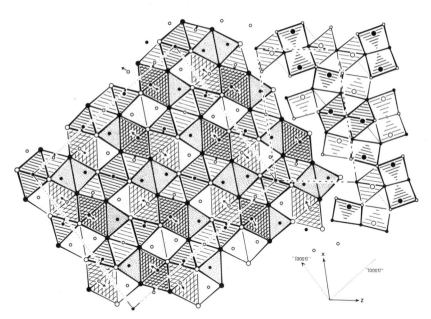

Figure 45.   B-type Eu$_2$O$_3$, projected as in Figure 44, but now depicted mainly in terms of *anion*-centered polyhedra, except for the right-hand side, which is identical to Figure 44. The atoms are represented as in Figure 44. The conventional unit cell is outlined with broken lines (on the right), and the corresponding axes as full lines. Dotted lines outline an alternative 1 × 3 × 3 × A-La$_2$O$_3$ unit cell and the corresponding axial directions. As in A-La$_2$O$_3$, the cations in B-Eu$_2$O$_3$ are seen to be in approximately hcp, with oxygens in octahedra and half-octahedra (both line shaded), tetrahedra (dotted), and trigonal bipyramids (double line shaded). The relation to A-La$_2$O$_3$ (Figure 7 of Chapter V) is very close, as is shown by the small arrows, which indicate the minor oxygen shifts needed to transform B-type to A-type. (At the same time, the monoclinic angle of the La$_2$O$_3$-related cell changes from ~86$\frac{1}{4}$° to 90°.)

Other Trigonal Prism Structures Derived from AlB_2 Derivatives by Reflection (Including Glide-Reflection) Twinning

Ni_4B_3

The structure of Ni_4B_3, shown in Figure 46, consists of $(\bar{1}10)$ FeB-type layers (cf. Figure 24) twinned by glide reflection. Structural data are given in Data Table 16. (Note that the axes in Figure 46 are for the alternative and common setting *Pbnm*.) By the simple shear or deformation process discussed earlier (Figure 4), the trigonal prisms "lying down" (pseudo three-fold axes in the plane of the drawing) can be readily transformed to half-octahedra. This changes the (monocapped) prismatic coordination to octahedral and the o-Ni_4B_3 structure to the $CaTi_2O_4$ type (described in Chapter VI). Note that the array of "standing up" prisms—pseudo-three-fold axes normal to the plane of the figure—is similar in both structures. It seems likely that $CaTi_2O_4 \rightarrow o$-Ni_4B_3 could be achieved by applying high pressure to the former.

Data Table 16 Ni_4B_3 (36)

Orthorhombic, space group *Pnma*, No. 62; $a = 11.954$, $b = 2.982$, $c = 6.568$ Å; $Z = 4$, $V = 234.10$ Å^3

Atomic positions
All atoms in 4(c): $\pm(x,\frac{1}{4},z; \frac{1}{2} + x,\frac{1}{4},\frac{1}{2} - z)$, with

	x	z
Ni(1)	0.1484	0.9912
Ni(2)	0.4497	0.7490
Ni(3)	0.2002	0.3788
Ni(4)	0.3756	0.1675
B(1)	0.4685	0.4238
B(2)	0.0356	0.4815
B(3)	0.2565	0.6792

There is also a monoclinic m-Ni_4B_3, which we will not discuss.

Many of the remaining structures in this section are twinned monoclinic YO(OH) or CrB-type layers with up to one-half of the prisms empty. As in the case of o-Ni_4B_3, the twinning operation generates further trigonal prisms in the composition plane. There appear to be a family of related structures (37), mainly mineral sulfosalts, although only a few members have had their structures confirmed by single-crystal X-ray diffraction methods. They are related to reflection-twinned ccp structures in the manner alluded to above.

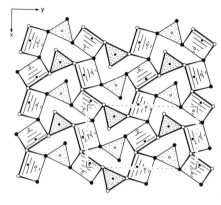

Figure 46. Orthorhombic Ni_4B_3 projected on (010) of *Pnma* (in the nonstandard setting *Pbnm*). Large circles are Ni; small ones are B; open at $y = \frac{1}{4}$, filled at $y = \frac{3}{4}$ (of *Pnma*).

Bi_2S_3

The structure of Bi_2S_3, shown in Figure 47, is the bismuthinite type (Strukturbericht symbol $D5_8$). Structural data are given in Data Table 17. Stibnite (Sb_2S_3) and α-U_2S_3 are isostructural. It is the high-pressure form of $GdYbS_3$ [whose 1-atm modification is $CaIrO_3$ type = twinned ccp $\cdots$ (*3*) $\cdots$,

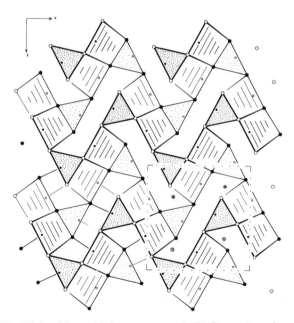

Figure 47. The Bi_2S_3, bismuthinite, structure (α-U_2S_3 type) projected on (010) of *Pnma*. Larger circles are S; smaller circles are Bi; open at $y = \frac{1}{4}$, filled at $y = \frac{3}{4}$. The doubly ringed positions within the outlined unit cell are available tetrahedral sites (open at $y = \frac{1}{4}$ and dotted at $y = \frac{3}{4}$), unoccupied in Bi_2S_3 but utilized for Cu in the aikinite series.

cf. Chapter VI]* and also of several lanthanide sesquisulfides (whose 1-atm polymorphs have the α-Ln_2S_3 types, to be discussed shortly). Hf_3P_2 is an antitype of bismuthinite.

Gd_3NiSi_2 is the similarly twinned CrB type (i.e., a filled Hf_3P_2) (39).

Data Table 17 Bi_2S_3 (38)

Orthorhombic, space group *Pnma*, No. 62; $a = 11.25$, $b = 3.97$, $c = 11.115$ Å; $Z = 4$, $V = 496._4$ Å^3

Atomic Positions
All atoms in 4(c): $\pm(x,\frac{1}{4},z; \frac{1}{2} + x,\frac{1}{4},\frac{1}{2} - z)$, with

	x	z
Bi(1)	0.0166	0.3257
Bi(2)	0.3406	0.5341
S(1)	0.0483	0.8715
S(2)	0.3750	0.9432
S(3)	0.2178	0.1938

Atomic Distances
Bi(1)–S = 2.67 Å (2×), 2.70 Å (cap), 3.02 Å, 3.05 Å (2×), 3.91 Å
Bi(2)–S = 2.56 Å (cap), 2.74 Å (2×), 2.96 Å (2×), 3.29 Å (2×)

The Bismuthinite–Aikinite Series

In the drawing of the bismuthinite structure (Figure 47), we have indicated some unoccupied tetrahedral sites within the unit cell. In some minerals a degree of substitution of (Cu^+ + Pb^{2+}) for Bi^{3+} has occurred, with the Cu^+ entering some or all of these sites. A series of ordered structures $Bi_{12-n}Pb_nCu_nS_{18}$ is known (40):

$n = 0$	$Bi_{12}S_{18}$ (= Bi_2S_3)	bismuthinite
$n = 1$	$Bi_{11}PbCuS_{18}$	pekoite
$n = 2$	$Bi_{10}Pb_2Cu_2S_{18}$ (= Bi_5PbCuS_9)	gladite
$n = 3$	$Bi_9Pb_3Cu_3S_{18}$ (= Bi_3PbCuS_6)	krupkaite
$n = 6$	$Bi_6Pb_6Cu_6S_{18}$ (= $BiPbCuS_3$)	aikinite

where $n = 6$ is the maximum possible.

Their unit cell volumes are respectively 1×, 3×, 3×, 1×, and 1× that of bismuthinite (cf. Figure 48). A number of compounds are isostructural with aikinite, examples are K_2CuCl_3, Ba_2FeS_3, and Ba_2ZnS_3.

Some of the members of another bismuthinite-related family also accom-

---

* The simple topological relation between the $CaIrO_3$ and Bi_2S_3 types (which is that of Figure 4) has been described in detail elsewhere (7).

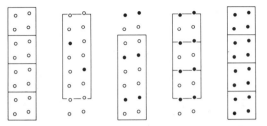

Figure 48. Pattern of unoccupied (open circles) and occupied (filled circles) tetrahedral sites (cf. Figure 47) used by Cu in the bismuthinite–aikinite series of structures, $Bi_{12-n}Pb_nCu_nS_{18}$. Unit cells are outlined. From left to right: bismuthinite, $n = 0$; pekoite, $n = 1$; gladite, $n = 2$; krupkaite, $n = 3$; aikinite, $n = 6$.

modate smaller cations, such as Cu, in the analogous tetrahedral sites, as follows.

A Homologous Series $Sb_2S_3 \cdot nPbS$

The first example is meneghinite, $CuPb_{13}Sb_7S_{24}$, the structural data for which are given in Data Table 18. There is some doubt about the exact multiplicity of the supercell, or true cell [it may be incommensurate with the subcell (42)], and so we ignore it and use only the subcell. [The true cell of the modulated structure, if there is one, reflects the Cu distribution, and probably the Pb/Sb ordering also. There is a Cu-free analogue, approximately $Sb_2S_3 \cdot 3PbS = Sb_2Pb_3S_6$, cf. $(Cu_{1/4})(Pb_{1/4}Sb_{7/4})_{\Sigma=2}Pb_3S_6$ for meneghinite.]

Data Table 18   Meneghinite, $CuPb_{13}Sb_7S_{24}$ (41)

True cell: Orthorhombic, space group $Pmn2_1$, No. 31; $a = 24 \times 4.128$
            $(= 99.07_2)$, $b = 11.363$, $c = 24.057$ Å; $Z = 24$, $V = 27,08_2$ Å^3

Subcell: Orthorhombic, space group $Pnma$, No. 62; $a = 24.057$, $b = 4.128$,
            $c = 11.363$ Å; $Z = 1$, $V = 1128.4$ Å^3

Atomic Positions (Subcell)
All atoms are in $4(c)$ of $Pnma$: $\pm(x,\frac{1}{4},z; \frac{1}{2} + x,\frac{1}{4},\frac{1}{2} - z)$; with

	x	z
$(Pb_{0.5},Sb_{0.5})$ (1)	0.026	0.150
$Pb(2)$	0.626	0.658
$(Pb_{0.5},Sb_{0.5})$ (3)	0.436	0.036
$Pb(4)$	0.716	0.985
$(Pb_{0.25},Sb_{0.75})$ (5)	0.849	0.780
$S(1)$	0.437	0.474
$S(2)$	0.110	0.406
$S(3)$	0.183	0.089
$S(4)$	0.763	0.664
$S(5)$	0.348	0.160
$S(6)$	0.521	0.783
Cu	0.256	0.240

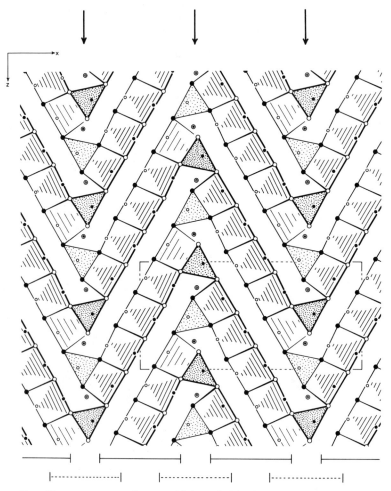

Figure 49. The structure of meneghinite, $CuPb_{13}Sb_7S_{24}$, projected on (010) of *Pnma*. Large circles are S; small circles are Pb or (Pb,Sn); open at $y = \frac{1}{4}$, filled at $y = \frac{3}{4}$. Double circles are Cu sites (occupancy $\frac{1}{4}$); open at $y = \frac{1}{4}$, filled center at $y = \frac{3}{4}$. Twin/composition planes are indicated by arrows (above), and (100) bands of SnS type and Sb_2S_3 type by solid and broken lines, respectively (below). One unit cell is outlined.

Its structure is shown in Figure 49. In Bi_2S_3 the slabs are YO(OH)-type and two trigonal prisms wide. Now these slabs are increased in width by inserting SnS-like lamellae (not YO(OH)], in the present instance an extra three prisms in width. Thus, the structure can be regarded as an intergrowth of the SnS (Figure 6) and Bi_2S_3 (Figure 47) types. (Note that the high-pressure form of PbS is SnS-type.)

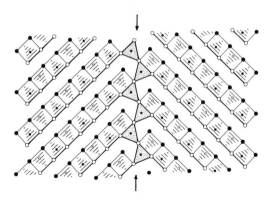

Figure 50. Glide-reflection twinning of the YO(OH) type. Compare Figure 30 of Chapter VI and Figure 29 of Chapter VIII.

Ignoring the tetrahedrally coordinated Cu, the general formula for a family of such **SnS** and **Bi$_2$S$_3$** intergrowths is Sb$_2$S$_3 \cdot n$PbS, where n is the width of the inserted SnS-type lamellae ($n = 3$ for meneghinite; $n = 0$ for bismuthinite).* Sb$_2$Pb$_2$S$_5$ is the $n = 2$ analogue, and Sb$_2$Sn$_2$S$_5$ is isostructural (35). Other n-valued homologues are unknown. [A different description, with wider connotations, has been given by Makovicky (43).]

Glide-Reflection Twinning of the YO(OH) Type

A second type of glide-reflection twinning of the YO(OH) type is shown in Figure 50. It yields the following structures.

Data Table 19 α-Gd$_2$S$_3$ (44)

Orthorhombic, space group *Pnma*, No. 62; $a = 7.339$, $b = 3.9318$, $c = 15.273$ Å; $Z = 4$, $V = 440.7_1$ Å^3

Atomic Positions
All atoms in 4(c): $\pm(x,\frac{1}{4},z; \frac{1}{2} + x,\frac{1}{4},\frac{1}{2} - z)$, with

	x	z
Gd(1)	0.14237	0.20418
Gd(2)	0.76637	0.54329
S(1)	0.0086	0.3914
S(2)	0.8747	0.9330
S(3)	0.6493	0.7186

---

* Boldface formulae refer to structure types, not chemical compounds.

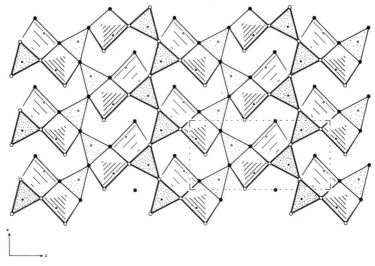

Figure 51. The α-Ln$_2$S$_3$-type structure of Gd$_2$S$_3$ projected on (010) of *Pnma*. Large circles are S; small circles are Gd; open at $y = \frac{1}{4}$, filled at $y = \frac{3}{4}$. Compare with Figure 47, and note the common building unit of four edge-shared trigonal prisms.

The type of unit cell twinning demonstrated by α-Gd$_2$S$_3$ (Data Table 19) has been dealt with earlier, although not applied to YO(OH). It yields PbCl$_2$-type trigonal prismatic layers in the twin/composition planes. This is the case for the α-type lanthanoid sesquisulfide structures, of which α-Gd$_2$S$_3$ is an example, shown in Figure 51. It occurs for all the lighter (larger) rare earths, from La to Dy. [The heavier lanthanoid sesquisulfides have the

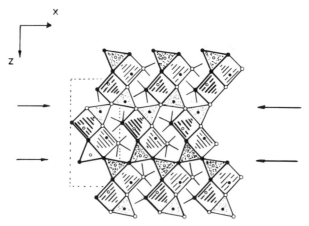

Figure 52. The structure of the Cr$_3$C$_2$ antitype of α-Ln$_2$S$_3$; compare with Figure 51. The trigonal prisms are of Cr atoms; they are centered by C.

Ho_2S_3 type structure and the heaviest, Yb and Lu, the corundum structure.] The antitype of α-Gd_2S_3 is the $D5_{10}$ structure of Cr_3C_2, (Figure 52).

Note that this structure and the α-U_2S_3 (or bismuthinite) type dealt with earlier are built of the same compact units of four edgeshared trigonal prisms—two "standing up" and two "lying down" (actually from columns with this cross section); compare Figures 47 and 51. For at least some rare earth sulfides, the latter structure transforms to the former under high pressure.

Except for the C23/$PbCl_2$ type (Chapter VIII, Figure 33), no other structure in this family is known. But K_2AgI_3 is a "stuffed" version of the α-Ln_2S_3 type, with the Ag atoms in available tetrahedral interstices (analogous to the Cu atoms in the bismuthinite type in aikinite). There are a number of isostructural compounds of the alkali halides and alkaline earth chalcogenides such as Ba_2CdS_3 and Ba_2MnSe_3.

References

1. E. J. Felten, *J. Am. Chem. Soc.* **78,** 5977 (1956).

2. B. T. Matthias, *Rev. Mod. Phys.* **35,** 1 (1963).

3. F. Hulliger and E. Mooser, *Prog. Solid State Chem.* **2,** 330 (1965).

4. P. I. Kripyakevich, *J. Struct. Chem. USSR* **4,** 257 (1963).

5. E. Parthé and V. Sadagopan, *Monatsh. Chem.* **93,** 263 (1962).

6. R. Kiessling, *Acta Chem. Scand.* **3,** 595 (1949).

7. B. G. Hyde, S. Andersson, M. Bakker, C. M. Plug, and M. O'Keeffe, *Prog. Solid State Chem.* **12,** 292 (1979).

8. M. Julien-Pouzol, S. Jaulmes, M. Guittard, and P. Laruelle, *J. Solid State Chem.* **26,** 185 (1978).

9. F. Hulliger, *Structural Chemistry of Layer-Type Phases.* Reidel, Dordrecht, 1976.

10. K. A. Agaev, V. A. Gasymov, and M. I. Ciragov, *Sov. Phys. Cryst.* **18,** 226 (1973).

11. R. Kiessling, *Acta Chem. Scand.* **3,** 603 (1949).

12. W. Rieger, H. Nowotny, and F. Benesovsky, *Monatsh. Chem.* **97,** 378 (1966).

13. Yu. B. Kuzma, P. I. Kripyakevich, and R. V. Skolozdra, *Struct. Rep.* **31A,** 27 (1966).

14. H. Jedlicka, F. Benesovsky, and H. Nowotny, *Monatsh. Chem.* **100,** 844 (1969).

15. R. Kiessling, *Acta Chem. Scand.* **1,** 893 (1947).

16. A. N. Christensen, *Acta Chem. Scand.* **20,** 2658 (1966).

17. A. N. Christensen, *Acta Chem. Scand.* **19,** 1391 (1965).

18. M. A. Kamermans, *Z. Kristallogr.* **101,** 406 (1939).

19. M. Dyke and R. L. Sass, *J. Phys. Chem.* **68,** 3259 (1964).

20. R. L. Sass, T. Brackett, and E. Brackett, *J. Phys. Chem.* **67,** 2862 (1963).

21. J. G. Smeggil and H. A. Eick, *Inorg. Chem.* **10,** 1458 (1971).

22. T. Bjurström, *Arkiv. Kemi Mineral. Geol.* **11A,** 12 (1933).

23. K. Klepp and E. Parthé, *Acta Cryst. B* **36,** 744 (1980).

24. E. Parthé, *Acta Cryst. B* **32,** 2813 (1976).

25. F. Merlo and M. L. Fornasini, *Acta Cryst. B* **37,** 500 (1981).

26. K. Klepp and E. Parthé, *Acta Cryst. B* **37,** 495 (1981).

27. O. I. Bodak, E. I. Gladysevskij, and O. I. Kharcenko, *Sov. Phys.—Crystallogr.* **19,** 45 (1974).

28. O. I. Bodak, E. I. Gladysevskij, and M. G. Mis'kiv, *Sov. Phys.—Crystallogr.* **17,** 439 (1972).

29. M. G. Mis'kiv, O. I. Bodak, and E. I. Gladysevskij, *Sov. Phys.—Crystallogr.* **18,** 450 (1973).

30. H. K. Chun and G. B. Carpenter, *Acta Cryst. B* **35,** 30 (1979).

31. K. Mariolacos, *Acta Cryst. B* **32,** 1947 (1976).

32. Ju. B. Kuzma, P. I. Kripyakevich, and N. F. Chaban, *Struct. Rep.* **39A,** 3 (1973).

33. A. D. Wadsley, A. F. Reid, and A. E. Ringwood, *Acta Cryst. B* **24,** 740 (1968).

34. A. E. Ringwood, *Composition and Petrology of the Earth's Mantle.* McGraw-Hill, New York, 1975.

35. H. L. Yakel, *Acta Cryst. B* **35,** 564 (1979).

36. S. Rundqvist and S. Pramatus, *Acta Chem. Scand.* **21,** 191 (1967).

37. P. P. K. Smith and B. G. Hyde, *Acta Cryst. C* **39,** 1498 (1983).

38. V. Kupčik and L. Veselá-Nováková, *Tschermaks Mineral. Petrogr. Mitt.* **14,** 55 (1970).

39. K. Klappe and E. Parthé, *Acta Cryst. B* **37,** 1500 (1981).

40. W. G. Mumme, E. Welin, and B. J. Wuensch, *Am. Mineral.* **61,** 15 (1976).

41. R. Euler and E. Hellner, *Z. Kristallogr.* **113,** 345 (1960).

42. P. P. K. Smith, unpublished work, 1982; private communication.

43. E. M. Makovicky, *Fortschr. Mineral.* **63,** 45 (1985).

44. C. T. Prewitt and A. W. Sleight, *Inorg. Chem.* **7,** 1090 (1968).

45a. R. Vincent and R. L. Withers, *Phil. Mag. Lett.* **56,** 57 (1987).

45b. R. L. Withers, G. L. Hua, T. R. Welberry, and R. Vincent, *J. Phys. C* **21,** 309 (1988).

45c. G. L. Hua, T. R. Welberry, and R. L. Withers, *J. Phys. C* (1988) in press.

45d. J. G. Thompson, A. D. Rae, R. L. Withers, T. R. Welberry, A. C. Willis, *J. Phys. C,* **21,** 4007 (1988).

46. F. Hulliger, *Struct. Bonding* **4,** 83 (1968).

CHAPTER X

The Stereochemistry of Valence and Lone Pair Electrons

The following structures are considered in this chapter: NH_3, H_2O, ClF_3, orthorhombic (yellow) PbO, α-PbO_2, tetragonal (red) PbO, o-Sb_2O_3, o- and m-Sb_2O_4, m-Sb_2O_5, Fe_3C, B-Nb_2O_5, $SbNbO_4$, $SbCl_3$, $BiCl_3$, SbF_3, XeO_3, HIO_3, $NaIO_3$, AsF_3, PCl_3, PBr_3, $AsCl_3$, AsF_3, NCl_3, $POBr_3$, Re_3B, α-$SbBr_3$, α-$BiBr_3$, LaF_3, BiOF, K_2MgF_4, L-SbOF, As_2S_3 (orpiment).

The Sidgwick–Powell and Gillespie–Nyholm Theories

In their theory of the stereochemistry of valence bonds, Sidgwick and Powell (1) assumed that bonding pairs and lone pairs of electrons are of equal importance and distribute themselves so as to minimize interelectron repulsion. For example, three pairs are arranged to form a triangle, four pairs a tetrahedron, five pairs a trigonal bipyramid, and six pairs an octahedron. The theory has proved successful but could not account for the deviations from the tetrahedral angle (109°28′) that exist in NH_3 (HN̂H = 106.6°) and H_2O (HÔH = 104.5°). Also, for the structure of the molecule ClF_3 it gave three different models.

Gillespie and Nyholm (2) modified the theory by assuming that lone pairs are larger than bonding pairs, so that the repulsion between electron pairs decreases in the order

lone pair–lone pair > lone pair–bonding pair > bonding pair–bonding pair

This was used to account for the bond angles in NH_3 and H_2O, and a satisfactory model of ClF_3 could also be given: the lone pairs are in two equatorial positions in a trigonal bipyramid, because the repulsion between them is then minimized. Their assumptions were that (1) the electron pairs are at the corners of a polyhedron with the nucleus at its center and (2) the lone pair is so much bigger than a bonding pair that there is a considerable distortion of the polyhedron (Figure 3a).

An Alternative Approach

In our attempts to analyze and systematize the structures of compounds with stereochemically active lone pairs of electrons, we used the space taken by a lone pair (and its cation) in a crystal to locate the centroid of its negative charge (3–5).

The volume of the lone pair was found to be of the same order as that of an oxygen or fluorine ion, as demonstrated in Table 1 and Figure 1. Hence the lone pair was located at the corner of a regular coordination polyhedron like the tetrahedron (Figure 1), the octahedron, or the trigonal bipyramid (Figure 2). The lone pair–nucleus distance could be calculated from structure determinations (3, 4) according to Figures 1 and 2. Such distances vary, of course, with the cation/anion size ratio and are given in Table 2 (4). Sometimes, in compounds like $SbCl_3$ (4), the lone pair volume is small com-

TABLE 1   Volume per Anion Site in Various Structures[a]

	V_1 (Å^3)	V_2 (Å^3)
TiO_2, rutile	15.6	15.6
$BaNiO_3$	21.8	16.3
$BaTiO_3$	21.4	16.0
$TiTe_3O_6$	20.5	15.0
α-$TeVO_4$	19.2	15.3
β-$TeVO_4$	20.1	16.1
L-SbOF	24.6	16.4
M-SbOF	25.1	16.7
β-Sb_2O_4	19.0	15.2
SbF_3	22.3	16.8
$KTeF_5$	24.4	17.4
α-PbO_2	20.2	20.2
PbO (orthorhombic)	38.8	19.4
PbO (tetragonal)	39.7	19.8
XeO_3	21.8	16.4
SnO	35.0	17.5

[a] V_1 = volume per anion. V_2 = volume when lone pairs or barium and potassium are included formally as anions in the corresponding calculations.

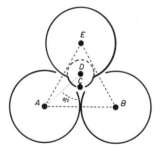

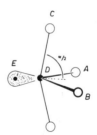

Figure 1. A, B, and C are anions with positions found experimentally. D is the position of a cation also found experimentally. E is the position of the lone pair, completing the close packing (a regular tetrahedron) with A, B, and C. It is now easy to calculate distances D–E, and corresponding angles. Compare with Figure 2.

Figure 2. Regular trigonal bipyramid of ABC_2E, where E stands for a lone pair of electrons. By swinging E in the plane CEC, one C is pushed away, and a regular tetrahedron ABCE is obtained instead.

pared with the anion volume, and the tetrahedron is not regular. In this case the structure is best described as containing trigonal prisms of Cl^-, with the lone pair in the center of the prism. This is also often the case for oxides or fluorides when the anion/cation stoichiometric ratio is high. Examples of this are SbF_3 and TeF_4, and the trigonal prisms are then distorted and enlarged.

Localization of lone pairs is always done with symmetry conservation; that means that the coordinates of the centroid(s) should agree with the available position(s) in the space group of the crystal.

The advantage of this approach is obvious and is demonstrated in Figure 3*b*. The polyhedron is regular, and bond angles and bond distances are easily calculated for various ions (4). There is no need to assume various magnitudes of repulsion between electron pairs. Instead, we conclude that nucleus–electron pair attraction brings the nucleus closer to a lone pair than to a bonding pair; after all, the bonding pairs are shared between two nuclei (5). The anions and the lone pair are of roughly the same size, and *the cation must therefore be off-center in the regular polyhedron*, as shown in Figure 3*b*.

TABLE 2 M⁺—Lone Pair Distances

Ga^{1+}	Ge^{2+}	As^{3+}	Se^{4+}	Br^{5+}	
(0.95)	1.05	1.26	1.22	1.47	
In^{1+}	Sn^{2+}	Sb^{3+}	Te^{4+}	I^{5+}	Xe^{6+}
(0.86)	0.95	1.06	1.25	1.23	1.49
Tl^{1+}	Pb^{2+}	Bi^{3+}	Po^{4+}		
0.69	0.86	0.98	(1.06)		

Figure 3. (*a*) Gillespie's model, illustrating the deviations from the ideal shapes of AX_4E molecules. The lone pair forms a corner, while the cation forms the center of a polyhedron. (*b*) Our model. The lone pair–X distance is determined from the ideal trigonal bipyramid. Angles and distances are simply a function of atom sizes. The lone pair plus attached cation form a corner of the polyhedron, the center of which is empty.

The Geometry of the ClF_3 Molecule

The three possible models for ClF_3 are shown in Figure 4. The two lone pairs in axial positions in model I, about 1 Å from the nucleus, would give F$\cdots$F distances ($\sim$1.2 Å) that are far too short; hence for steric reasons the structure is impossible. Model II is impossible because of the great differences in bond lengths. Model III is the only plausible one, and it represents the correct structure of the molecule ClF_3, both in the gas phase (6) and in the crystal (7).

Crystal Structures

Yellow PbO and α-PbO_2

Figure 5 shows the structure of yellow PbO* (crystallographic data are in Data Table 1), and we shall use it here to demonstrate how lone pairs are located. Circles are lead atoms on 0 or $\frac{1}{2}$, the projection axis being about 5 Å. All the oxygens are also on 0 or $\frac{1}{2}$ and are always on one side of the cation. If the lone pairs are plotted so that regular or almost regular triangles are obtained in each plane, Figure 6 is obtained. (The lone pairs are in the mirror

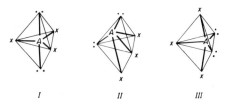

I   *II*   *III*

Figure 4. Possible structures of ClF_3. If the A–X distance is to be kept constant, model I is impossible for steric reasons, model II has two very different A–X distances, and the only possible model is III, which also corresponds to the real structure of ClF_3.

* Note that it is red PbO (Figure 8) twinned on $\{101\}_{red} \equiv (100)_{yellow}$.

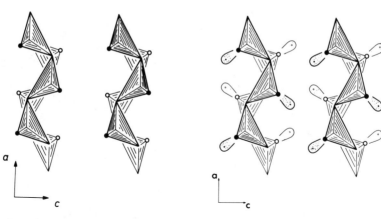

Figure 5. The real structure of "hexagonal" or yellow PbO. Circles are Pb^{2+}; polyhedral corners are oxygens. All atoms on 0 or $\frac{1}{2}$.

Figure 6. Lone pairs (the centroids of the volume lone pairs take) localized in PbO.

planes, according to the rule of conservation of symmetry.) In Figure 7 the positions of the oxygens and lone pairs are plotted, and it is obvious that an almost regular hcp arrangement is obtained. The cation–lone pair distance becomes about 1 Å. When PbO is oxidized, one can imagine O atoms entering the crystal and taking up the positions of the (now nonexistent) lone pairs; divalent lead is oxidized to tetravalent and "walks" into the O_6 octahedra so that the structure of α-PbO_2 is obtained. (Recall that the oxygens in α-PbO_2 are approximately in hcp, Chapter III.)

Red PbO and LiOH

Figure 8 gives the structure of the tetragonal red PbO. (Crystallographic data are given in Data Table 2.) The anion (O) plus lone pair (E) array is distorted ccp (ccp I projection), and the structure is thus related to NaCl. But, as is frequently the case, simple structures can be described in more than one way. Another obvious way is to relate red PbO to the CsCl structure. The Pb^{2+} ion is close to one square of oxygens, and its lone pair is then in the center of an elongated cube of anions. (Compare the description of fcc structures via a body-centered tetragonal cell in Chapter VI.)

[Note that LiOH is an antitype, with Li in place of O, O in place of Pb, and H in place of E: $l(O–H) = 0.94$ Å, cf. $l(Pb–E) = 0.86$ Å (Data Table 2).]

Structures of Some Compounds of Antimony (III)

Orthorhombic Sb_2O_3 [$= Sb_2(O_3E_2)_{\Sigma=5}$], orthorhombic and monoclinic Sb_2O_4 [$= Sb_2(O_4E)_{\Sigma=5}$], and monoclinic Sb_2O_5 have structural relationships similar

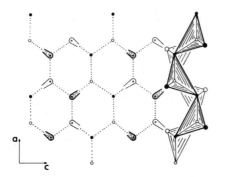

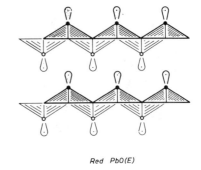

Red PbO(E)

Figure 7. The hcp (HCP I) arrangement of lone pairs and oxygens in PbO.

Figure 8. The real structure of red or tetragonal PbO.

to those between yellow PbO and α-PbO_2. The real structure of Sb_2O_3 is shown in Figure 9. The oxygens and lone pairs form a twinned hcp array of the same type as the iron atoms form in Fe_3C (8) (Chapter IV). The twin blocks here are larger than we have seen previously: . . . , 5,5,5, . . . ; but it is easy to distinguish them in the idealized drawing in Figure 10a. In orthorhombic Sb_2O_4 (= $Sb^{3+}Sb^{5+}O_4^{2-}$), every second Sb is pentavalent, and in the structure of Figure 10b the twin blocks now contain real cation-centered octahedra of oxygen (9). (Half of the lone pairs of electrons have been replaced by O atoms, and the oxidized Sb atoms have moved from

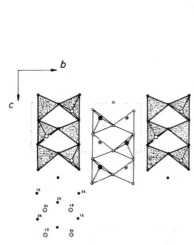

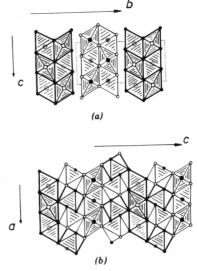

Figure 9. The real structure of orthorhombic Sb_2O_3. Larger circles are Sb^{3+}.

Figure 10. (a) Idealized structure of Sb_2O_3. Large circles are lone pairs. (b) Idealized structure of orthorhombic Sb_2O_4 (the α form). Larger circles are lone pairs.

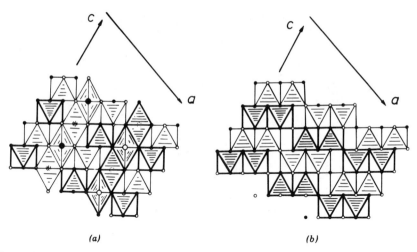

(a) *(b)*

Figure 11. (*a*) Idealized structure of monoclinic Sb$_2$O$_4$ (the β form). (*b*) Idealized structure of B-Nb$_2$O$_5$.

trigonal bipyramids to octahedra.) Untwinned (monoclinic) Sb$_2$O$_4$ also exists (BiSbO$_4$ is isostructural), and this structure is consequently very similar to the structures of Sb$_2$O$_5$ and B-Nb$_2$O$_5$ (3). Idealized drawings of these last two structures are shown in Figures 11*a* and *b*. Crystallographic data for Sb$_2$O$_3$ and α- and β-Sb$_2$O$_4$ are given in Data Tables 3–5.

Unit-cell dimensions of all these related pairs of compounds are compared in Table 3.* The sometimes remarkable similarity in structures is also demonstrated in Table 4,* where the X-ray powder patterns of the orthorhombic structures of Sb$_2$O$_3$ and SbNbO$_4$ (which is isostructural with α-Sb$_2$O$_4$) are compared.

TABLE 3 Unit-Cell Dimensions of Related Lone Pair–Anion Structures

Compound	Unit Cell	a (Å)	b (Å)	c (Å)	β	Volume (Å^3)
Yellow PbO(E)	Orthorhombic	5.489	4.775	5.891	—	154
α-PbO$_2$	Orthorhombic	5.497	4.947	5.951	—	162
Sb$_2$O$_3$(E$_2$)	Orthorhombic	5.42	4.92	12.46	—	333
α-Sb$_2$O$_4$(E)	Orthorhombic	5.456	4.814	11.787	—	310
SbNbO$_4$(E)	Orthorhombic	5.561	4.939	11.810	—	324
β-Sb$_2$O$_4$(E)	Monoclinic	12.060	4.834	5.383	104.58°	304
BiSbO$_4$(E)	Monoclinic	11.8	4.88	5.46	101.0°	309
Sb$_2$O$_5$	Monoclinic	12.646	4.7820	5.4247	103.91°	318
B-Nb$_2$O$_5$	Monoclinic	12.73	4.88	5.56	105.1°	334

* Where necessary, the unit cell axes have been permuted so that equivalent axes have the same label.

TABLE 4 Comparison of X-Ray Powder Patterns of Sb_2O_3 and $SbNbO_4$

Sb$_2$O$_3$			SbNbO$_4$ (α-Sb$_2$O$_4$ type)		
d (Å)	Int	hkl	d (Å)	Int	hkl
4.56	17	110	4.54	8	110
			3.68	2	101
3.49	25	111	3.52	18	111
3.17	20	130	3.07	6	130
3.14	100	121	3.13	100	121
3.12	80	040	2.95	30	040
2.73	10	131	2.69	6	131
2.71	10	002	2.78	6	002

Data Table 1 Yellow PbO* (10)

Orthorhombic, space group, *Pbma*, No. 57; $a = 5.489$, $b = 4.755$, $c = 5.891$ Å; $Z = 4$, $V = 38.44$ Å^3

Atomic Positions
Pb in 4(d): $\pm(x,\frac{1}{4},z; \bar{x},\frac{1}{4},\frac{1}{2} + z)$; $x = -0.0208 \pm 0.0027$, $z = 0.2309 \pm 0.0020$
O in 4(d): $x = 0.0886 \pm 0.0025$, $z = -0.1309 \pm 0.0034$
Atomic Distances
Pb–O: 2.21, 2.22, 2.49 Å

* The structure determination was done with neutron powder diffraction methods.

Data Table 2 Red PbO (11) and LiOH* (12)

Tetragonal, space group $P4/nmm$, No. 129; $a = 3.976$ (3.549), $c = 5.023$ (4.334) Å; $Z = 2$, $V = 79.41$ (54.59) Å^3; $\frac{1}{2}V = 39.70$ Å^3 [$c/a = 1.263$ (1.222), compared with $\sqrt{2} = 1.414$ for **NaCl**, 1.000 for **CsCl**]

Atomic Positions
Pb(O) in 2(c): $0,\frac{1}{2},z; \frac{1}{2},0,\bar{z}$; $z = 0.237 \pm 0.002$ (0.1938)
O(Li) in 2(a): $0,0,0; \frac{1}{2},\frac{1}{2},0$
(H in 2(c): $z = 0.410$)
Atomic Distances
Pb–O: 2.317 Å
(O–Li: 1.963 Å)

* The structure determination was done with neutron powder (single-crystal) diffraction. Data for LiOH are given in parentheses.

Data Table 3 α-Sb₃O₃ (Valentinite) (13)

Orthorhombic, space group, *Pccn*, No. 56; $a = 4.911$, $b = 12.464$, $c = 5.412$ Å;
$Z = 4$, $V = 331.27$ Å³

Atomic Positions

Sb in 8(*e*): $\pm(x,y,z; \frac{1}{2} - x, \frac{1}{2} - y, z; \frac{1}{2} + x, \bar{y}, \frac{1}{2} - z; \bar{x}, \frac{1}{2} + y, \frac{1}{2} - z)$; $x = 0.0418$,
  $y = 0.1277$, $z = 0.1786$

O in 8(*e*): $x = 0.1557$, $y = 0.0601$, $z = 0.1416$

O in 4(*b*): $\pm(\frac{1}{4},\frac{1}{4},\frac{1}{2} + z)$; $z = 0.0244$

Atomic Distances

Sb–O: 1.98 Å, 2.02 Å (2×)

Data Table 4 α-Sb₂O₄ (14, 15)

Orthorhombic, space group *Pna2₁*, No. 33; $a = 5.436$, $b = 4.810$, $c = 11.760$ Å;
$Z = 4$, $V = 324$ Å³

Atomic Positions

See refs. 14 and 15.

Data Table 5 β-Sb₂O₄ (16, 17)

Monoclinic, space group *C2/c*, No. 15; $a = 12.060$, $b = 4.834$, $c = 5.383$ Å,
$\beta = 104.58°$; $Z = 4$, $V = 303.71$ Å³

Atomic Positions

See refs. 16 and 17.

Fe₃C-Like Structures Containing Lone Pairs

When the anions are larger—that is, when the anion/cation radius ratio is high—the lone pair is in the center of a relatively larger anion polyhedron. One example of this has been described earlier, namely, $PbCl_2$; and here we shall discuss $SbCl_3$ (Data Table 6), which is the antiform of Fe_3C, cementite. Sb^{3+} is off-center in a trigonal prism, with the three chlorines of one triangular face as nearest neighbors forming the $SbCl_3$ molecule. The lone pair of the Sb^{3+} ion is situated in the center of the trigonal prism, analogous to the carbon position in Fe_3C. This is shown in Figure 12.

$BiCl_3$ is isostructural with $SbCl_3$, but the lone pair is closer to the nucleus of Bi^{3+} (see Table 2), and the surrounding trigonal prism is therefore somewhat smaller. This geometry can still accommodate a somewhat longer Bi–Cl distance, and consequently the unit cell volume of $BiCl_3$ (473.0 Å³) is smaller than that of $SbCl_3$ (489.8 Å³) although the Bi–Cl bonds are longer than Sb–Cl. Such an inverted volume relation seems to be a common phenomenon for lone-pair elements belonging to these two periods of the periodic system (4).

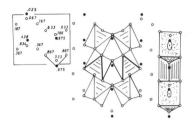

Figure 12. The real structure of $SbCl_3$. Unit cell and $SbCl_3$ molecule to the left, the empty octahedra in the middle, and the trigonal prism containing Sb^{3+} and attached lone pair to the right.

In SbF_3 the trigonal prism is elongated to provide space for the Sb^{3+} ion and its lone pair. According to Table 2, Xe^{6+} has the largest lone pair–cation distance. It is obvious from the structure of XeO_3 that this is the reason that Xe^{6+} now is on the *outside* of the trigonal prism and in one of the two capping square pyramids (5). XeO_3 is therefore a distorted Fe_3C structure, as shown in Figure 13. (Its structural data are given in Data Table 7.)

XeO_3 is isoelectronic and isostructural with HIO_3, while $NaIO_3$ is a normal $GdFeO_3$ (stuffed anticementite) type (Chapter IV, Figure 29). The I^{5+}–lone pair distance is calculated to be 1.23 Å, and I^{5+} is just outside a "square" face of the trigonal prism in HIO_3 just as Xe is in XeO_3.* However, with Na in the normally empty octahedron in the cementite structure, there is a considerable expansion of the unit-cell dimensions in the crystal, and the trigonal prism is now big enough to accommodate I^{5+} and its lone pair. AsF_3, again with a rather large cation–lone pair distance, also has a structure very similar to that of XeO_3.

For a general description of similar structures such as PCl_3, PBr_3, $AsCl_3$, AsF_3, and NCl_3, see various reports by Galy et al. (18, 19). In these it is again interesting to see that oxygenation often leads to the substitution of E by O, as in $PBr_3 \rightarrow POBr_3$ (18).

The structures of the (main) Group V trihalides form an extremely interesting collection from more than one viewpoint. For example, some of them are to be regarded as Re_3B- rather than Fe_3C-related, that is, as twinned ccp rather than twinned hcp anions (α-$SbBr_3$ and α-$BiBr_3$); or perhaps also LaF_3/$IrAl_3$ types with various stereochemical effects due to a lone pair on the cation.

Some Other Structures with Lone Pairs

The BiOF structure (Figure 38, Chapter III) consists of what is usually described as $Bi_2O_2^{2+}$ layers separated by F^- ions. The Bi_2O_2 layers have the same structure as the layers of red PbO. If the lone pairs complete the geometry of BiOF, a structure is obtained that has the same anion plus lone pair arrangement as the fluorine plus potassium array in K_2MgF_4 (Figure 32,

---

* The H atom is in a tetrahedron of the twinned hcp array of O atoms.

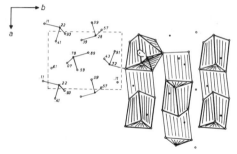

Figure 13. The real structure of XeO_3. The structure is very similar to those of Fe_3C and $SbCl_3$. Xe is outside the trigonal prism and in the capping square pyramid. Its lone pair is in the center of the prism. The unit cell and XeO_3 molecules are shown on the left.

Chapter II). Formally the structure is obtained by a slip in ccp, giving cubes in the slip plane. This is demonstrated in Figures 14 and 15. (Compare the latter with Figure 32 of Chapter II.) For BiOF, see also Chapter III, including Data Table 25.

A more complicated structure is that of L-SbOF, shown in Figure 16. (Structure data are in Data Table 8.) Trigonal bipyramids (SbX_4E) share edges along b to form an elegant structure. If the OFE array is drawn in the form of a net, as in Figure 17, it is easy to see that this net is a twinned version of $PbCl_2$ (which was discussed in Chapter VI). It is interesting that the lone pair is sometimes at a corner of a polyhedron as in Sb_2O_3, and

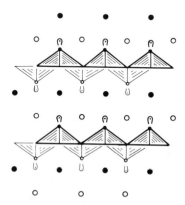

Figure 14. The real structure of BiOF. Larger circles are F^-, smaller circles are Bi^{3+}, and the oxygens are at the polyhedral corners.

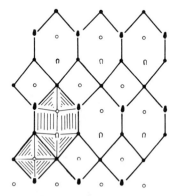

Figure 15. The OFE array of BiOF. This is the same as the K_2F_4 array in K_2MgF_4; K and lone pairs are in the same positions. O and F of BiOF are in the same positions as fluorines are in K_2MgF_4.

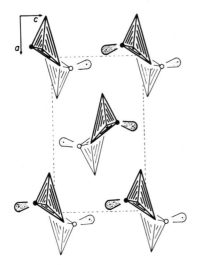

Figure 16. The real structure of L-SbOF.

sometimes in the center of an elongated prism as in SbF_3, reflecting the intermediate composition of SbOF.

Another complex structure is that of As_2S_3. This is a cyclic intergrowth of units of the AlB_2 structure type. The real structure is shown in Figure 18 (with structure data in Data Table 9), and the lone pairs are in the centers of trigonal prisms (forcing the As atoms into the faces of the prisms). The cyclic intergrowth principle on which this structure is based is shown schematically in Figure 19. It is not difficult to see that this structure can also be derived by twinning the Mo_2BC structure.

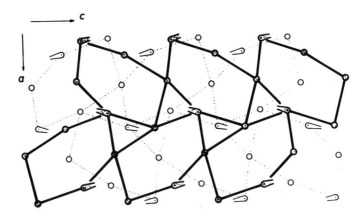

Figure 17. The OFE array of L-SbOF. For explanation, see text.

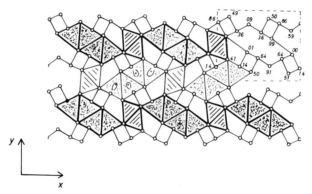

Figure 18. The real structure of As_2S_3.

Data Table 6   SbCl₃ (20)

Orthorhombic, space group *Pbnm*, No. 62; $a = 6.37$, $b = 8.12$, $c = 9.47$ Å;
$Z = 4$, $V = 490$ Å³

Atomic Positions
Sb in 4(c): $x = 0.025$, $y = 0.992$
Cl in 4(c): $x = 0.667$, $y = 0.075$
Cl in 8(d): $x = 0.133$, $y = 0.183$, $z = 0.067$
Atomic Distances
Sb–Cl = 2.36 Å

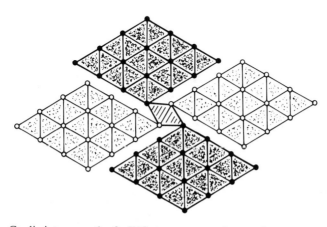

Figure 19. Cyclic intergrowth of a WC structure, to give an element of the structure of As_2S_3.

Data Table 7   XeO$_3$ (21)

Orthorhombic, space group $P2_12_12_1$, No. 19; $a = 6.163$, $b = 8.115$, $c = 5.234$ Å; $Z = 4$, $V = 262$ Å^3

Atomic Positions
All atoms in 4(a): $x,y,z; \frac{1}{2} - x,\bar{y},\frac{1}{2} + z; \frac{1}{2} + x,\frac{1}{2} - y,\bar{z}; \bar{x},\frac{1}{2} + y,\frac{1}{2} - z$

	x	y	z
Xe	0.944(2)	0.150(2)	0.219(2)
O(1)	0.54(3)	0.27(3)	0.07(3)
O(2)	0.17(3)	0.10(3)	0.41(3)
O(3)	0.14(3)	0.45(3)	0.39(3)

Atomic Distances
Xe–O = 1.74 Å, 1.76 Å, 1.77 Å

Data Table 8   L-SbOF (22)

Orthorhombic, space group *Pnma*, No. 62; $a = 8.873$, $b = 4.099$, $c = 5.403$ Å; $Z = 4$, $V = 196.50$ Å^3

Atomic Positions
All atoms in 4(c): $\pm(x,\frac{1}{4},z; \frac{1}{2} - x,\frac{3}{4},\frac{1}{2} + z)$

	x	y	z
Sb	0.0707	$\frac{1}{4}$	0.2124
O	0.054	$\frac{3}{4}$	0.090
F	0.266	$\frac{1}{4}$	0.047

Atomic Distances
Sb–O: 2.162 Å (2×), 1.991 Å (1×)
Sb–F: 1.956 Å (1×), 3.01–3.11 Å (3×)

Data Table 9   As$_2$S$_3$, Orpiment (23)

Monoclinic, space group $P2_1/c$, No. 14; $a = 11.46$, $b = 9.57$, $c = 4.22$ Å, $\beta = 90.5°$; $Z = 4$, $V = 462.79$ Å^3

Atomic Positions
All atoms in 4(e): $\pm(x,y,z; \frac{1}{2} - x,\frac{1}{2} + y,\frac{1}{2} - z)$

	As I	As II	S I	S II	S III
x	0.267	0.484	0.395	0.355	0.125
y	0.190	0.323	0.120	0.397	0.293
z	0.143	0.643	0.500	0.013	0.410

Atomic Distances
As–S = 2.21–2.28 Å (3×)

References

1. N. V. Sidgwick and H. M. Powell, *Proc. Roy. Soc. Lond.* **A176**, 153 (1940).
2. R. J. Gillespie and R. S. Nyholm, *Quart. Rev. Chem. Soc.* **11**, 339 (1957).
3. S. Andersson and A. Åström, *Nat. Bur. Standards, 5th Mat. Symp.* 1972, p. 3.
4. J. Galy, G. Meunier, S. Andersson, and A. Åström, *J. Solid State Chem.* **9**, 92 (1975).
5. S. Andersson, *Acta Cryst. B* **35**, 1321 (1979).
6. D. F. Smith, *J. Chem. Phys.* **21**, 609 (1953).
7. R. D. Burbank and F. N. Bensey, *J. Chem. Phys.* **21**, 602 (1953).
8. J.-O. Bovin, The hydrolysis of antimony(III) and the hydrogen coordination of antimony(III) in the solid state, thesis, Lund, 1975.
9. S. Andersson and B. G. Hyde, *J. Solid State Chem.* **9**, 92 (1974).
10. M. J. Kay, *Acta Cryst.* **14**, 80 (1961).
11. J. Leciejewicz, *Acta Cryst.* **14**, 1304 (1961).
12. S. L. Mair, *Acta Cryst. A* **34**, 542 (1978).
13. C. Svensson, *Acta Cryst. B* **30**, 458 (1974).
14. A. C. Skapski and D. Rogers, *Chem. Commun.* **1965**, 611.
15. P. S. Gopala Krishnan and H. Manohar, *Cryst. Struct. Commun.* **4**, 203 (1975).
16. B. Aurivillius, *Ark. Kemi.* **3**, 153 (1951).
17. D. Rogers and A. C. Skapski, *Proc. Chem. Soc.* **1964**, 400.
18. R. Enjalbert and J. Galy, *Acta Cryst. B* **35**, 546 (1979).
19. R. Enjalbert and J. Galy, *Acta Cryst. B* **36**, 914 (1980).
20. I. Lindqvist and A. Niggli, *J. Inorg. Nucl. Chem.* **2**, 345 (1956).
21. D. H. Templeton, A. Zalkin, J. D. Forrester, and S. M. Williamson, *J. Am. Chem. Soc.* **85**, 817 (1963).
22. A. Åström and S. Andersson, *J. Solid State Chem.* **6**, 191 (1973).
23. N. Morimoto, *Miner. J. (Japan)* **1**, 160 (1954).

CHAPTER XI

Topological Transformations

In this chapter we consider various plane nets, polyhedra, and (three-dimensional) crystal structures and ways of relating them: 4^4, 3^6, $3^3 \cdot 4^2$, $3^2 \cdot 4 \cdot 3 \cdot 4$, $3 \cdot 6 \cdot 3 \cdot 6$; rhombohedron, cube, octahedron, tricapped trigonal prism, bisdisphenoid, tetrahedron, square antiprism, tetraederstern; NiAs, NaCl, WC, CsCl, rutile, $CdCl_2$, α-PbO_2, fluorite, baddeleyite (ZrO_2), $Cd(OH)_2$, PdF_3, ReO_3, zircon ($ZrSiO_4$), scheelite ($CaWO_4$), $AgClO_4$, $CaSO_4$, U_3Si_2, $CuAl_2$, $CoGa_3$, ThB_4, $NbTe_4$, $PtPb_4$, TlSe, KHF_2, KN_3, $PdGa_5$, $NH_4Pb_2Br_5$, cubic and lower symmetry "perovskites" $CaCu_3Ge_4O_{12}$, $CaCu_3Mn_4O_{12}$, $ThCu_3Mn_4O_{12}$, $La_3Fe_4P_{12}$, skutterudite ($CoAs_3$), $Sc(OH)_3$, $LiNbO_3$, calcite ($CaCO_3$), MBO_3 (M = Al, Fe, Lu), $NaNO_3$, $GdFeO_3$, perovskite ($CaTiO_3$), $UCrS_3$, $MgSeO_3$, cementite (Fe_3C), YF_3.

Introduction

So far our approach to structure has been mainly a commonplace geometrical one—static and utilizing regular arrays of atoms such as close-packing and primitive cubic packing and various coordination polyhedra. The polyhedra are often idealized as being perfectly regular, and they are connected together in various ways to form units that are then repeated by (discontinuous) symmetry operations (translation, rotation, reflection) to generate a structure. This process is essentially metrical and is basically a very ancient geometry.

A younger branch of geometry ignores the metrical aspects and concentrates on the topology or connectedness of points (in our case atoms); it is sometimes called "rubber-sheet geometry." We need only the simplest aspects of this for our purpose of relating crystal structures, but even so it gives a new insight, revealing relations that may not otherwise be obvious. Essentially, a topological transformation involves a (continuous) nonrigid motion that changes one figure into another so that each point in the first has a corresponding point in the second. No points (atoms) are lost or generated in the transformation, although distances and orientations change (not by very much in the cases of interest to us). Difficult to describe abstractly and in words, what is meant becomes obvious (at least in simple cases) when actual examples are described. We deal first with two-dimensional figures— that is, two-dimensional aspects of three-dimensional structures—and discuss only cases of direct interest in crystal chemistry. [Broader aspects are discussed in simple terms in various references; see refs. 1–3 for the more mathematical aspects, including maps, Möbius strips, and Klein bottles, and ref. 4 for more general and especially biological aspects, including reference to Dürer's intuitive utilization (in 1613!) of topology in describing animal and human appearances and facial expressions.]

Polygonal and Plane Net Transformations

A particularly simple example is the transformation of a square to a rhombus (Figure 1). We notice that in this continuous transformation there is a change of connectedness at the point where the angle of the rhombus, initially 90°, becomes 120°, for at that point the distance $d(1-3)$ becomes equal to the distances $d(1-2)$, $d(2-3)$, and so on. In the topological spirit, such identity is not demanded in crystal chemistry. (Coordination polyhedra in crystals are often not as exactly regular as are the "regular" or "semiregular" polyhedra of the geometer.)

Extending the same operation into infinite two-dimensional nets (5), a 4^4 (square) net can be readily transformed into a 3^6 (triangular) net, as shown in Figure 2. And note that Figure $2a \rightarrow c$ is a twinning operation (proceeding via Figure $2b$). Regular twinning of this sort, on the finest scale, can change the shape of a (finite) 3^6 net (Figure 3); or it can minimize the shape change in $4^4 \rightarrow 3^6$ (Figure $2b \rightarrow$ Figure $3b$). All these transformations may be regarded as simple shear operations.

If only some of the squares in a 4^4 net are transformed by the shear shown in Figure 1, then one obtains nets that consist of intergrowths of squares and triangles, that is, intergrowths of elements of 4^4 and 3^6. One example is shown in Figure 4: $3^6 \rightarrow 3^3 \cdot 4^2$. Clearly the $3^3 \cdot 4^2$ net is readily derived from 4^4

Figure 1. Topological transformation of a square to a 60° rhombus by shear. The arrows indicate the shear involved.

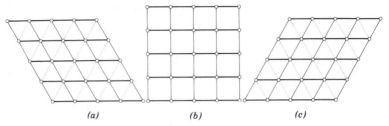

Figure 2. Topological transformation of the plane net $4^4 \rightarrow 3^6$; two possibilities: $(b) \rightarrow (a)$ and $(b) \rightarrow (c)$.

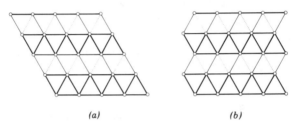

Figure 3. Change of shape of 3^6 by twinning.

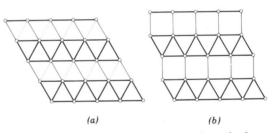

Figure 4. Shear transformation of $3^6 \rightarrow 3^3 \cdot 4^2$.

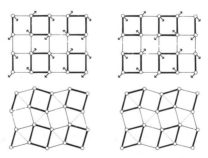

Figure 5. Transformation of $4^4 \rightarrow 3^2 \cdot 4 \cdot 3 \cdot 4$.

by carrying out the shear operation on alternate rows of squares in Figure 2b; i.e., $3^3 \cdot 4^2$ (Figure 4b) is a possible intermediate between 4^4 (Figure 2b) and 3^6 (Figures 2a and 4a). Another possibility, in which half the squares in 4^4 are transformed, is shown in Figure 5: in the right-hand part, squares (above) $\rightarrow$ pairs of triangles (below) by shear, exactly as in Figure 1; but in the left-hand part, the same operation is described as a rotation of those squares that are *not* transformed to diamonds.

Of different interest are those net transformations that appear to change the number of nodes in the repeat unit. They would seem to be most relevant to processes involving changes of stoichiometry and to diffusion mechanisms. As an example we show, in Figure 6, the transformation of $3^6 \rightarrow 3 \cdot 6 \cdot 3 \cdot 6$. In the figure the change is a topological one; clearly this is an alternative to the, at first sight more obvious, method that involves omitting one-fourth of the nodes (atoms). (Cf., in Chapter III, the section on the production of tri-capped trigonal prisms from hcp.) Another possibility, corresponding to the omission of one-thirteenth of the nodes, is shown in Figure 7, where again the change is only topological. Similar transformations can be envisioned for square nets, as shown in Figure 8.

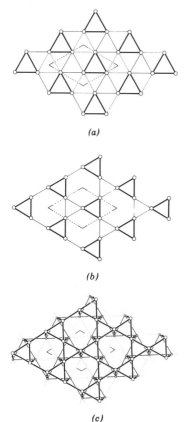

(a)

(b)

(c)

Figure 6.   Transformation of $3^6 \rightarrow 3 \cdot 6 \cdot 3 \cdot 6 = 3^6$ with 25% "vacancies."

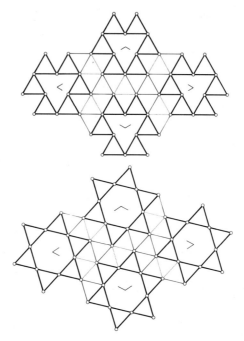

Figure 7. Transformation of $3^6 \rightarrow 3^6$ with one-thirteenth "vacancies."

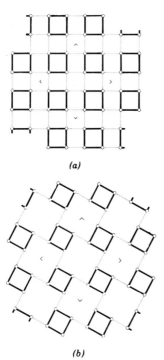

(a)

(b)

Figure 8. Transformation of $4^4 \rightarrow 4 \cdot 8^2 = 4^4$ with 20% "vacancies."

Transformations in Three Dimensions

All the nets, 4^4, 3^6, $3^3 \cdot 4^2$, and $3^2 \cdot 4 \cdot 3 \cdot 4$, are common atomic layers in various crystal structures, and so the operations described above can be used to describe transformations between such structures *in three dimensions* (6, 7), although transformations in three dimensions are not so easily depicted in two-dimensional drawings. A few examples follow, starting with $3^6 \to 4^4$.

NiAs, NaCl, or WC → CsCl

The transformation of a layer of NiAs, NaCl, or WC, etc., to a layer of CsCl is shown in Figure 9. All the "reactant" layers are identical: a triangular 3^6 net of anions plus another 3^6 net of cations. These are (0001) (composite) layers of NiAs or WC or (111) layers of NaCl. The shear operation of Figure $2a \to b$ transforms each into (composite) 4^4 layers of anions and cations. In the third (projection) direction, the operations are not identical. In NiAs the anion layer stacking is hcp, in WC it is ph, while in NaCl it is ccp; i.e., $\cdots a\, \gamma\, b\, \gamma\, a\, \gamma\, b \cdots$, $\cdots a\, \beta\, a\, \beta\, a\, \beta\, a \cdots$, and $\cdots a\, \gamma\, b\, a\, c\, \beta\, a\, \gamma\, b\, a\, c \cdots$, respectively. In CsCl the 4^4 nets of anions (and cations) superimpose along this same direction. It is not too difficult to visualize both the NiAs and NaCl transformations as involving only $3^6 \to 4^4$ in all three dimensions. For NaCl → CsCl, the transformation involves only the primitive (rhombohedral) unit cell (without any reorientation being necessary), as we shall see later. One can imagine this happening as the central cation increases in size ($Na^+ \to Cs^+$). Although in principle continuous, this transformation is, in fact, discontinuous (first-order). But this is an energetic, not a geometric, necessity, and there is no doubt that it sometimes does proceed by this simple mechanism. Other, closely related geometrical mechanisms may be envisaged (8, 9), but they do not differ in principle, only in detail (shape change).

Rutile, CdCl$_2$, α-PbO$_2$, etc. → CaF$_2$, Baddeleyite, etc.

If the appropriate half of the octahedral sites are occupied in a cp anion array, a similar transformation can lead to the fluorite (CaF_2)-type structure (with a pc anion array, as in CsCl, but only half the cubes occupied by cations), as already discussed in Chapter III. Thus the operation of Figure $2a \to b$ can transform the rutile type to fluorite type, as shown in Figure 10. An alternative continuous transformation is that shown in Figure $4a \to b$, followed by $3^3 \cdot 4^2 \to 4^4$. These are two distinct possibilities (there are obviously many others), the former a homogeneous shear, the latter having an intermediate stage (Figure $4b$) that corresponds reasonably well to the (010) layers of the baddeleyite structure of monoclinic ZrO_2. (In this and in the rutile structure, the models idealize the anion layers to *plane* nets. In fact, of course, they are slightly puckered. However, in projection their appearance is very close to that shown in the figures.) There is no reported observation

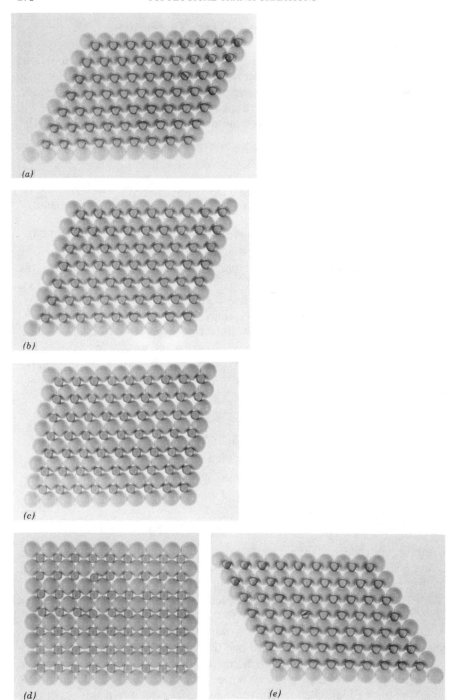

Figure 9. Transformation of a NiAs or NaCl layer in (a) to a layer of CsCl in (d). Continuing the operation leads to (e), a twin of (a) (cf. Figure 2).

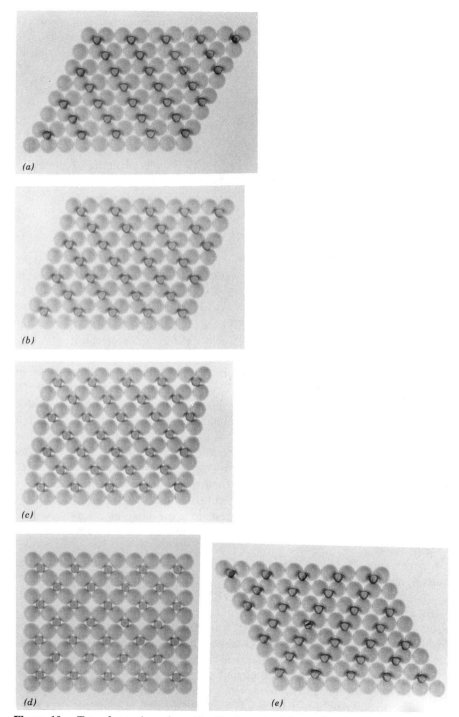

Figure 10.   Transformation of a rutile-like layer in (a) to a fluorite-like layer in (d).

of the rutile type to the baddeleyite type (10), but there are several reports of rutile → "distorted fluorite," especially for transition metal difluorides.

Figure 10 also shows clearly (as Figure 2 implies) that a continuation of the homogeneous shear that transforms rutile to fluorite (Figure $10a \rightarrow d$) will regenerate the rutile type, but in a twin orientation (in Figure $10e$). If a rutile type is transformed to a fluorite type by the application of high pressure, then when the pressure is released it will transform back. (The reverse transformation is spontaneous and cannot be prevented.) But Figure $2b \rightarrow a$ and $b \rightarrow c$ are equally probable paths. A likely criterion will be minimum shape change, which can be achieved by fine-scale twinning of the rutile product, i.e., equal amounts of Figure $10d \rightarrow a$ and $d \rightarrow e$. Twinning on the finest possible scale can occur and is depicted in Figure $11a \rightarrow d$; the product (Figure $11d$) is the α-PbO_2 type. Again it is noteworthy that high-pressure fluorites do revert to this orthorhombic structure (rather than the rutile type) when the pressure is released (11).

Alternate (0001) layers of $Cd(OH)_2$ type or (111) layers of $CdCl_2$ type are like the (111) layers of NaCl [or the (0001) layers of NiAs or WC]; intervening layers contain no cations. Transformation analogous to NiAs/NaCl → CsCl will therefore give in this case stacked CsCl-like layers with no intervening cations, i.e., $MoSi_2$ or α-$FeSi_2$ types (Chapter VIII). (However, Figure $10a \rightarrow d$ can also represent the high-pressure transformation of $CdCl_2$ type to CaF_2 type.)

Similar "games" can be played with various other stoichiometries and structures. For example, Figure 6 corresponds to the transformation (in three dimensions, by stacking the nets) of the anion array in PdF_3 to that in ReO_3 type (cf. Figure 12). (Many trifluorides have structures intermediate between the two extremes.) The point is that simple topological transformations are, in principle at least and often in practice, capable of changing one structure continuously into another, the coordination numbers being different in the two cases. In some cases, for example, CaF_2 → rutile, there is a reduction of symmetry that allows the possibility of twinning (often on a very fine scale). This is consistent with many experimental observations and also accounts for some others in which twinning has not been looked for but seems likely.

An example is the reported NiAs-like high-pressure phase produced when GeO_2 and SiO_2 glasses are compressed at high temperature and then "quenched" (cooled to room temperature before the pressure is released). The result is a small volume *increase* (over that of the rutile form), and it seems likely that the high-pressure transformation is to the fluorite type, which is unquenchable and reverts to very highly and finely twinned α-PbO_2 and/or rutile type (when the pressure is released). Both the latter are approximately hcp anions with an average half-occupancy of the octahedral sites by cations; disorder due to twinning, intergrowth, etc., will yield an average of half a cation per octahedron. X-ray powder diffraction will not reveal the fine-scale effects suggested, and electron microscopy (which will) has not

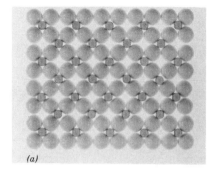

(a)

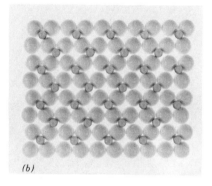

(b)

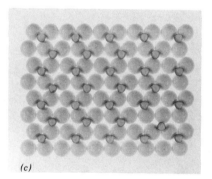

(c)

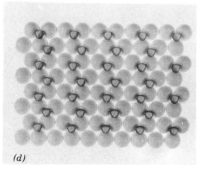

(d)

Figure 11. Repeated lamellar twinning on the finest possible scale in the fluorite → rutile transformation. The product is α-PbO$_2$ type.

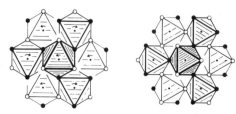

Figure 12.   Transformation of the octahedral framework structure of PdF_3 (left, hcp anions) to that of ReO_3 (right, anions form 75% of a ccp array) by rotation of the octahedra as indicated by the arrows. (Both structures are viewed down their three-fold axes, [0001] and [111], respectively.)

yet been applied. The resulting disorder may very well account for the small increase in volume.

A few examples involving similar two-dimensional net transformations, but other than 3^6, 4^4, and their combinations, have also been described (5).

Transformations of Polyhedra

A proper consideration of three-dimensional nets and their transformations is a complex matter, and we shall not attempt it here (see ref. 12). But it is aided by considering transformations of polyhedra by topological distortion, which is also of great help in relating crystal structures. A particularly simple case is the distortion of a 60° rhombohedron into a 90° rhombohedron—a cube. This is depicted in Figure 13. It is a complete description of the $NaCl \rightarrow CsCl$ transformation discussed earlier (and also of $CdCl_2 \rightarrow CaF_2$), since the 60° rhombohedron is the primitive (though not the conventional) unit cell of the NaCl type (Figure 14), while the cube is the unit cell of the CsCl type (also primitive). The former is, of course, the basic, space-filling unit of ccp consisting of one octahedron plus two tetrahedra.

Another useful transformation is the distortion of a pair of face-shared octahedra to a tricapped trigonal prism depicted in Figure 15. For a given

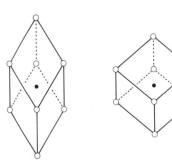

Figure 13.   Transformation of a primitive (rhombohedral) unit cell of NaCl to the corresponding unit cell of CsCl, with persistence of the [111] axis. The former is a 60° rhombohedron; the latter a 90° rhombohedron.

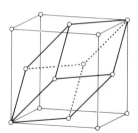

(a) (b)

Figure 14. Relation between the fcc unit cell ($Z = 4$, light lines) and the corresponding primitive unit cell ($Z = 1$, heavy lines), which is a 60° rhombohedron. Circles represent lattice points.

Figure 15. (a) A pair of face-sharing octahedra (an element of hcp); (b) a tricapped trigonal prism. Their topological identity is obvious.

edge length (the same in the two cases), the main difference is between the heights of the two polyhedra: from $\sqrt{8/3} = 1.633$ for the former to 1 for a perfectly regular trigonal prism or, more plausibly, ~ 1.264 for a tricapped trigonal prism (13). This is the basis of a group of formal structural relationships already discussed in Chapter III, for example, rutile $\rightarrow$ PbFCl, α-PbO$_2$ $\rightarrow$ PbCl$_2$. It is also a plausible reaction mechanism for some high-pressure transformations and for their reversal when the pressure is released. It is obvious that in the latter case (as in CaF$_2$ $\rightarrow$ α-PbO$_2$, rutile etc., mentioned above) the high-pressure structure is "nonquenchable"—that is, it cannot be retained when the pressure is released. This provides an alternative model for the production of "half-filled NiAs type" when MX$_2$ structures are compressed and then quenched. Compare the earlier discussion in which the high-pressure form is the CaF$_2$ type; here it is rutile $\rightleftharpoons$ PbFCl, α-PbO$_2$ $\rightleftharpoons$ PbCl$_2$, etc.

The "nonquenchability" is an important characteristic of such continuous, topological transformations. The reason is that they are nondiffusive; such transformations have also been given various other names, such as displacive, martensitic, and, in certain cases, ferroelastic. The last is restricted to *small distortions* with a reduction of symmetry, so that more than one orientation of the product phase, and therefore a domain texture in the product, must result. Relative to its surrounding atoms, no atom has to move as much as a bond length (say no more than about 1 Å) in such transformations. Often the shifts ("shuffles") will be only a few tenths or even a few hundredths of an ångström, and extensive, repeated twinning is an inevitable result. For small displacements the domains (of different orientations of the product phase) will be coherent but perhaps strained, which makes this an important mechanism for making strong solids.

For larger displacements the strain can be great enough to shatter all but the smallest single crystals. In such cases the transformation superficially appears to be *reconstructive*, but this is not the case. It is *displacive,* but the

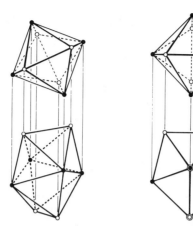

Figure 16. Three different views of a bisdisphenoid: Top, "plan" projected along its 4 axis; below left, perspective view; below right, in elevation.

displacements, and the resulting strain, are too large to be accommodated; that is, the strength of the crystal is insufficient. An example that has been carefully studied (14) is the high- to low-temperature transformation of MnCoGe (Ni$_2$In → PbCl$_2$ type). Another is possibly β- → γ-Ca$_2$SiO$_4$ (15). Descriptions of structures and structure relations, of the sort given here, provide a simple intuitive approach to such phenomena (which are often obscured by the complicated mathematics of elasticity and group theories, which, however, are essential for anything more than a simple, naive appreciation of what is occurring).

In Chapter XIII we shall show how the rather complicated anion arrangements in zircon (ZrSiO$_4$), scheelite (CaWO$_4$), and several other structures can be derived from much simpler ones: bcc and pc arrays. The characteristic coordination polyhedron in these structures is the bisdisphenoid or dodeca(delta)hedron, CN = 8. Here we show how this can be readily derived from more familiar polyhedra: octahedra or tetrahedra (16). (It is even simpler to derive it by distorting a cube.)

Three different views of a bisdisphenoid are shown in Figure 16. It has 8 vertices, 18 edges, and 12 faces. The last accounts for the name dodecahedron.* This dodecahedron has triangular faces and is therefore a deltahedron: a dodecadeltahedron. The name bisdisphenoid arises in the following way: a sphenoid is a wedge; a disphenoid is a tetrahedron—two wedges, accounting for its four faces; a bisdisphenoid is two (interpenetrating) tetrahedra (cf. the cube), as shown in elevation in Figure 17. One of these tetrahedra (lightly dotted) is elongated, and the other (heavily dotted) is squat; the deformations being tetragonal distortions (in opposite senses) parallel to the $\bar{4}$ axis in both cases. The centroids and $\bar{4}$ axes of both tetrahedra coincide,

* The pentagonal dodecahedron is a "regular" polyhedron that has 12 pentagonal faces.

Figure 17.   A bisdisphenoid depicted as two interpenetrating tetrahedra.

and hence an atom at the center of the bisdisphenoid has two sets each of four equidistant coordinating atoms (at its eight vertices). Invariably, in crystals, the bonds to the corners of the squashed tetrahedron are slightly shorter than those to the corners of the elongated one.

Figure 18 shows, in plan and in elevation, a bisdisphenoid edge-connected to a tetrahedron, a pair of edge-shared octahedra, and a chain of four edge-shared tetrahedra. It is not difficult to see that each group consists of 10 vertices and that each may be transformed into the others by small topological distortions. We will now see how such distortions relate several types of structures that are not at first sight similar.

Some Further Examples of Structure Transformations

An elegant example is the relation between the rutile and zircon structures, both of which are depicted (in equivalent projections) in Figure 19. A comparison of the two structures in this figure reveals that the chains of edge-sharing TiO_6 octahedra in the rutile type have become chains of alternating ZrO_8 bisdisphenoids and SiO_4 tetrahedra in the zircon type. The relation is

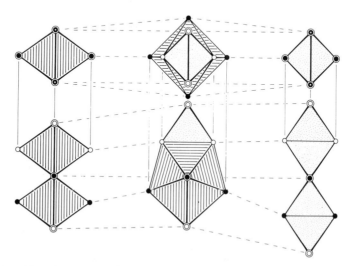

Figure 18.   Elevation views of, center, a bisdisphenoid connected to a tetrahedron by a common edge; right, a chain of four edge-shared tetrahedra; left, a pair of edge-shared octahedra. Above, the same groups are shown in plan.

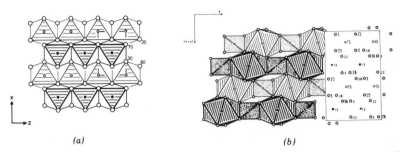

(a)   (b)

Figure 19. The rutile (*a*) and zircon (*b*) structures projected on (010) and ($1\bar{1}0$), respectively.

exactly that between a pair of octahedra and a bisdisphenoid plus tetrahedron shown in Figure 18.

It is therefore interesting that zircon was originally reported (17) as having the rutile-type structure [although this was subsequently corrected (18)]. It is perhaps even more interesting that such eminent authorities as Bragg and Claringbull (19) have commented that "zircon, $ZrSiO_4$, was once referred to the rutile group because of similarity in cell dimensions but X-ray analysis has shown this to be *fortuitous, as there is no relation between the two structures*" [our italics]. The relation is, in fact, quite simple; it corresponds to replacing Ti atoms, in the chains of edge-sharing octahedra in TiO_2, by alternately larger (Zr) and smaller (Si) atoms. The larger cations have a longer bond length to oxygen and consequently a higher CN, and the smaller have a shorter bond length and consequently a lower CN: $^{VI}Ti^{VI}Ti(^{III}O_4) \rightarrow$ $^{VIII}Zr^{IV}Si(^{III}O_4)$. This is shown schematically in Figure 20.

Similar shifts generate the related structures (also with large and small cations) of anhydrite ($CaSO_4$) and $AgClO_4$ (16). The differences between these three ABX_4 structures lies simply in the relative directions of the small shuffles of the cations that are necessary when Ti_2 is replaced by (large) A and (small) B. Figure 21 shows that if the shifts in adjacent rows (of edge-sharing octahedra in the rutile parent) are all parallel, then the resulting bisdisphenoids share edges. If these shifts are antiparallel, then they share corners. In zircon the bisdisphenoids share edges in both (100) and (010) layers, and so all shifts are parallel throughout the structure (which is therefore tetragonal). In anhydrite they share edges in (100) layers but corners in (001) layers, so shifts are parallel in the former but antiparallel in the latter. (The structure is therefore orthorhombic.) In $AgClO_4$ the bisdisphenoids

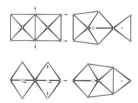

Figure 20. Two projections of the transformation of a pair of edge-shared octahedra, on the left, to a bisdisphenoid and tetrahedron (also sharing an edge), on the right.

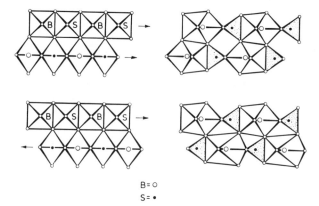

B = ○
S = •

Figure 21. Transformation of chains of edge-sharing octahedra (left) to chains of alternating bisdisphenoids and tetrahedra (right). B and S are "big" (A) and "small" (B) cations, respectively. The arrows represent the directions of cation shifts.

share only corners, so the shifts in each row are antiparallel with respect to those in all immediately adjacent rows. (This structure is therefore also tetragonal.) These shifts are shown schematically in Figure 22, and Figure 23 shows the two types of layers involved. Figure 24 shows the complete structure of CaSO₄ in two projections. The zircon structure is depicted in two bounded (corresponding) projections in Figures 5a and b of Chapter XIII.

Crystallographic data for these structures are given in Data Tables 1–4.

Data Table 1   Rutile, TiO₂ (20)

Tetragonal, space group $P4_2/mnm$, No. 136; $a = 4.59366$, $c = 2.95868$ Å, $c/a = 0.64408 = 0.91087/\sqrt{2}$; $Z = 2$, $V = 62.433 = 249.73/4$ Å³

Atomic Positions
Ti in 2(a): $0,0,0; \frac{1}{2},\frac{1}{2},\frac{1}{2}$
O in 4(f): $\pm(x,x,0; \frac{1}{2}+x,\frac{1}{2}-x,\frac{1}{2}); x = 0.3048$
Bond Lengths
l(Ti–O) = 1.980 Å (2×), 1.948 Å (4×)

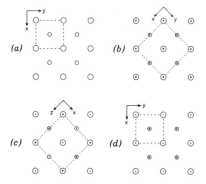

Figure 22. Projections of equivalent cation rows in (a) rutile, (b) zircon, (c) anhydrite, and (d) AgClO₄. The projection directions are, respectively, [001], [001], [010], and [001]. The plus and dot symbols indicate concerted shifts (from rutile to ABX₄) in opposite senses. Large circles are cations at height 0; small circles are cations at height $\frac{1}{2}$ in rutile.

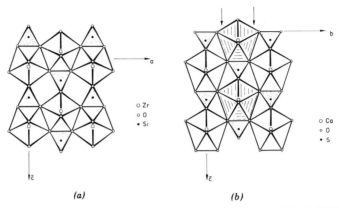

Figure 23. The two layer types in various ABX_4 structures. (a) A (010) layer of zircon $\equiv$ (100) zircon layer $\equiv$ (100) layer of $CaSO_4$; (b) a (001) layer of $CaSO_4$ $\equiv$ (100) $AgClO_4$ $\equiv$ (010) $AgClO_4$.

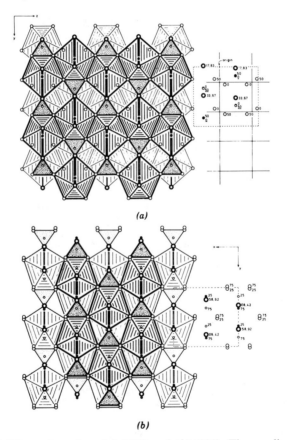

Figure 24. $CaSO_4$ projected on (a) (100) and (b) (001). The smaller circles are S (open) and Ca (filled); larger circles are oxygens. Heights are in hundredths of the projection axis. [S atom heights are underlined in (a).]

Data Table 2 Zircon, ZrSiO$_4$ (21)

Tetragonal, space group $I4_1/amd$, No. 141; $a = 6.607$, $c = 5.982$ Å, $c/a = 0.9054$; $Z = 4$, $V = 261.13$ Å^3

Atomic Positions (1st choice of origin, at $\overline{4}m2$): all atoms at $(0,0,0; \frac{1}{2},\frac{1}{2},\frac{1}{2})+$
Zr in $4(a)$: $0,0,0; 0,\frac{1}{2},\frac{1}{4}$
Si in $4(b)$: $0,0,\frac{1}{2}; 0,\frac{1}{2},\frac{3}{4}$
O in $16(h)$: $0,x,z; x,0,\overline{z}; 0,\overline{x},z; \overline{x},0,\overline{z}; 0,\frac{1}{2}+x,\frac{1}{4}-z; x,\frac{1}{2},\frac{1}{4}+z; 0,\frac{1}{2}-x,\frac{1}{4}-z;$
  $\overline{x},\frac{1}{2},\frac{1}{4}+z; x = 0.8161, z = 0.3203$
Bond Lengths
$l(\text{Si–O}) = 1.622$ Å ($4\times$)
$l(\text{Zr–O}) = 2.130$ Å ($4\times$), 2.269 Å ($4\times$)

Data Table 3 Anhydrite, CaSO$_4$ (22)

Orthorhombic, space group $Cmcm$, No. 63; $a = 6.998$, $b = 6.245$, $c = 7.006$ Å, $b/a = 0.8924$, $b/c = 0.8914$; $Z = 4$, $V = 306.2$ Å^3

Atomic Positions: $(0,0,0; \frac{1}{2},\frac{1}{2},0)+$
Ca in $4(c)$: $\pm(0,y,\frac{1}{4}); y = 0.6524$
S in $4(c)$: $y = 0.1556$
O(1) in $8(g)$: $\pm(x,y,\frac{1}{4}; \overline{x},y,\frac{1}{4}); x = 0.1695, y = 0.0155$
O(2) in $8(f)$: $\pm(0,y,z; 0,y,\frac{1}{2}-z); y = 0.2976, z = 0.0817$
Bond Lengths
$l(\text{S–O}) = 1.474$ Å ($2\times$), 1.475 Å ($2\times$)
$l(\text{Ca–O}) = 2.466$ Å ($4\times$), 2.510 Å ($4\times$)

Data Table 4 AgClO$_4$ (23)

Tetragonal, space group $I\overline{4}2m$, No. 121; $a = 4.976$, $c = 6.746$ Å, $c/a = 1.3557 = 0.9586 \times \sqrt{2}$; $Z = 2$, $V = 167.0 = 334.1/2$ Å^3

Atomic Positions: $(0,0,0; \frac{1}{2},\frac{1}{2},\frac{1}{2})+$
Ag in $2(b)$: $0,0,\frac{1}{2}$
Cl in $2(a)$: $0,0,0$
O in $8(i)$: $x,x,z; \overline{x},\overline{x},z; x,\overline{x},\overline{z}; \overline{x},x,\overline{z}; x = 0.1653, z = 0.1259$
Bond Lengths
$l(\text{Cl–O}) = 1.440$ Å ($4\times$)
$l(\text{Ag–O}) = 2.504$ Å ($4\times$), 2.779 Å ($4\times$)

[In Chapter XIII we will reconsider the complications of the anion arrays in rutile and zircon and the relationships between them and show that they fall into perspective if we concentrate on the cation arrays, which provide a fixed, almost unaltered, framework: bct A in rutile and (ordered) AB in zircon. Then one simply changes (in a very simple and obvious way) the anion coordinates to accommodate two different bond lengths in the ABO$_4$ compounds (cf. ref 24).]

The other topological relation in Figure 18 is that between a bisdisphenoid plus a tetrahedron, on the one hand, and a chain of four edge-shared tetrahedra on the other. The magnitude of the distortion (three tetrahedra to a bisdisphenoid) is somewhat greater in this case but is still not too large to be useful. It is manifested in the relation between some "stuffed cristobalites" ABX_2 such as chalcopyrite and β-$KCoO_2$ or $CaGeN_2$ (25), as well as between the ABX_4 compounds just discussed and various tetrahedral structures (16) such as $ZrSiO_4$ and Cu_2O. We mention this in passing; we but will not go into further detail here.

There are many other distortions that relate various types of polyhedra. Frequently these incorporate transformations relating polygons, some of which we have already considered. If a pair of opposite faces on a cube are rotated by 45° relative to each other, then the cube obviously becomes a square antiprism (Figure 25). If this is done repeatedly on a pair of stacked square nets, so that in each of them $4^4 \rightarrow 3^2 \cdot 4 \cdot 3 \cdot 4$ (in which case the rotation is 30° rather than the full 45°), then a layer of cubes can be transformed into a layer of cubes and rhombic prisms (= two triangular prisms sharing a square face) or a layer of approximately square antiprisms sharing edges, plus intervening tetracapped tetrahedra (tetraederstern, see Chapter XIV). These two possibilities are shown in Figure 26a and b. The two products are, in fact, projections of well-known structures: Figure 26a shows part of the U array in U_3Si_2, and Figure 26b, the Al array in $CuAl_2$.

In the former the U_8 cubes are centered by additional U atoms, and the U_6 trigonal prisms are centered by Si atoms. The centered cubes form columns of bcc U (the structure of one polymorph of U metal), and the centered trigonal prisms form columns of AlB_2 type (the structure of USi_2). A comparison of Figure 26a with Figure 6 of Chapter VIII shows that the U_3 part of U_3Si_2 is a collapsed $PtHg_2$ type [$Pt_2Hg = (Pt,Hg)_3$]. There are many compounds isostructural with U_3Si_2 and also many related structure types with partial occupancy of the interstitial sites. In $CoGa_3$, for example, half the trigonal prisms are emptied (in a strictly ordered way that doubles the c axis) so that $U_3Si_2 \rightarrow Ga_3Co$. ThB_4 has a structure related to the U_3Si_2 type; the framework atoms (Figure 26a) are Th, and B atoms occupy the trigonal prisms, but the cubes are now centered by B_6 octahedra; thus $U_2USi_2 \rightarrow Th_2B_6B_2 = 2\ ThB_4$. The crystallographic data are given in Data Tables 5–7.

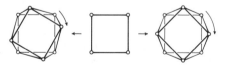

Figure 25. Transformation of a cube (center) to a regular square antiprism (on the right; rotation of top face = 45°) and a distorted square antiprism, of the sort occurring in $CuAl_2$, etc. (on the left; rotation = 30°).

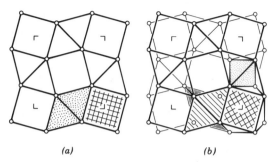

(a)　　　　　　　　*(b)*

Figure 26. Transformation (cf. Figure 5) of two adjacent 4^4 nets each to $3^2 \cdot 4 \cdot 3 \cdot 4$ (*a*) superimposed, (*b*) in antipositions. An example is shown of each of the polyhedra thus generated.

Data Table 5　U_3Si_2 (26)

Tetragonal, space group $P4/mbm$, No. 127; $a = 7.330$, $c = 3.900$ Å, $c/a = 0.5321$; $Z = 2$, $V = 209.5$ Å^3

Atomic Positions
U(1) in 2(*b*):　　$0,0,\frac{1}{2}; \frac{1}{2},\frac{1}{2},\frac{1}{2}$
U(2) in 4(*g*):　　$\pm(x,\frac{1}{2} + x,0; \frac{1}{2} + x,\bar{x},0); x = 0.181$
Si in 4(*h*):　　$\pm(x,\frac{1}{2} + x,\frac{1}{2}; \frac{1}{2} + x,\bar{x},\frac{1}{2}); x = 0.389$
Atomic Distances
U(1)–4Si = Si–2U(1) = 2.96 Å
U(2)–6Si = Si–6U(2) = 2.92 Å
Si–Si = 2.30 Å

Compounds isostructural with U_3Si_2 include Nb_3Be_2, Ta_3B_2, Mo_2CoB_2, Th_3Al_2, Th_3Si_2, and Th_3Ge_2.

Data Table 6　$CoGa_3$ (27)

Tetragonal, space group $P\bar{4}n2$, No. 118; $a = 6.26$, $c = 6.45$ Å, $c/a = 1.03 = 0.5_2 \times 2$; $Z = 4$, $V = 251$ Å^3

Atomic Positions
Co in 4(*f*):　　$x,\frac{1}{2} - x,\frac{1}{4}; \bar{x},\frac{1}{2} + x,\frac{1}{4}; \frac{1}{2} + x,x,\frac{3}{4}; \frac{1}{2} - x,\bar{x},\frac{3}{4}; x = 0.35(3)$
Ga(1) in 4(*e*):　　$\pm(0,0,z; \frac{1}{2},\frac{1}{2},\frac{1}{2} + z); z = 0.25(0)$
Ga(2) in 8(*i*):　　$x,y,z; \bar{x},\bar{y},z; \bar{y},x,\bar{z}; y,\bar{x},\bar{z}; \frac{1}{2} - x,\frac{1}{2} + y,\frac{1}{2} + z; \frac{1}{2} + x,\frac{1}{2} - y,\frac{1}{2} + z;$
　　　　　　$\frac{1}{2} + y,\frac{1}{2} + x,\frac{1}{2} - z; \frac{1}{2} - y,\frac{1}{2} - x,\frac{1}{2} - z; x = 0.15(3), y = 0.34(7),$
　　　　　　$z = 0.00(0)$
Atomic Distances
Co–Ga(2) = 2.36 Å (2×), 2.82 Å (4×) (trigonal prism)
Co–Ga(1) = 2.38 Å (2×) (caps)

Several other transition metal trigallides and tri-indides are isostructural with $CoGa_3$, and many lanthanide and actinide tetraborides are isostructural with ThB_4.

Data Table 7 ThB_4 (28)

Tetragonal, space group $P4/mbm$, No. 127; $a = 7.256$, $c = 4.113$ Å, $c/a = 0.567$; $Z = 4$, $V = 216.5$ Å^3

Atomic Positions
Th in $4(g)$: $\pm(x,\frac{1}{2} + x,0; \frac{1}{2} + x,\bar{x},0)$; $x = 0.313(2)$
B(1) in $4(e)$: $\pm(0,0,z; \frac{1}{2},\frac{1}{2},z)$; $z = 0.212$
B(2) in $4(h)$: $\pm(x,\frac{1}{2} + x,\frac{1}{2}; \frac{1}{2} + x,\bar{x},\frac{1}{2})$; $x = 0.087$
B(3) in $8(j)$: $\pm(x,y,\frac{1}{2}; \bar{y},x,\frac{1}{2}; \frac{1}{2} + x,\frac{1}{2} - y,\frac{1}{2}; \frac{1}{2} + y,\frac{1}{2} + x,\frac{1}{2})$; $x = 0.170$, $y = 0.042$

Atomic Distances
B(2)–Th = 3.10 Å (2×), 2.96 Å (4×) (trigonal prism)
B(2)–B(2) = 1.79 Å (B_2 pairs)
B(2)–B(3) = 1.79 Å (between B_2 and B_6 octahedron)
B(3)–B(3) = 1.80 Å; B(1)–B(3) 1.74 Å (both within B_6 octahedra)
B(1)–B(1) = 1.74 Å (between B_6 octahedra along **c**)

In the latter (the Al array in $CuAl_2$, Figure 26b), the Al_8 square antiprisms are centered by Cu atoms. It may also be derived from the $PtHg_2$ type (cf. U_3Si_2 above, and Figure 6 of Chapter VIII), but now it is the filled cubes that collapse (to square antiprisms, Figure 26b) rather than the empty ones. Again there are many related structures. In $NbTe_4$ the framework (Figure 26b) atoms are Te, but only alternate columns of square antiprisms are occupied by Nb (but see the discussion on p. 310); the others are empty. The diagonals of the rhombuses are slightly short due to Te–Te bonding. In $PtPb_4$ the framework atoms are Pb, and Pt atoms occupy only alternate (001) layers of square antiprisms. In still other structures the tetrahedra are used as well as the square antiprisms. For example, in "TlSe" the latter are occupied by the (larger) Tl^+ and the former by the (smaller) Tl^{3+}, so that the compound is $Tl^+Tl^{3+}Se_2^{2-}$. In a group of related structures the additional atoms are in the edges rather than the centers of the tetrahedra; examples are KHF_2 and $NaHF_2$, with the large cations in the square antiprisms and F–H–F in the diagonal of each rhombus [parallel to (001)]. Alkali metal azides, such as KN_3, have N–N–N groups in place of F–H–F. (Compare the Te–Te bonds in the same positions in $NbTe_4$.)

The structure of $PdGa_5$ is an elegant variant, with alternate framework layers of the U_3Si_2 and $CuAl_2$ types. The columns are then cubes alternating with square antiprisms, and tetraedersterns (capped tetrahedra) alternating with rhombic prisms. The Ga_8 cubes are centered by other Ga atoms, and the square antiprisms by Pd atoms. $NH_4Pb_2Br_5$ (Figure 27) is similar, with Br atoms forming the framework; the Br_8 cubes are each centered by an additional Br, and the square antiprisms by NH_4 (so that $NH_4Br_5^{4-} \equiv PdGa_5$), and

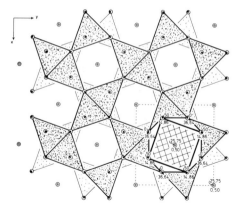

Figure 27. The tetragonal ($I4/mcm$) structure of $NH_4Pb_2Br_5$ projected on (001). Large circles are Br atoms, small open circles are NH_4, and small filled circles are Pb atoms; heights are in units of $c/100$. NH_4Br_8 square antiprisms are cross-hatched; $PbBr_6$ trigonal prisms are shaded. (Tetrahedra of Br in the tetraederstern are unoccupied.)

the Pb atoms center all the triangular prisms. Hence alternate (001) layers are $(NH_4)Br_2^-$ ($\equiv CuAl_2$) and $Br_3Pb_2^+$ ($\equiv U_3Si_2$).

The crystallographic data for several of these structures are given in Data Tables 8–13.

Data Table 8   NbTe$_4$* (29)

Tetragonal, space group $P4cc$, No. 103; $a = 6.499$, $c = 6.837$ Å, $c/a = 1.052$; $Z = 2$, $V = 288.8$ Å^3

Atomic Positions
Nb in 2(a): $0,0,z; 0,0,\frac{1}{2} + z; z = 0.250$
Te in 8(d): $x,y,z; \bar{y},x,z; \bar{x},\bar{y},z; y,\bar{x},z; \bar{x},y,\frac{1}{2} + z; y,x,\frac{1}{2} + z; x,\bar{y},\frac{1}{2} + z; \bar{y},\bar{x},\frac{1}{2} + z;$
        $x = 0.1425, y = 0.3316, z = 0$

Atomic Distances
Nb–Te = 2.902 Å (8×)
Nb$\cdots$Nb = 3.419 Å (2×)†

* This is an average, nonmodulated substructure (cf. p. 310).
† Cf. 2.858 and 3.300 Å in bcc Nb metal.

Data Table 9   PtPb$_4$ (30)

Tetragonal, space group $P4/nbm$, No. 125; $a = 6.66$, $c = 5.97$ Å, $c/a = 0.896$; $Z = 2$, $V = 264.8$ Å^3

Atomic Positions
Pt in 2(a): $0,0,0; \frac{1}{2},\frac{1}{2},0$
Pb in 8(m): $x,\frac{1}{2} + x,z; x,\frac{1}{2} - x,\bar{z}; \bar{x},\frac{1}{2} - x,z; \bar{x},\frac{1}{2} + x,\bar{z}; \frac{1}{2} + x,x,\bar{z}; \frac{1}{2} + x,\bar{x},z;$
        $\frac{1}{2} - x,\bar{x},\bar{z}; \frac{1}{2} - x,x,z; x = 0.175(5), z = 0.0255(5)$

Data Table 10   TlSe (31)

Tetragonal, space group $I4/mcm$, No. 140; $a = 8.02(1)$, $c = 7.00(2)$ Å, $c/a = 0.87_3$; $Z = 4$, $V = 450$ Å^3

Atomic Positions $(0,0,0; \frac{1}{2},\frac{1}{2},\frac{1}{2})+$
Tl(1) in 4(a): $\pm(0,0,\frac{1}{4})$
Tl(2) in 4(b): $0,\frac{1}{2},\frac{1}{4}; \frac{1}{2},0,\frac{1}{4}$
Se in 8(h): $\pm(x,\frac{1}{2}+x,0; \frac{1}{2}+x,\bar{x},0)$; $x = 0.179(3)$
Bond Lengths
l(Tl(1)–Se) = 3.42 Å (8×)
l(Tl(2)–Se) = 2.68 Å (4×)

Data Table 11   KHF$_2$ (32)

Tetragonal, space group $I4/mcm$, No. 140; $a = 5.67$, $c = 6.81$ Å, $c/a = 1.201$; $Z = 4$, $V = 219$ Å^3

Atomic Positions: $(0,0,0; \frac{1}{2},\frac{1}{2},\frac{1}{2})+$
K in 4(a): $\pm(0,0,\frac{1}{4})$
H in 4(d): $0,\frac{1}{2},0; \frac{1}{2},0,0$
F in 8(h): $\pm(x,\frac{1}{2}+x,0; \frac{1}{2}+x,\bar{x},0)$; $x = 0.1420$
Atomic Distances
l(K–F) = 2.769 Å (8×)
d(F–H–F) = 2.277 Å (linear, symmetrical)

Data Table 12   Potassium Azide, KN$_3$ (33)

Tetragonal, space group $I4/mcm$, No. 140; $a = 6.119$, $c = 7.102$ Å, $c/a = 1.1606$; $Z = 4$, $V = 265.9$ Å^3

Atomic Positions: $(0,0,0; \frac{1}{2},\frac{1}{2},\frac{1}{2})+$
K in 4(a): $\pm(0,0,\frac{1}{4})$
N(2) in 4(d): $0,\frac{1}{2},0; \frac{1}{2},0,0$
N(1) in 8(h): $\pm(x,\frac{1}{2}+x,0; \frac{1}{2}+x,\bar{x},0)$; $x = 0.136$
Atomic Distances
l(K–N) = 2.967 Å (8×)
d(N–N–N) = 2.354 Å (linear and symmetrical)

Since a cube is readily described as two concentric tetrahedra, it is obvious how it may be deformed to a bisdisphenoid. This is the basis of the fluorite → scheelite transformation. It is only slightly less obvious that a bisdisphenoid is easily transformed into a bicapped trigonal prism or a square antiprism. It soon becomes apparent that, in topological terms, all coordination polyhedra with the same number of vertices are readily transformed from one to another (35).

Earlier we showed that even the number of vertices may be changed quasi-continuously; for example, in rutile → fluorite. Another way of doing

Data Table 13   NH₄Pb₂Br₅ (34)

Tetragonal, space group $I4/mcm$, No. 140; $a = 8.430$, $c = 14.470$ Å, $c/a = 1.716 = 0.858 \times 2$; $Z = 4$, $V = 1028$ Å³

Atomic Positions: $(0,0,0; \frac{1}{2},\frac{1}{2},\frac{1}{2})+$

N(H₄) in $4(a)$: $\pm(0,0,\frac{1}{4})$

Pb in $8(h)$: $\pm(x,\frac{1}{2}+x,0; \frac{1}{2}+x,\bar{x},0)$; $x = 0.158$

Br(1) in $16(l)$: $\pm(x,\frac{1}{2}+x,z; x,\frac{1}{2}+x,\bar{z}; \frac{1}{2}+x,\bar{x},z; \frac{1}{2}+x,\bar{x},\bar{z})$;
           $x = 0.163$, $z = 0.363$

Br(2) in $4(c)$: $0,0,0; 0,0,\frac{1}{2}$

Bond Lengths

l(NH₄–Br(1)) = 3.554 Å (8×) = sap; 3.618 Å (2×) = caps

l(Pb–Br) = 2.913 Å (2×) plus 3.355 Å (4×) = tp, 3.176 Å (2×) = caps

this is considered in some detail in the following section. [We mention in passing that, particularly for coordination numbers greater than 6, it is often difficult to identify the coordination polyhedron or indeed to determine the coordination number. A method for doing the latter has been proposed by O'Keeffe (36).]

Topological Distortions of the Cubic Perovskite Structure (13)

The parent [aristotype (37)] structure has been considered in Chapter II. It is ABX₃, primitive cubic, consisting of a regular array of corner-connected BX₆ octahedra with A atoms in the cuboctahedral, X₁₂, interstices. Two projections of the structure, [001] and [1̄10], are shown in Figure 28. All the

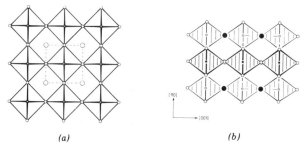

(a)   *(b)*

Figure 28.   The "ideal" cubic perovskite structure, ABX₃, projected along (a) [001], (b) [1̄10]. In (a) the large circles are A atoms at $z = 0$, the medium circles are X atoms at $z = \frac{1}{2}$, and the very small circles are X atoms at $z = 0$ and B atoms at $z = \frac{1}{2}$. The BX₆ octahedra are outlined, and the broken line indicates the unit cell. In (b) the largest circles are A atoms at height $\frac{1}{2}$ (open, partly obscured) and height 0 (filled); medium circles are X atoms, some at 0, some at $\frac{1}{2}$; and the smallest circles are B atoms at $\frac{1}{2}$ (open) and 0 (filled). (These heights are along [1̄10].) The BX₆ octahedra are emphasized.

TABLE 1   Patterns of Rotation Axes of Octahedra Centered at Corners of Ideal Perovskite Unit Cell[a]

System	First Layer		Second Layer	
(i)	[11$\bar{1}$]	[$\bar{1}$1$\bar{1}$]	[$\bar{1}$1$\bar{1}$]	[1$\bar{1}$1]
	[1$\bar{1}$1]	[$\bar{1}$11]	[$\bar{1}$11]	[111]
(ii)	[111]	[1$\bar{1}$1]	[1$\bar{1}$1]	[111]
	[1$\bar{1}$1]	[111]	[111]	[1$\bar{1}$1]
(iii)	[11$\bar{1}$]	[$\bar{1}$11]	[$\bar{1}$1$\bar{1}$]	[111]
	[1$\bar{1}$1]	[111]	[111]	[111]

[a] The x axis is vertical, pointing down the page; the y axis is horizontal, pointing from left to right. First and second layers refer to octahedra with centers at $z = 0$ and $z = 1$, respectively, in the aristotype.

deformations we will consider involve a partial collapse of the BX_3 framework; the bond angles B$\hat{X}$B being reduced from their 180° value in the aristotype. They arise because the appropriate (correct) A–X bond length is less than that for the cubic structure. In the cubic structure, $l(B–X) = a_c/2$ (a_c = cubic unit cell edge) and $l(A–X) = a_c/\sqrt{2}$. Therefore $l(A–X) = \sqrt{2}$ $l(B–X)$ can be regarded as being fixed by $l(B–X)$. Collapse of the structure makes some $l(A–X) < \sqrt{2}\, l(B–X)$, and so smaller A cations are readily accommodated without stretching (and weakening) the A–X bonds or compressing the B–X bonds.

The collapse is achieved (in the cases discussed here) by tilting the BX_6 octahedra about one of their $\bar{3}$ axes and restricting the number of possibilities by considering only those cases that double each unit-cell edge (increase the cell volume to about $8a_c^3$). Each octahedron has eight $\bar{3}$ axes (if we allow positive and negative directions), all parallel to $\langle 111 \rangle_c$ in the cubic aristotype. Since the BX_3 framework is retained in the collapsed structures (hettotypes (37)], only appropriate combinations of the various $\langle 111 \rangle$ axes are allowed; otherwise cornersharing is destroyed, i.e., B–X bonds will be broken. For a $2a_c \times 2a_c \times 2a_c$ unit cell, it transpires that only three different combinations of tilt axes are possible, and we deal with all three hettotype families, assuming that the octahedra remain regular.* The patterns of rotation axes in each case are given in Table 1.

Note that the $2 \times 2 \times 2$ cell used is 200/020/002 for system (i), but 002/200/020 for systems (ii) and (iii).

System (i). Structures with Cubic $Im3$ Symmetry

Two different types of A sites are generated, so that the new structure is $A'A''_3B_4X_{12}$. The crystallographic data are given in Data Table 14, in which ϕ

---

* A fourth tetragonal, $I4/mmm$, family described previously (13) is omitted here because the net tilts have to be about $\langle 110 \rangle$ rather than $\langle 111 \rangle$ axes, and also there are no known ABX_3 examples, only deficit structures.

is the angle through which each octahedron is tilted, and $l = l(B-X)$. The "ideal" collapsed structure with $\phi = \cos^{-1}[(3 + \sqrt{5})/\sqrt{32}] = 22.24°$ so that $y = 0.3010$ and $z = 0.1860$, in which the next-nearest $X \cdots X$ distance is reduced from a_c to the edge length of the BX_6 octahedron, is shown in Figure 29: Figure 29a emphasizes the BX_3 array of octahedra, 29b the $A'X_{12}$ icosahedra, and 29c the (not quite regular) $A''X_8$ rhombic prisms (with A'' in their mirror planes), which are, in fact, tetracapped so that the coordination is $A''X_{4+4+4}$. The bond angle $\theta(B\hat{X}B) = 143.8°$ in this ideal case.

Data Table 14 $A'A''_3B_4X_{12}$

Space group $Im3$, No. 204; $Z = 2$; $a = l(8 \cos \phi + 4)/3$,
$V_\phi = V_0 (2 \cos \phi + 1)^3/27$

Atomic Positions: $(0,0,0; \frac{1}{2},\frac{1}{2},\frac{1}{2})+$
A' in 2(a): $0,0,0$
A'' in 6(b): $0,\frac{1}{2},\frac{1}{2}; \frac{1}{2},0,\frac{1}{2}; \frac{1}{2},\frac{1}{2},0$
B in 8(c): $\frac{1}{4},\frac{1}{4},\frac{1}{4}; \frac{1}{4},\frac{3}{4},\frac{3}{4}; \frac{3}{4},\frac{1}{4},\frac{3}{4}; \frac{3}{4},\frac{3}{4},\frac{1}{4}$
X in 24(g): $\pm(0,y,z; 0,y,\bar{z}; z,0,y; \bar{z},0,y; y,z,0; y,\bar{z},0);$
    $y = (3 \cos \phi + \sqrt{3} \sin \phi)/(8 \cos \phi + 4),$
    $z = (3 \cos \phi - \sqrt{3} \sin \phi)/(8 \cos \phi + 4)$

The A' site is suitable for a large cation; and the A'' site, for a (smaller) "Jahn–Teller cation," such as Mn^{3+} and Cu^{2+}, which have a preference for square planar coordination. [The "square" is, in fact, a rectangle with edges of $\sqrt{2}l = 1.414l$ and $(3 + \sqrt{32} - \sqrt{5})l/\sqrt{18} = 1.513l$ in the "ideal" collapsed structure.] It is not surprising to find the following compounds with this structure:

$CaCu_3Ge_4O_{12}$ with $y = 0.3012$, $z = 0.1859$ (almost exactly "ideal"), and
    $a = 7.202$ Å (38)
$CaCu_3Mn_4O_{12}$ with $y = 0.3033$, $z = 0.1822$, and $a = 7.241$ Å (39)
$ThCu_3Mn_4O_{12}$ with $y = 0.2988$, $z = 0.1771$, and $a = 7.359$ Å (40)

The departures of y and z from the ideal values are small (approximately zero for the germanate), indicating small distortions of the octahedra.
    In addition, "deficit" structures are known as follows ($\Box$ = unoccupied site):

$A'\Box_3^{A''}\Box_4^B Al_{12}$ with $A' = $ W, Mo, Re, or Tc, and $y = 0.3083$, $z = 0.1854$, $a = 7.582$ Å ($\phi = 23.32°$) for $MoAl_{12}$
$La\Box_3^{A''}Fe_4P_{12}$ with $y = 0.3534$, $z = 0.1504$, and $a = 7.8316$ Å ($\phi \approx 40°$); as well as a series $Ln\Box_3^{A''}B_3P_{12}$, with $Ln = $ La–Eu and $B = $ Fe, Ru, Os (41)

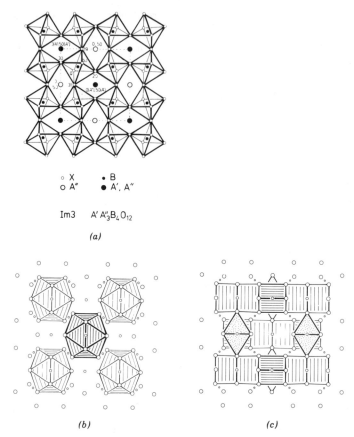

Figure 29. (*a*) The collapsed ideal perovskite structure with space group *Im*3 (ϕ = 22.24°), $A'A''_3B_4O_{12}$, projected along [001]. (Compare with Figure 28*a*.) Large filled circles are A' atoms at 0,0,0, and $\frac{1}{2},\frac{1}{2},\frac{1}{2}$ and A'' atoms at 0,0,$\frac{1}{2}$ and $\frac{1}{2},\frac{1}{2}$,0. Large, open circles are A'' atoms; small filled circles are B atoms; small open circles are X atoms. BX_6 octahedra in the lower half of the unit cell are outlined; those in the upper half are generated by a mirror plane at $z = \frac{1}{2}$. (Note also the mirror planes at $x = \frac{1}{2}$ and $y = \frac{1}{2}$.) (*b*) As (*a*), but the $A'X_{12}$ coordination icosahedra are now emphasized. (*c*) As (*a*), but now the tetracapped rhombic prisms around the A'' sites are emphasized. [In (*b*) and (*c*) the large, medium, and small circles are respectively X, A, and B atoms.]

$\square^{A'}\square^{A''}_3Co_4As_{12}$ = CoAs₃ (skutterudite) and a number of isostructural phosphides, arsenides, and antimonides, with $y \approx 0.34$, $z \approx 0.15$, $\phi \approx 35°$ (42)

$\square^{A'}\square^{A''}_3B_4(OH)_{12}$ = B(OH)₃ with B = Sc, Ga, In, and $y = 0.307$, $z = 0.182$, $\phi \approx 24°$ for Sc

In the last three deficit structures there is bonding along the edges of the "square" array of anion positions coordinating the (empty) A'' sites: P–P bonding, As–As bonding, and O–H $\cdots$ O hydrogen bonding, respectively. (In the last, one could also say that each A'' site is occupied by four hydrogen atoms, a 4H group.) The departures from the ideal structure are therefore a little larger than in the previous group but are still not very large. (Notice the larger ϕ in the second and third examples.)

System (ii). Structures with Trigonal, $R\bar{3}c$ Symmetry

This is perhaps the best known family of collapsed perovskites (43, 44). It involves the same operation as that introduced earlier to relate the **ReO₃** and **PdF₃** structures (45). The rotation axes involved are all parallel or antiparallel. The transformation matrix for the resulting hexagonal cell is $\bar{1}01/1\bar{1}0/222$. (The data in Data Table 15 are again for regular BX_6 octahedra.)

Data Table 15   ABX₃

Rhombohedral, space group $R\bar{3}c$, No. 167; $a_{\text{hex}} = \sqrt{8}\, l \cos \phi$, $c_{\text{hex}} = \sqrt{48}\, l$; $V_\phi = V_0 \cos^2 \phi$

Atomic Positions: $(0,0,0;\ \frac{1}{3},\frac{2}{3},\frac{2}{3};\ \frac{2}{3},\frac{1}{3},\frac{1}{3})+$
A in 6(a): $\pm(0,0,\frac{1}{4})$
B in 6(b): $0,0,0;\ 0,0,\frac{1}{2}$
X in 18(e): $\pm(x,0,\frac{1}{4};\ 0,x,\frac{1}{4};\ x,x,\frac{3}{4});\ x = (\sqrt{3} - \tan \phi)/\sqrt{12}$

Next-nearest distances $d(X \cdots X)$ become equal to the octahedron edge length at $\phi = 30°$, at which point the incomplete (75%) ccp array of X atoms in the aristotype has transformed to a perfect (complete) hcp array in the hettotype. The BX_3 array in the former is **ReO₃**; in the latter, **PdF₃**. At intermediate angles, $13° \leq \phi \leq 27.5°$, various MF_3 structures (M = Ti, Fe, Co, V, Cr, Ga, Mo, Ru) occur (45).

Figure 30 shows the aristotype and the "ideal" hettotype ($\phi = 30°$), the

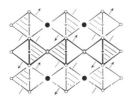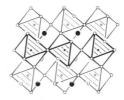

Figure 30. The transformation of the ideal, cubic perovskite type (on the left) to the collapsed ABX₃ structure with space group $R\bar{3}c$ and $\phi = 30°$ (on the right). (Compare Figure 28b.) Rotation axes are shown, and corresponding faces of the BX_6 octahedra are emphasized in the two cases. Note the small shuffles of the A atoms; also note that in the collapsed structure the AX_3 and BX_3 arrays are identical. The B atoms outline one unit cell in each case; on the left it is C-centered tetragonal (**c** is horizontal); on the right it has become monoclinic.

latter in an unusual projection that facilitates comparison with system (iii). (Compare also Figure 12.) One can, for this hettotype, pick out an obvious monoclinic unit cell with the transformation matrix $0\bar{1}1/200/0\bar{1}1$ and monoclinic angle $\gamma = 97.75°$. It will be noted that the A atoms are also in X_6 octahedra and that they have been shuffled by $c_{hex}/12$ to get there. This is the low-temperature structure of $LiNbO_3$. If they remain in their original positions, the A atoms are in triangular coordination, AX_3; which is the structure of high-temperature, paraelectric $LiNbO_3$. There are a series of isostructural compounds with AX_3 "complex ions"; calcite, $CaCO_3$, with $\phi \approx 40°$ is the best known, but there are also $AlBO_3$, $FeBO_3$, $LuBO_3$, and $NaNO_3$ with $33.5° < \phi < 42.0°$.

The cuboctahedron X_{12} around the A site in the aristotype has, at $\phi = 30°$, transformed to a pair of face-sharing octahedra (X_9) with three trigonal bipyramids on the reentrant faces (i.e., three capping X atoms). There is only one angle $\theta(B\hat{X}B) = \cos^{-1}[(1 - 4\cos^2 \phi)/3] = 131.8°$ at $\phi = 30°$.

System (iii). Structures with Orthorhombic, *Pnma* Symmetry

All the tilt axes are in the $(0\bar{1}1)$ plane of the parent structure = (001) plane of the orthorhombic structure. Perfect regularity of the octahedra requires a small tilt of the tilt axes; a primary rotation of ϕ demands a secondary tilt (of the primary tilt axes and the octahedra) of $\pm\psi$ about $[001]_{ortho}$, which is normal to the first tilt axis; $\psi = \tan^{-1}[\sqrt{2}(1 - \cos \phi)/(2 + \cos \phi)]$ for regular octahedra, so that $\phi = 15°$, $30°$ means $\psi = 0.93°$, $3.78°$, respectively.

Crystallographic parameters are given in Data Table 16. For regular octahedra, ϕ can be deduced from the unit-cell parameters by $\phi = \cos^{-1}[\sqrt{2}c^2/(ab)]$. However, a, b, and c are very sensitive to distortion of the octahedra, and ϕ is best calculated from $y(X'')$: $\phi = \tan^{-1}(-\sqrt{48}\, y)$. The ideal hettotype, with $\phi = 30°$ and regular octahedra, is shown in Figure 31. Comparing this with Figure 30 quickly reveals that the *Pnma* structure is a unit-cell-twinned derivative of the rhombohedral structure. The monoclinic cell of the latter becomes orthorhombic by the reflection in (002).

The tilt angle at which next-nearest distances $d(X \cdots X)$ become equal to the octahedron edge length is less than 30°; it is $\phi = \tan^{-1}(\sqrt{3}/4) = 23.41°$ (when $\psi = 2.29°$). The distance $d(X'' \cdots X'') = \sqrt{2}(\sqrt{3} - \tan \phi) \times (1 + 2\sec^2 \phi)^{-1/2}$, which is $(8/11)^{1/2} = 0.853$ times the edge length of the octahedron when $\phi = 30°$.

The coordination polyhedron about A is, in the collapsed structure, a slightly distorted bicapped trigonal prism. At $\phi = 30°$ it is a right prism [with two short edges, $d(X'' \cdots X'')$, see above]. The bond lengths $l(A-X)$ become more nearly equal if the x and z parameters for A become slightly positive (instead of zero).

This is the well-known **GdFeO₃** = "orthorhombic perovskite" structure, of which there are many examples including the mineral perovskite, $CaTiO_3$. The lanthanoid orthoferrites provide a well-known sequence with ϕ ranging

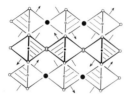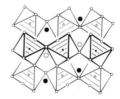

Figure 31. Derivation of the ideal ($\phi = 30°$) collapsed *Pnma* structure from perovskite (analogous to Figure 30) projected on (001). Again the B atoms outline one unit cell; the (primary) rotation axes are shown; and the small shuffles of the A atoms should be noted.

from 15° for $LaFeO_3$ to 23° for $LuFeO_3$.* Many isostructural compounds have Al, Sc, Cr, Ga, Co, V, or Rh in place of Fe. There are also sulfides such as $UCrS_3$, $BaUS_3$, and $YScS_3$ and fluorides such as $NaMgF_3$, and even hydrides such as $CsCaH_3$ (cf. $SrLiH_3$, cubic perovskite). $MgSiO_3$ adopts this structure (with SiO_6 octahedra) at high pressure (46), which may be important for understanding the earth's mantle (47, 48). In compounds like $SeMgO_3$ and $TeCoO_3$ the A atoms are displaced from the centers of the trigonal prisms (to lower values of x and z) by their stereochemically active lone pairs of electrons.

Fe_3C has the analogous $A\square^BX_3$ structure (derived from the aristotype Cu_3Au), with $\phi = 25°$. Our previous description of this structure was as twinned hcp Fe (Chapter IV). The picture is now clear: operation (ii) transforms the Cu_3 array of Cu_3Au from incomplete (75%) ccp to (ideally, at $\phi = 30°$) hcp; in operation (iii) this product is twinned on (002), and so the X array becomes twinned hcp. In fact, in Fe_3C the Fe array is not perfect (twinned) hcp because the tilt angle $\phi = 25°$ instead of the ideal 30°. This is quite clear from the (001) projection of its structure (Figure 32). Projections of this sort for several perovskites and a slightly deformed antistructure of Fe_3C, namely, YF_3, are shown in Figure 33. (Compare Figure 24*d* of Chapter IV.)

Other compounds isostructural with Fe_3C have been considered in Chapters IV and X.

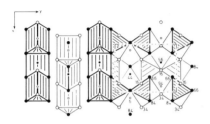

Figure 32. Cementite, Fe_3C ($\phi = 25°$), projected on (001).

---

* Here, and in cases (*i*) and (*ii*), larger tilt angles for the octahedra mean smaller values of $l(A-O)/l(B-O)$, of course.

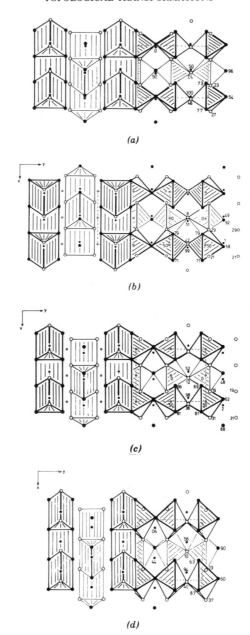

Figure 33. Projections on (001) of the *Pnma* (or equivalent) unit cell, analogous to Figure 32 but with smaller rotation angles ϕ. (*a*) $CaTiO_3$ ($\phi = 10°$); (*b*) $PrFeO_3$ ($\phi = 17°$); (*c*) $LuFeO_3$ ($\phi = 23°$); (*d*) YF_3 ($\phi \approx 23°$).

References

1. R. Courant and H. Robbins, *What Is Mathematics*. Oxford University Press, London, 1967, Chap. V, Sec. 2.

2. M. Kline, *Mathematical Thought from Ancient to Modern Times*. Oxford University Press, New York, 1972, Chap. 50.

3. H. S. M. Coxeter, *Introduction to Geometry*. Wiley, New York, 1961, Chap. 21.

4. D'A. W. Thompson, *On Growth and Form*. Cambridge University Press, Cambridge, 1942.

5. M. O'Keeffe and B. G. Hyde, *Phil. Trans. Roy. Soc. Lond.* **A295,** 553 (1980).

6. B. G. Hyde, L. A. Bursill, M. O'Keeffe, and S. Andersson, *Nature Phys. Sci.* **237,** 35 (1972).

7. B. G. Hyde, in *Reactivity of Solids,* J. S. Anderson et al., Eds. Chapman and Hall, London, 1972, p. 23.

8. B. G. Hyde and M. O'Keeffe, in *Phase Transitions—1973,* L. E. Cross, Ed. Pergamon, New York, 1973, p. 345.

9. W. L. Fraser and S. W. Kennedy, *Acta Cryst. A* **33,** 13 (1974).

10. L.-G. Liu, private communication, 1985.

11. L.-G. Liu, *High-Pressure Sci. Technol.* **2,** 17 (1979); K. Kusaba, M. Kikuchi, M. Fukuoka, and Y. Syono, *Phys. Chem. Miner.* **15,** 238 (1988).

12. A. F. Wells, *Three-Dimensional Nets and Polyhedra*. Wiley, New York, 1977.

13. M. O'Keeffe and B. G. Hyde, *Acta Cryst. B* **33,** 3802 (1977).

14. W. Jeitschko, *Acta Cryst. B* **31,** 1187 (1975).

15. J. Barbier and B. G. Hyde, *Acta Cryst. B* **41,** 383 (1985).

16. H. Nyman, B. G. Hyde, and S. Andersson, *Acta Cryst. B* **40,** 441 (1984).

17. L. Vegard, *Phil. Mag. Ser. 6,* **32,** 65 (1916).

18. L. Vegard, *Phil. Mag. Ser. 7,* **1,** 1151 (1926).

19. W. L. Bragg and G. F. Claringbull, *Crystal Structures of Minerals*. Bell, London, 1965, p. 113.

20. S. C. Abrahams and J. L. Bernstein, *J. Chem. Phys.* **55,** 3206 (1971).

21. K. Robinson, G. V. Gibbs, and P. H. Ribbe, *Am. Mineral.* **56,** 782 (1971).

22. A. Kirfel and G. Will, *Acta Cryst. B* **36,** 2881 (1980).

23. H. J. Berthold, W. Ludwig, and R. Wartchow, *Z. Kristallogr.* **149,** 327 (1979).

24. M. O'Keeffe and B. G. Hyde, *Struct. Bonding* **61,** 77 (1985).

25. M. O'Keeffe and B. G. Hyde, *Acta Cryst. B* **32,** 2923 (1976).

26. W. H. Zachariasen, *Acta Cryst.* **1,** 265 (1948).

27. K. Schubert, H. L. Lukas, H.-G. Meissner, and S. Bhan, *Z. Metallk.* **50,** 534 (1959).

28. A. Zalkin and D. H. Templeton, *Acta Cryst.* **6,** 269 (1953).

29. K. Selte and A. Kjekshus, *Acta Chem. Scand.* **18,** 690 (1964).

30. U. Rösler and K. Schubert, *Z. Metallk.* **4,** 395 (1951).

31. J. A. A. Ketelaar, W. H. t'Hart, M. Moerel, and D. Polder, *Z. Kristallogr. A* **101,** 396 (1939).

32. S. W. Peterson and H. A. Levy, *J. Chem. Phys.* **20,** 704 (1951); J. A. Ibers, *J. Chem. Phys.* **40,** 402 (1964); H. L. Carrell and J. Donohue, *Israel J. Chem.* **10,** 195 (1972).

33. E. D. Stevens, *Acta Cryst. A* **33,** 580 (1977).

34. F. Ras, University of Leiden, private communication, 1975.

35. D. L. Kepert, *Inorganic Stereochemistry.* Springer-Verlag, Berlin, 1982.

36. M. O'Keeffe, *Acta Cryst. A* **35,** 772 (1979).

37. H. D. Megaw, *Crystal Structures: A Working Approach.* W. B. Saunders, Philadelphia, 1973.

38. Y. Ozaki, M. Ghedira, J. Chenavas, J. C. Joubert, and M. Marezio, *Acta Cryst. B* **33,** 3615 (1977).

39. J. Chenavas, J. C. Joubert, M. Marezio, and B. Bochu, *J. Solid State Chem.* **14,** 25 (1975).

40. M. N. Deschizeaux, J. C. Joubert, A. Vegas, A. Collomb, J. Chenavas, and M. Marezio, *J. Solid State Chem.* **19,** 45 (1976).

41. W. Jeitschko and D. Braun, *Acta Cryst. B* **33,** 3401 (1977).

42. A. Kjekshus and T. Rakke, *Acta Chem. Scand. A* **28,** 99 (1974).

43. H. D. Megaw and C. N. W. Darlington, *Acta Cryst. A* **31,** 161 (1975).

44. C. Michel, J. M. Moreau, and W. J. James, *Acta Cryst. B* **27,** 501 (1971).

45. M. A. Hepworth, K. H. Jack, R. D. Peacock, and G. J. Westland, *Acta Cryst.* **10,** 63 (1957).

46. L.-G. Liu, *Phys. Earth Planet. Interiors* **11,** 289 (1976); H. Horiuchi, E. Ito, and D. J. Weidner, *Am. Mineral.* **72,** 357 (1987); Y. Kudoh, E. Ito, and H. Takeda, *Phys. Chem. Miner.* **14,** 350 (1987).

47. J. P. Poirier, J. Peyronneau, J. Y. Gesland, and G. Brebec, *Phys. Earth Planet. Interiors* **32,** 273 (1983).

48. R. W. Cahn, *Nature* **308,** 493 (1984).

CHAPTER XII

Noncommensurate Vernier (or Nonius) Structures

The following structures are discussed in this chapter: Nowotny's "chimney-ladder" phases $TX_{2-\delta}$ (based on $TiSi_2$), α- and β-$BaFe_2S_4$, $Ba_p(Fe_2S_4)_q$, $Hg_{2.86}AsF_6$, $Eu_{1-p}Cr_2Se_{4-p}$, CaF_2 and ZrO_2 (baddeleyite), $Y_6O_5F_8$, $Y_7O_6F_9$, $Zr_{108}N_{98}F_{138}$, cannizzarite $Pb_{46}Bi_{54}S_{127}$ (= $46PbS \cdot 27Bi_2S_3$), cylindrite $\sim FePb_3Sn_4Sb_2S_{14}$, "$LaCrS_3$," Dy_5Cl_{11}, $Nb_2Zr_6O_{17}$, $Ta_2Zr_8O_{21}$, and various sulfosalts.

We often find it convenient to describe structures as consisting of two interpenetrating substructures: lamellar intergrowths are simple examples (e.g., CS and unit-cell-twinned structures); less simple are three-dimensionally interpenetrating cases (such as pyrochlore and W_3Fe_3C), which we will consider in Chapter XIII. In this chapter we will be concerned with special examples of such cases, in which the two substructures are not commensurate with each other. *Commensurate* (C) structures are those in which the unit cells of each substructure have axes that are parallel and equal (or perhaps one is a *small* multiple of the other). *Incommensurate* (I) structures have two unit cells with at least one of the corresponding axes of each being incommensurate (having no common multiple), the others being commensurate. In *semicommensurate* (S) structures, at least one axis of one substructure is a relatively large multiple (say more than 3) of the corresponding axis in the other substructure, the others again being commensurate. This classification is somewhat subjective and not readily defined. It will become clearer when we deal with examples.

The term *vernier* (or *nonius*) was coined (1) for semicommensurate cases in which there are obvious subcell axes that have a multiplicity n in one substructure and $n + 1$ in the other. (In more complex cases—multiple verniers—n and $n + p$, where both n and p are integers, $n >> p$.)

There are one- and two-dimensional examples.

One-Dimensional, Columnar Misfit Structures

The Nowotny "Chimney-Ladder" Structures (2–4)

These occur among compounds of transition metals (T = V, Cr, Mo, Ti, Ru, Rh, Os, Ir) and group III and IV elements (X = Ga, Si, Ge). They are based on the $TiSi_2$ structure type (5), which consists of pseudohexagonal layers of the sort depicted on the left-hand side of Figure 1. (Identical layers are stacked, with offset, along the **c** direction so as to give a four-layer repeat; we show only one layer.) The structure is orthorhombic (space group *Fddd*, with $a = 8.2671$, $b = 4.8000$, $c = 8.5505$ Å) but almost tetragonal.

In each layer there are [010] rows of T (= Ti) at $x/a = 0$ and at $x/a = \frac{1}{2}$. Between these there are zigzag [010] rows of X (= Si) of twice the density. In the (noncommensurate) chimney-ladder structures, the former remain virtually unchanged,* but the latter have a reduced density, so that the stoichiometry becomes $TX_{2-\delta}$, $\delta > 0$. That component of the average interatomic spacing $d(X–X)$ along $[010]_{TiSi_2} \equiv [001]_{c-1}$ is reduced by a factor of $(2 + \delta)/2$. In general it is no longer commensurate with c ($\equiv b_{TiSi_2}$). It is as if the X lattice and the T lattice became independent along that direction. The X rows (the "ladders"), with variable interatomic spacing from one compound to another, are simply inserted into channels ("chimneys") in the regular T array.

In the simplest cases (semicommensurate = "superstructures" of $TiSi_2$), the repeat unit along the X rows is a simple multiple of that along the T rows. The structures can then be termed semicommensurate. Some examples are shown in the remaining part of Figure 1, with multiplicities of 2, 3, 4, and 4, respectively, and $\delta = \frac{1}{2}, \frac{1}{3}, \frac{1}{4}, \frac{3}{4}$. In principle, it seems entirely possible that incommensurate cases must also occur in which no multiple of $\mathbf{c}(T)$ is commensurate with any multiple of $\mathbf{c}(X)$. (The crystal then has no unit cell!) The range of δ is from ~0 to 0.75; **c** axes range up to at least 83 Å in length ($a \approx 6$ Å).

For the structures that have been accurately determined, it has been pointed out that the X atoms (often) lie on helices (cf. lower part of Figure 1). It seems likely that this is simply the result of minimizing interatomic repulsions. If this is so, then neither $d(T–T)$ nor $d(X–X)$ spacings will repeat accurately all the way along **c**.

---

* However, the symmetry becomes tetragonal instead of pseudotetragonal.

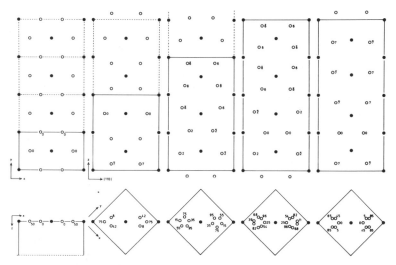

Figure 1. Structures of (approximately) planar *sections* in (left to right) TiSi$_2$, Ru$_2$Sn$_3$, Ir$_3$Ga$_5$, Mn$_4$Si$_7$, and Ir$_4$Ge$_5$, projected in two directions. Open circles are X atoms; filled circles are T atoms. Projected unit cells are outlined. Note that in the first the X and T arrays are commensurate but that in the others they become semicommensurate.

Ba$_p$(Fe$_2$S$_4$)$_q$

The basic structure here can be regarded as the α-BaFe$_2$S$_4$ type (6) ($p = q = 1$), although the β-BaFe$_2$S$_4$ type (7), which is readily derived from the α-form, is equally relevant. The first is shown in Figure 2. It is related to (Chapter XI) TlSe or NaAlSe$_2$ and to CuAl$_2$, consisting of strings of edge-shared FeS$_4$ tetrahedra parallel to the **c** axis (at $\frac{1}{2},0,z$ and $0,\frac{1}{2},z$),* with Ba occupying half the square antiprism, S$_8$, sites between them (along $0,0,z$ and $\frac{1}{2},\frac{1}{2},z$). The Ba occupancy is said to be random, but this seems unlikely; it is more plausible that alternate antiprisms in each [001] row are filled but that the correlation between adjacent rows is imperfect. [Random occupancy implies some distances d(Ba–Ba) $= c/2 = 2.80$ Å, which is very short.] If this is so, it may well be that the structure of α-BaFe$_2$S$_4$ is derived from the PtHg$_4$ type (Figure 7 of Chapter VIII) in the same way as that of TlSe is derived from the PtHg$_2$ type (Figure 6 of Chapter VIII; cf. also Chapter XI).

The related β-BaFe$_2$S$_4$ structure is shown in Figure 3. It is a topological distortion of the α form; the FeS$_2$ strings parallel to [001] being rotated slightly. At the same time the Ba atoms have jumped from heights $0,0,\pm\frac{1}{4}$ and $\frac{1}{2},\frac{1}{2},\pm\frac{1}{4}$ (both sets half-occupied) to $0,0,0$ and $\frac{1}{2},\frac{1}{2},\frac{1}{2}$ (both fully occupied). This shifts them from square antiprisms to tetracapped square prisms [basal

---

* Compare the structure of SiS$_2$.

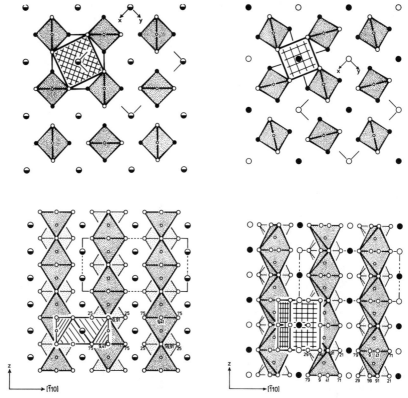

Figure 2. The structure of α-BaFe$_2$S$_4$ projected on (001) above and (110) below. Columns of face-sharing Ba$_{1/2}$S$_8$ square antiprisms (50% occupancy of Ba sites) are joined by edge-sharing to rods of edge-shared FeS$_4$ tetrahedra.

Figure 3. The structure of β-BaFe$_2$S$_4$ projected on (001) above and (110) below. Compare Figure 2. The rods of FeS$_4$ tetrahedra are similar to those in the α form but are tilted about **c**. The Ba atoms are strictly ordered and now in tetra-capped square prisms, S$_{12}$.

edges = 3.52 Å (8×), height 5.29 Å (4×)]. The Ba atoms are now necessarily strictly ordered along z.*

Data for these two structures are given in Data Tables 1 and 2. (The remaining structures in this chapter have large unit cells; crystallographic data are not listed for them. The data can be obtained from the appropriate listed reference.)

---

* Note that the Ba + S$_4$ array is rather close to ccp, a not unusual arrangement of anions and larger cations.

Data Table 1 α-BaFe$_2$S$_4$ (6)

Tetragonal, space group $I4/mcm$, No. 140; $a = 8.111$, $c = 5.590$ Å; $Z = 2$,
$V = 367.8$ Å^3

Atomic Positions: $(0,0,0; \frac{1}{2},\frac{1}{2},\frac{1}{2})+$
Ba$_{1/2}$ in 4(*a*): $\pm(0,0,\frac{1}{4})$
Fe in 4(*b*): $0,\frac{1}{2},\frac{1}{4}; \frac{1}{2},0,\frac{1}{4}$
S in 8(*h*): $\pm(x,\frac{1}{2}+x,0; \frac{1}{2}+x,\bar{x},0); x = 0.161$
Atomic Distances
Ba–S = 3.35 Å (8×)
Fe–S = 2.32 Å (4×)

Data Table 2 β-BaFe$_2$S$_4$ (7)

Tetragonal, space group $I4/m$, No. 87; $a = 7.678$, $c = 5.292$ Å; $Z = 2$,
$V = 312.0$ Å^3

Atomic Positions: $(0,0,0; \frac{1}{2},\frac{1}{2},\frac{1}{2})+$
Ba in 2(*a*): $0,0,0$
Fe in 4(*d*): $0,\frac{1}{2},\frac{1}{4}; \frac{1}{2},0,\frac{1}{4}$
S in 8(*h*): $\pm(x,y,0; \bar{y},x,0); x = 0.6196, y = 0.1986$
Atomic Distances
Ba–S = 3.29 Å (4×), 3.63 Å (8×)
Fe–S = 2.22 Å (4×)

   The two structures (α- and β-BaFe$_2$S$_4$) can be described as consisting of two alternating types of chains, Fe$_2$S$_4$ and Ba, which are commensurate. In the noncommensurate structures, Ba$_p$(Fe$_2$S$_4$)$_q$, the ratio $p/q > 1$; examples are Ba$_{10}$(Fe$_2$S$_4$)$_9$ (8) and Ba$_9$(Fe$_2$S$_4$)$_8$ (9). That is, relative to the FeS$_2$ rods the density of atoms in the Ba row is greater than it is in BaFe$_2$S$_4$. The coordination of Ba by S probably varies between that in α-BaFe$_2$S$_4$ (square antiprism) and that in β-BaFe$_2$S$_4$ (tetracapped square prism), *including intermediate stages.* Presumably (although they have not been so described), the orientation of the FeS$_2$ chains will vary continuously, in step with the Ba positions, between that in the α form and that in the β form of BaFe$_2$S$_4$. Again we have the chimney (FeS$_2$ array) and ladder (Ba rod) combination.
   In addition to the two semicommensurate ("superstructure") cases cited ($p/q = 10/9$ and $9/8$), one presumes that, in principle at least, truly incommensurate examples must also occur (again with no unit cell, i.e., no translation repeat unit).
   [Since this was written, an impressive analysis of a related structure problem has appeared. The conclusions are not dissimilar from those reached here (in a more qualitative way) for Ba$_p$(Fe$_2$S$_4$)$_q$. Reference 43 concerns the structure of NbTe$_4$, which is also simply related to α-BaFe$_2$S$_4$, etc. Basically (cf. Chapter XI), it is CuAl$_2$ type with alternate columns of square antiprisms (parallel to **c**) filled and empty. That is, compared with α-

$BaFe_2S_4$, the Te atoms occupy the S positions, the Fe atom positions are unoccupied, and the Nb atoms occupy all square antiprisms in half the columns (e.g., those at $00z$, but not those at $\frac{1}{2},\frac{1}{2},z$) instead of alternate square antiprisms in all such columns.

However, there is a complication, due to Nb–Nb bonding. Each Te in one column is bonded to its nearest-neighbor Te in an adjacent column, so that the structure is composed of $(Te_2)^{2-}$ anions, not Te^{2-} anions. Formally, therefore, the metal atoms are Nb^{4+}, i.e., d^1; which allows and leads to Nb–Nb bonding, so that there are longer and shorter Nb–Nb distances parallel to **c**. Mostly, it is said, this leads to triplets, Nb_3, but occasionally to Nb–Nb pairs; so the resulting "superstructure" has an approximately (instead of exactly) tripled c axis; it is a "modulated" structure with a periodicity of $\sim 3.20 \times$ along **c**.

The variation of Nb–Nb distances inevitably leads to shifts of some Te atoms (so as to maintain appropriate Nb–Te bond lengths), and this is achieved by rotation of some Te_4 squares (faces of square antiprisms), rather as described for α- $\rightarrow$ β-$BaFe_2S_4$, etc., above, and also some changes in the edge lengths of these squares.

The original paper (43) should be referred to for a great deal of information on the important details of the modulated structure of $NbTe_4$ and the way in which it was solved. But new observations are still being published.]

Mercury Hexafluoroarsenate ("Alchemist's Gold"), $Hg_{2.86}AsF_6$ (10)

This compound's structure is an interesting evolution of the "chimney ladder" type. Isolated AsF_6 octahedra are arrayed so as to leave *two* nonintersecting sets of "chimneys" (channels) that are orthogonal. (The structure is tetragonal.) Each contains an infinite row of Hg atoms with metallic character. The distance Hg–Hg = 2.64 Å is incommensurate with the lattice ($a = b = 7.538$ Å). [In α-Hg the distance Hg–Hg is 2.993 Å ($6\times$) and 3.465 Å ($6\times$), i.e., a mean of 3.23 Å for 12-coordination. In $Hg_{2.86}AsF_6$ the Hg–Hg coordination is only 2, hence the much smaller Hg–Hg separation.] Another report (11) suggests that the structure is stoichiometric, Hg_3AsF_6, though still incommensurate as described here.

$A_{1-p}Cr_2X_{4-p}$, with A = Ba, Sr, Eu, Pb and X = S, Se; $p \approx 0.29$ (12, 13)

These hexagonal structures were rather difficult to determine and are of great interest. They each consist of an approximately constant framework $Cr_{21}X_{36}$ of triply twinned $\{11\bar{1}\}$ NaCl-like [or $CdCl_2$-like, or (0001) NiAs- or $Cd(OH)_2$-like] strips of edge-connected CrX_6 octahedra with $a \approx 21.5$ Å, $c_0 \approx 3.45$ Å for X = S (22.5 and 3.63 Å for X = Se). The trilling framework contains two types of (rather wide) tunnels—hexagonal and triangular (see Figure 4). The hexagonal tunnels contain a repeat unit $A_6Cr_2X_6$ with $c_6 \approx$

5.7 Å (6.0 Å), and the triangular tunnels, A_3X with $c_3 \approx 4.2$ Å (4.6 Å). The compositions and unit cells of the structures are determined by the ratios c_6/c_0 and c_3/c_0. Note that there are now two noncommensurabilities: both the hexagonal and triangular column fillings are noncommensurate with the framework *and also with each other*.

The correlations between the positions (z parameters) of the three units (framework, hexagonal tunnel filling, and triangular tunnel filling) vary from one compound to another. In $Ba_{1-p}Cr_2Se_{4-p}$ they are correlated (and there is a semicommensurate superstructure, with $c = 5c_0 = 4c_3 = 3c_6$). In $Pb_{1-p}Cr_2S_{4-p}$ they are uncorrelated. In other compounds the hexagonal tunnel units are uncorrelated, but the triangular tunnel units are correlated with the framework. Detailed structural data have been published for $Eu_{1-p}Cr_2Se_{4-p}$ (12, 13).

The six-fold units are chains of face-sharing CrX_6 octahedra (as in hcp) surrounded by A atoms: stoichiometry $A_6Cr_2X_6$. The A atoms form triangular caps (cf. Chapter XIV) on those X_2 edges (of the Cr-centered octahedra in the chain) that are parallel to the basal plane (Figures 4 and 5, cf. the structure of Mn_5Si_3). They are therefore coordinated to four X atoms in the chain (all on one side of the A atom) and to (three to five) additional X atoms in the framework (on the other side of A). The c_0 repeat is a CrX_6 (framework) octahedral edge, and the c_6 repeat is twice the distance between opposite faces of a (tunnel-filling) CrX_6 octahedron. The ideal ratio is therefore equal to c/a for hcp $= c_6/c_0 = \sqrt{8/3} = 1.633$. The actual values are not far from ideal, ~ 1.64–1.69.

The three-fold units are $\sim A_3X$ per repeat distance c_3. They consist of (triangular) face-sharing trigonal prisms of A, centered by X and forming a single column. But this arrangement is regularly distorted (modulated) by the $\sim 4/5$ ratio of c_0/c_3 and also by the attempt of the A atoms to fit into the pattern of neighboring X atoms in the framework. (Furthermore, some of the centering X atoms, up to one-third, may be absent.) The ideal ratio c_3/c_0 will now depend on A and X, for c_3 is a trigonal prism edge, $d(A–A)$, while c_0 is an octahedral edge, $d(X–X)$.

Layer Misfit Structures

These have been reviewed and discussed in considerable detail (14). They cover a wide range in types of compounds but not many component structure types. The noncommensurability may be one- or two-dimensional. We first consider some one-dimensional examples.

Figure 6 shows schematic, semicommensurate examples: (*a*) a "unit vernier," in which six subcells in one layer is commensurate with seven subcells in the other ($n = 6$, $p = 1$); and (*b*) a "multiple vernier," in which 23 subcells in one layer is commensurate with 27 subcells in the other ($n = 23$, $p = 4$; $n' = n/p = 5.75$). Ideally at least, the structures consist of two

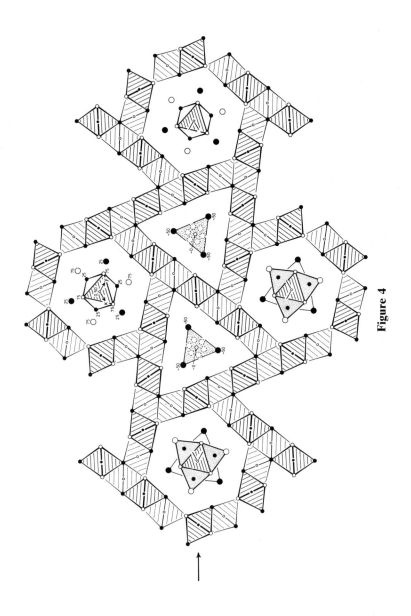

Figure 4

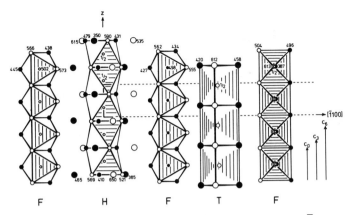

Figure 5. A bounded projection of $Eu_{1-p}Cr_2S_{4-p}$ ($p = 0.284$) on $(11\bar{2}0) = (100)_{\text{orthohex}}$ centered on $x_{\text{orthohex}} = \frac{1}{2}$; atomic heights in units of $a_{\text{orthohex}}/1000$. (Atoms with $x < a/2$ are shown as filled circles, which, in order of decreasing size, are Eu, S, and Cr, as in Figure 4.) The columns F, H, and T are, respectively, part of the framework, the hexagonal tunnel filling, and the triangular tunnel filling. Note the different lengths of their respective c axes, c_0, c_6, and c_3, the relations between the origins of which are taken arbitrarily, so that the structure shown is schematic.

layer types alternating strictly in a direction orthogonal to the layers (or nearly so).

$Y_nO_{n-1}F_{n+2}$ and Related Compounds (1, 14–19)

These form one class of "anion-excess fluorite-related structures" already alluded to in Chapter VIII. Similar series exist for rare earth oxide fluorides (Ln = Sm, Gd, Er, and Lu) and for zirconium or uranium nitride fluorides. All the structures can be most simply described and understood on the basis

---

Figure 4. The structure of $Eu_{1-p}Cr_2S_{4-p}$ ($p = 0.284$) projected onto (0001). Large circles are Eu, medium circles are S, and small ones are Cr. In the framework of edge-connected CrS_6 octahedra, open circles are at $z_0 = 0$, filled circles at $z_0 = \frac{1}{2}$. In the hexagonal tunnels, filled circles are at $z_6 = \frac{1}{4}$ and open circles at $z_6 = \frac{3}{4}$ (for Eu and S) or filled at 0 and open at $\frac{1}{2}$ (for Cr). In the triangular tunnels, open circles (S) are at $z_3 \approx 0$, filled circles (Eu) at $z_3 \approx \frac{1}{2}$, and the broken circles (Eu) have no defined z parameter (and the site occupancy is only 0.04, compared with 0.68 for Eu at $\sim\frac{1}{2}$ and 0.86 for S at $\sim$0. Note that while $c_0 = 3.446$ Å ($a = 21.41$ Å), $c_3 = 1.224c_0 = 4.218$ Å and $c_6 = 1.646c_0 = 5.672$ Å (cf. Figure 5).

S-centered Eu_6 trigonal prisms are shown in the triangular tunnels. In the hexagonal tunnels there are columns of face-sharing CrS_6 octahedra plus Eu atoms (in two tunnels), which, in the other two tunnels, are shown as triangular "caps" on the edges of the shared faces of the octahedra. [The heavy arrow on the left shows the midplane of the bounded projection on $(11\bar{2}0)$ in Figure 5.]

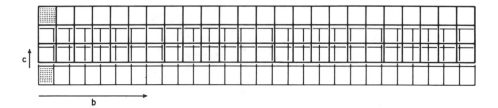

(a) n = 6 , p = 1 .

(b) n' = 23 , p = 4 , "n" = 5·75 .

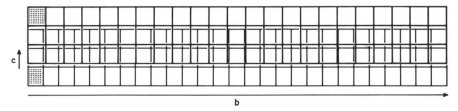

Figure 6.   Schematic illustrations of two types of double-layer, semicommensurate structures: (*a*) a unit vernier (*p* = 1), (*b*) a multiple vernier (*p* = 4). The two submeshes are shown by light and heavy lines, respectively. The unit meshes (unit subcell projections) are stippled.

of a description of the fluorite structure in terms of anion-centered rather than cation-centered coordination polyhedra. In these terms, fluorite itself consists of XM_4 tetrahedra sharing all their edges with similar tetrahedra. However, for present purposes this is best reduced to (tetragonal) layers of tetrahedra one tetrahedron thick, stoichiometry XM, alternating with 4^4 layers of anions (identical to those in the tetrahedral layers), stoichiometry X (see Figure 7).

Additional anions (in excess of two per cation) are accommodated by transforming these notional 4^4 anion-only layers to 3^6. (They may be supplied by, for example, adding some YF_3 to YOF, the valence of the yttrium atoms remaining at 3. YOF is fluorite-type or a small tetragonal distortion thereof; cf. Chapter VIII.) The anions in the tetrahedra remain virtually unaffected. Ideally, for perfectly planar, regular 4^4 and 3^6 anion layers (Figure 2, Chapter XI), the anion-row spacing then changes from $a/2$ to $(\sqrt{3}/2)(a/2)^*$, and the corresponding increase in density is $2/\sqrt{3} = 1.1547$. Hence the ideal stoichiometry of the anion layer changes from $X_{1.0000}$ to $X_{1.1547}$, and the overall

---

* $a = b = c$ are the edges of the fcc unit cell of the fluorite parent.

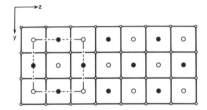

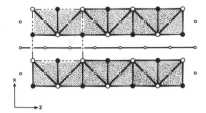

Figure 7. Representation of the fluorite structure (a = 5.463 Å) as alternating layers of anion-centered, edge-sharing tetrahedra and 4^4 layers of anions. Atoms at 0 or $\frac{1}{2}$. (Compare with Figures 8–10, 15–18 below.)

stoichiometry for the structure changes from MX_2 = $MX + X$ to $MX + X_{1.1547}$ = $MX_{2.1547}$. In practice, the observed composition range is $2.13 \lesssim X/M \lesssim 2.22$, e.g., $Y_nO_{n-1}F_{n+2}$ = $Y_n(O,F)_{2n+1}$ with $7.69 \gtrsim n \gtrsim 4.55$. [Note that $X/M = (2n + 1)/n = 2 + 1/n$ decreases as n increases.]

In this composition range every crystal appears to have its own structure. Semicommensurate structures that have been observed include unit verniers with n = 5, 6, and 7 and multiple verniers with very long unit cell axes—up to "a" = 317 Å. (Compare $a_{\text{subcell}} \approx 5.5$ Å, b_5 = 27.7 Å, c_6 = 33.1 Å, b_7 = 38.6 Å.*) Note that the ideal $MX_{2.1547}$ structure would be incommensurate (I). The determined (semicommensurate, S) structures for n = 6 and 7 are shown in Figures 8 and 9. They are "unit verniers." Figure 10 shows a related "multiple vernier" structure in a related system: $Zr_{108}N_{98}F_{138}$, in which 27 subcells of the tetragonal layer is commensurate with 32 subcells of the $\sim3^6$ anion net; a fivefold vernier with n = 27, p = 5, and n' = n/p = 5.40 (20), cf. Figure 6b.

These figures reveal departures from the ideal picture, but the departures are small—mainly some puckering of both types of anion nets, particularly the denser one. Note also that the multiple vernier structure in Figure 10 is not a simple intergrowth of five single-vernier substructures. The vernier effect is distributed over the whole cell length (b = 145.10 Å). Examination of Figures 8 and 9 reveals gradual transformation of the local structure (most obviously in terms of cation-centered polyhedra) from $\sim ZrO_2$ (baddeleyite) type at $y/b = 0, \frac{1}{2}$ to (Figure 11) the YF_3 type (cf. Figure 24d of Chapter IV or Figure 33d of Chapter XI) at $y/b = \pm\frac{1}{4}$—a quite reasonable result, since the

---

* The symmetries and space groups appear to be different for even and odd values of n. Hence the long axes are $\mathbf{b}_{\text{odd}} \equiv \mathbf{c}_{\text{even}}$.

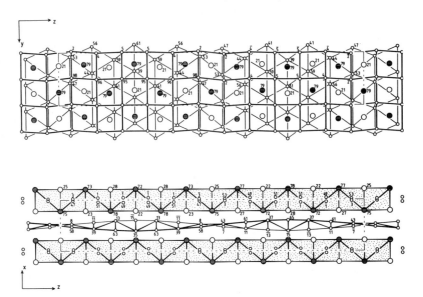

Figure 8. The vernier structure of $Y_6O_5F_8$ ($n = 6$, $p = 1$) projected on (100), above, and (010), below. Cf. Figure 7. Larger circles are cations; smaller circles are anions. Atom heights are in units of $1/100$ of the projection axis. In the upper drawing the (approximately) 4^4 anion net is shown by heavy lines and the (approximately) 3^6 anion net by light lines. In the lower drawing the atoms in the former center the cation tetrahedra. (The space group is *Pbcm*, No. 57; $a = 5.4154$, $b = 5.5303$, $c = 33.133$ Å.)

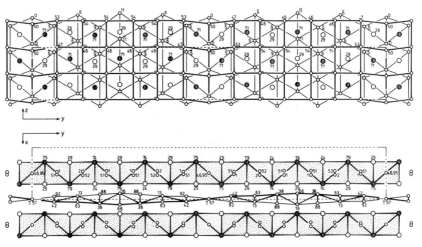

Figure 9. The vernier structure of $Y_7O_6F_9$ ($n = 7$, $p = 1$) in two projections. Compare Figures 8 and 7. (The space group is *Abm2*, No. 39; $a = 5.423$, $b = 38.624$, $c = 5.527$ Å.)

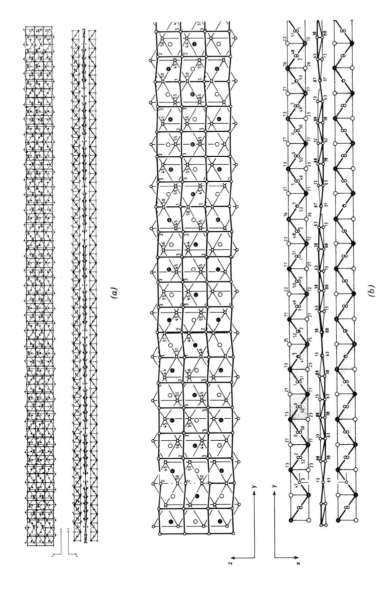

Figure 10. (a) The multiple-vernier (n = 27, p = 5) structure of $Zr_{108}N_{98}F_{138}$ in two projections. Compare Figures 7–9. (The space group is again Abm2, No. 39; a = 5.186, b = 145.1, c = 5.368 Å.) (b) Part of (a) at the usual scale, 1 cm = 4 Å.

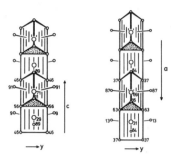

Figure 11. Comparison of the anion environment of (left) the Y(4) atoms (at $y/b = \pm\frac{1}{4}$) in $Y_7O_6F_9$ with (right) that of Y in YF_3. Larger circles are Y; smaller circles are anions. In both cases the cation coordination is a bicapped trigonal prism of anions. (Heights are in units of $10^{-2} \times$ the projection axis in each case. *Correction:* On the left, 56 should read 54.)

$Y_nO_{n-1}F_{n+2}$ structures can be regarded as resulting from the dissolution of YF_3 in YOF. (The ZrO_2 type is a small deformation of the fluorite type.)

Strictly speaking, most of the structures discussed in the remainder of this chapter are not "verniers." But they are layer misfit structures, and it is convenient and appropriate to include them here.

Cannizzarite, $mPbS \cdot nBi_2S_3$ with $m/n \approx 1.70–1.71$ (21)

This is the first example of a common and simple building principle in the complex crystal chemistry of mineral sulfosalts, i.e., basically $mAS \cdot nB_2S_3$ compounds with A = Sn or Pb and B = As, Sb, or Bi,* m and n being integers. The same principle is also important in complex sulfides of the rare earth elements, especially those with transition metals (cf. below). The principle is that blocks of NaCl-type structure with two different orientations are fitted together so that the interfaces are between {100} planes of cations and anions (tetragonal, or approximately so) and {111} planes of anions only (hexagonal, or approximately so) of the fcc unit cell.

Cannizzarite, shown in Figure 12, is perhaps the most straightforward example, although it was an extremely difficult structure determination. (It is monoclinic, space group $P2_1/m$; $a = 74.06$, $b = 4.09$, $c = 189.8$ Å, $\beta = 11.93°$; $Pb_{46}Bi_{54}S_{127}$, $Z = 2$!) As the figure shows, it consists of two layer types in strict alternation along $\mathbf{c}^*$. Both are NaCl-like, but the orientation of the NaCl-type fcc subcell is different for the two. One (A) is a two-atom-thick layer in the {100}$_{NaCl}$ orientation, $(Pb,Bi)_{46}S_{46}$; the other (B) is a five-atom-thick layer in the {111}$_{NaCl}$ orientation, $(Pb,Bi)_{54}S_{81}$, i.e., 46(Pb,Bi)S and 27(Pb,Bi)$_2S_3$. (The ratio 46/27 = 1.704; cf. the ideal value = $\sqrt{3}$ = 1.732.) These component substructures are commensurate except in the **a**

---

* They often contain minor amounts of other elements also, especially Cu or Ag. Compare the aikinite series of Chapter IX.

direction, along which 46 subcells of the first is (approximately?) commensurate with 27 subcells of the second.*

Along the interfaces between the layer types, cations occur only in the pseudotetragonal (100) layer (and their coordination within that layer is a square pyramid/half octahedron, S_5). Their total coordination by sulfur varies, more or less smoothly, between a bicapped trigonal prism, a monocapped trigonal prism, and an octahedron (taking due account of the fact that stereochemically active lone pairs of s^2 electrons will, as we have seen in Chapter X, tend to displace the cations from the centers of their coordination polyhedra of anions). Clearly, this sort of structure is likely to be restricted to those cases where the coordination polyhedra are readily distorted without any appreciable increase of energy, that is, where the cations are large and their coordination numbers high. (Compare the yttrium oxyfluorides above.)

Small changes in stoichiometry could lead to small distortions of either or both of the component substructures, which could easily change the length of the overall unit-cell c axis substantially. Another possibility (for massively varying the stoichiometry) is to vary the thicknesses of either or both layer types: the pseudotetragonal layer will always have a stoichiometric ratio, anion/cation = 1/1; the pseudohexagonal layer (bounded by anion planes) will have a stoichiometric ratio, anion/cation = $(n + 1)/n = 1 + 1/n$, with n an integer equal to the number of octahedra in the layer thickness. {If $n = 0$ the structure would resemble those of the $Y_nO_{n-1}F_{n+2}$ series (above), except that now the tetragonal layers are MX with half-octahedral/square-pyramidal coordination of the anions instead of MX with tetrahedral coordination [in $Y(O,F)_x$]}. Nonstoichiometry and stacking disorder are obviously possible.

There are several other mineral and synthetic cannizzarites; see the references and discussion in ref. 14. It is a striking fact that all their stoichiometries lie between those of lillianite or "phase III," $3PbS \cdot Bi_2S_3$ (22), on the one hand, and the "phases V," V-2 = $\frac{2}{3}PbS \cdot Bi_2S_3$, V-3 = $\frac{3}{5}PbS \cdot Bi_2S_3$, V-1 = $\frac{1}{2}PbS \cdot Bi_2S_3$ (23), on the other. All these latter structures (phases III and V) are based on a common structural principle—unit-cell twinning of NaCl-type PbS (see Chapter VI)—which (at least at first sight) is quite different from that involved in the intervening cannizzarite. (This difference is not due to "impurity" cations such as Cu or Ag in either. All have been synthesized from pure PbS plus Bi_2S_3).

Other sulfosalts with noncommensurate layer structures based on exactly the same structural principle include cylindrite, $\sim FePb_3Sn_4Sb_2S_{14}$ (but really a series of minerals with a range of noncommensurability) (24, 25). In this structure the layers are semicommensurate in one direction and incommen-

---

* In the projection direction, both have $\mathbf{b} = \frac{1}{2}\langle 110 \rangle_{NaCl}$: compare $b = 4.09$ Å with $a/\sqrt{2} = 4.20$ Å for PbS. In the $\mathbf{c}^*$ direction, which is the layer-stacking direction, they are necessarily commensurate.

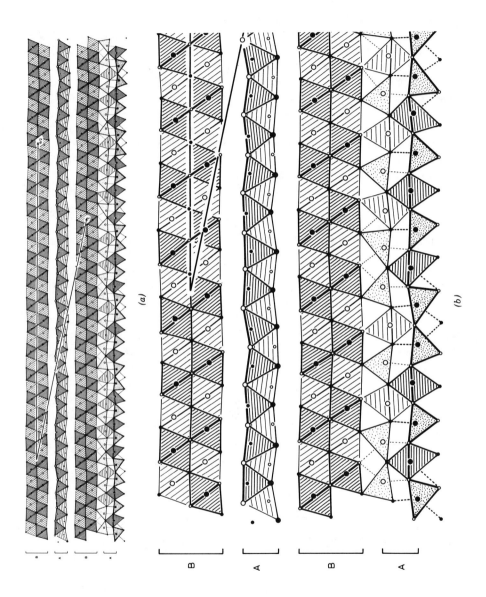

(a)

(b)

surate in the other. (But the pseudohexagonal layer is now only one octahedron, i.e., three atoms, thick.) The layer stack is curved into a cylindrical spiral (like a roll of paper or a "brandy snap"), rather like chrysotile asbestos but with a much bigger radius of curvature—several millimeters, compared with ~100 Å in asbestos. Other examples are franckeite, incaite, and lengenbachite (25) [cf. ref. 14].

"LaCrS$_3$" (26)

This is another example, similar to cannizzarite but with trigonal layers one octahedron (three atoms) thick, instead of two octahedra (five atoms) thick. There are several other compounds more or less isostructural, each with different lanthanoid and/or transition metal atoms and, sometimes, Se in place of S (27a–27c). As with cannizzarite, these are also noncommensurate in only one direction.

At present there is still some doubt about the exact stoichiometry of "LaCrS$_3$." The structure [which has not been *completely* solved, (26)] is shown schematically in Figure 13. If the pseudotetragonal layers are pure LaS and the trigonal layers pure CrS$_2$ (respectively, positively and negatively charged) and the structures of both are perfectly ordered, then the ideal stoichiometry (unit-cell content) is not La$_{64}$Cr$_{64}$S$_{192}$ (= 64LaCrS$_3$) but La$_{72}$Cr$_{60}$S$_{192}$; i.e., the stoichiometric cation ratio is not La/Cr = 1/1 but La/Cr = 1.20/1.00. To conform to the former, considerable disorder was proposed (26): four of the 72 La sites empty and four others occupied by Cr. Such gross disorder in a structure elegantly constructed so as to achieve La/Cr = 1.20/1.00 seems unlikely, and, indeed, microprobe analysis of single crystals has yielded a ratio La/Cr = 1.201 ± 0.006. Density measurements and structure factor calculations also support the more complex stoichiometry (28), but the question is not yet unambiguously resolved.

Accepting the structure shown in Figure 13, a comparison with that of cannizzarite (Figure 12) is interesting. The broad structural principle involved [stacking of alternate (100) and (111) layers of NaCl type] is the same in both structures, but there are considerable differences:

Figure 12. (*a*) The structure of cannizzarite, 46PbS · 27Bi$_2$S$_3$ = Pb$_{46}$Bi$_{54}$S$_{127}$, projected onto (010), after Matzat (21). Large circles are cations; small circles are S atoms: open at $y = \frac{1}{4}$, filled at $y = \frac{3}{4}$. The A layers are two-atom-thick (100) slices of **NaCl** type, stoichiometry MX, and the B layers are five-atom-thick (111) slices of **NaCl** type, stoichiometry M$_2$X$_3$. In the upper part of the figure these are shown as S-centered M$_5$ square pyramids and M-centered S$_6$ octahedra, respectively, so that the noncommensurability of the interface is obvious. At the bottom the A-layer cations are shown as centering polyhedra of anions—octahedra and trigonal prisms. The unit cell axes **c** and **a**/2 are shown. (The length of the drawing is ~80% of **a**.) (*b*) Part of *a* at the usual scale, 1 cm = 4 Å.

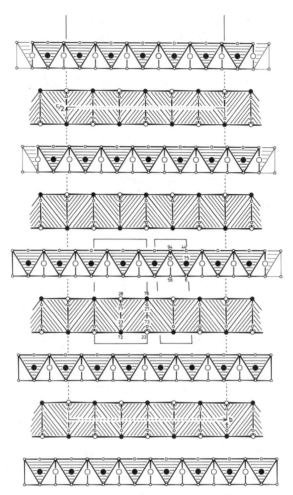

Figure 13. The approximate solution of the structure of "LaCrS$_3$" (probably La$_{72}$Cr$_{60}$S$_{192}$ = 12 La$_6$Cr$_5$S$_{16}$) projected along the **a** axis of its triclinic cell (26); a = 5.936, b = 17.257, c = 66.215 Å, α = 90.39°, β = 95.3°, γ = 90.02°. The largest circles are Cr, medium circles are La (open at $x \approx \frac{1}{4}$, filled at $x \approx \frac{3}{4}$ in both cases), and smallest circles are S atoms. Atom heights are in units of $a/100$ above (100). Half a unit cell is outlined with broken and dotted lines: Full lines near the center of the figure outline the approximate (ideal) subcells of the La$_4$S$_4$ layers (on the left) and the Cr$_2$S$_4$ layers (on the right). (Note the different stacking vectors, which give rise to the long **c** axis.) The structure is triclinic, so the length of $b = b \sin \gamma$ (γ = 90.02°) and that of $c/2$ is actually ($c \sin \beta)/2$ (β = 95.3°): the angle α = 90.39° (90.37° in projection). The shaded polyhedra are SLa$_5$ square pyramids (about S at $x \approx \frac{3}{4}$) and CrS$_6$ octahedra (mainly about Cr at $x = \frac{3}{4}$). The exact planarity of the atom layers parallel to (001) suggests that the structure solution is approximate. (Contrast Figure 12.)

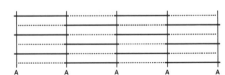

Figure 14. Schematic illustration of an incommensurate (two-) layer structure (above) that becomes commensurate by the introduction of regular APBs (below).

1. The orientation of the (100) layers [relative to the (111) type] differs by 45°.
2. The misfit direction differs by 90° for the (111) layers.

[An additional, very obvious, difference is that the (111) layers are twice as thick in cannizzarite.]

Different examples of noncommensurate structures consisting of two misfitting layer types are some graphite intercalates (e.g., with $MoCl_5$), metal chloride/hydroxide layers (e.g., koenenite), metal hydroxide/hydroxide layers, metal sulfide/hydroxide layers (e.g., valleriites and tochilinites), and silicates (e.g., greenalite); see ref. 14.

Related Commensurate Structures

A number of other structures are derived from the two-layer noncommensurate types just discussed. They are made commensurate by the introduction of antiphase boundaries (APBs) normal (or approximately normal) to the layers, so that any noncommensurability is canceled out (cf. Figure 14). That is, they consist of infinite slabs of a noncommensurate structure type repeated by translation.

For example, there are a large number of structures thus derived from the $Y_nO_{n-1}F_{n+2}$ types, including various rare earth halides, mixed oxides of Zr and Nb or Ta, and zirconium and related oxide fluorides $Zr_n(O,F)_{2n+1} = Zr_nO_{2n-1}F_2$ ($5 \lesssim n \lesssim 9$)* (29a) and $UZr_6O_{14}F$ (29b). Several examples are shown in Figures 15–18 (30–33). In each case they are members of more extensive families with varying degrees of misfit. In these cases the cation coordination midway between the APBs is YF_3-like (as in $Y_nO_{n-1}F_{n+2}$), and toward the APBs it gradually transforms to that of Zr in baddeleyite (monoclinic ZrO_2). However, now it goes even further, to the α-PbO_2 type at the

* Note: n is not always an integer.

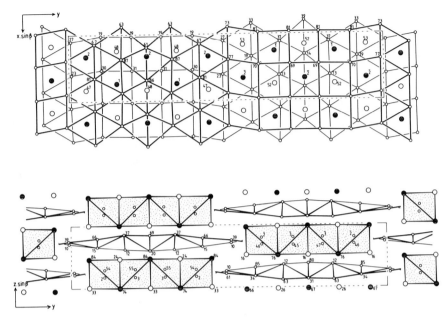

Figure 15. The structure of Dy_5Cl_{11} in two projections. Larger circles are cations, smaller circles are anions; heights are in hundredths of the projection axis. ($P2_1/m$; $a = 7.110$, $b = 34.68$, $c = 6.634$ Å, $\beta = 90.23°$.)

APBs. [This last is also a simple topological distortion of fluorite-type (and baddeleyite), as shown in ref. 34 and in Chapters III and XI.]

Makovicky (35) has published an impressive analysis of a very large number of mineral and synthetic sulfosalt structures of extreme complexity and range, all based on exactly the same principle as in cannizzarite, but with galena-like (NaCl) columns or slabs *each* of which presents both (111) and (100) surfaces to its neighbors, again with the common principle (111)‖(100). In this way structures such as those of cosalite, jamesonite, boulangerite, bournonite, weibullite, nuffieldite, and $Pb_4In_9S_{17}$, and even the lillianite series, are readily understood. (See also ref. 14.) The same principle clarifies a very large number of ternary chalcogenides of the rare earths (36). Of the many structures that can be simply described in this way, we depict only that of cosalite, $2PbS \cdot Bi_2S_3$, (37), shown in Figure 19. (Notice that its stoichiometry is close to that of cannizzarite, lying between it and that of lillianite; cf. above.)

Another family of sulfosalt structures $nMS \cdot Sb_2S_3$, including stibnite, Sb_2S_3, itself, can be generated in a rather similar way. But now all the structures consist of columns of CrB (or TlI) bounded on (010) and (021), or the closely similar SnS type bounded on (100) and (210) (*Pnma* settings in both instances). These columns are fitted together so that, as far as possible, these two types of boundary planes face each other. As a result, additional

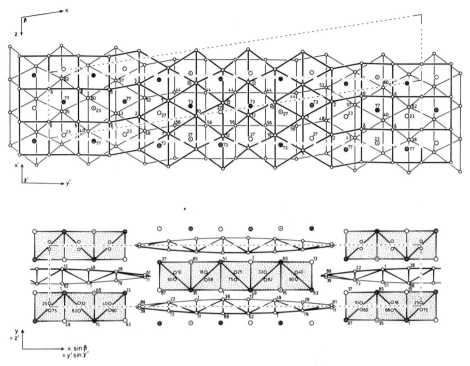

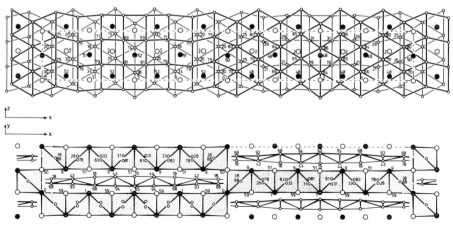

Figure 16. The structure of Yb_5ErCl_{13}. ($C2/c$; $a = 41.44$, $b = 6.53_7$, $c = 7.00_4$ Å, $\beta = 98.5°$. The primed axes correspond to an alternative monoclinic cell, equivalent to the unit cells in Figures 8, 9, 10, 15, 17, and 18.) Compare Figure 15.

Figure 17. The structure of $Nb_2Zr_6O_{17}$ in two projections. ($Ima2$; $a = 40.92$, $b = 4.93$, $c = 5.27$ Å.) Compare with Figures 15, 16, 8, and 9.

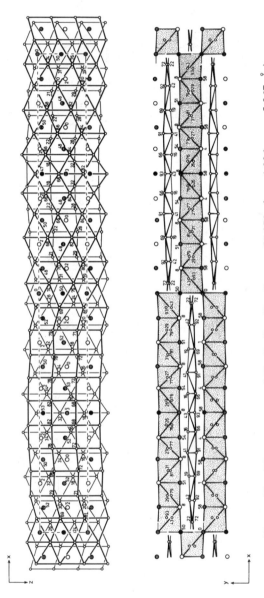

Figure 18. The structure of $Ta_2Zr_8O_{21}$ ($Ima2$; $a = 50.978$, $b = 4.996$, $c = 5.267$ Å.) Compare Figure 17. (Large circles with vertical hatching are Ta; those without are Zr.)

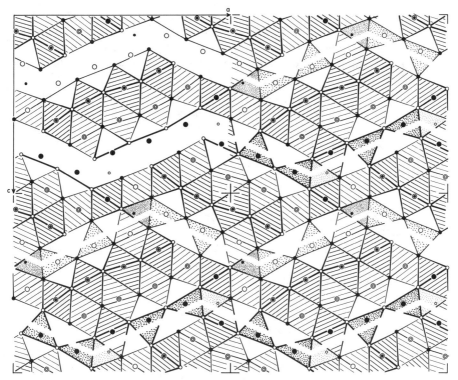

Figure 19. Four unit cells of cosalite, $2PbS \cdot Bi_2S_3 = Pb_2Bi_2S_5$, projected along [010], the short axis for unit-cell setting *Pnma*. ($a = 23.89$, $b = 4.057$, $c = 19.10$ Å.) Large circles are Pb or Bi (the latter being double circles), medium circles are S, and small circles are Cu (site occupancy 0.12); open at $y = \frac{1}{4}$, filled at $y = \frac{3}{4}$. The structure consists of corrugated layers of **NaCl** type [parallel to (001)] with $(100)/(111)_{NaCl}$ interfaces between adjacent layers. The interfaces are emphasized by white (un-hatched) zigzag bands, especially in the top left-hand unit cell. The distortions of the octahedra, particularly BiS_6, are clearly to accommodate stereochemically active lone pairs of electrons.

trigonal prisms (in two orthogonal orientations—"standing up" and "lying down") are generated across the interfaces.

In this chapter we have so far entirely ignored a quite different type of noncommensurate structure that is becoming increasingly common. [But see the discussion on p. 310 (43).] In this, some atomic parameters and/or the (partial) occupancy of some sites vary in a periodic fashion through the structure. The periodicity may or may not be commensurate with the unit cell of the basic structure (semi- and incommensurate cases, respectively). Examples include the γ form of sodium carbonate, pyrrhotite (FeS_x), etc., and important mineral species, for example the feldspars.* Their single-

---

* And, in a narrow temperature range, even quartz (38).

crystal diffraction patterns contain weak satellite reflections in addition to the main (strong) reflections from the substructure. Collectively, they are often termed *modulated structures* (39). An increasing number of nonstoichiometric compounds exhibit such modulations—for example, FeS_x (40) and $\sim Yb_3S_4$ (41a, 41b)—which further reduces the number of examples still believed to contain the classical "random point defects" and raises important questions about why modulated *stoichiometric* compounds (see, e.g., refs. 42a and 42b) are stable.

References

1. B. G. Hyde, A. N. Bagshaw, S. Andersson, and M. O'Keeffe, *Ann. Rev. Mater. Sci.* **4**, 43 (1974).

2. H. Nowotny, in *The Chemistry of Extended Defects in Non-metallic Solids*, L. Eyring and M. O'Keeffe, Eds. North-Holland, Amsterdam, 1970, p. 223.

3. W. Jeitschko and E. Parthé, *Acta Cryst.* **22**, 417 (1967).

4. H. Q. Ye and S. Amelinckx, in *Modulated Structure Materials*, T. Tsakalakos, Ed. Martinus Nijhoff, Dordrecht, 1984, p. 173.

5. W. Jeitschko, *Acta Cryst. B* **33**, 2347 (1977).

6. H. Boller, *Monatsh. Chem.* **109**, 975 (1978).

7. J. S. Swinnea and H. Steinfink, *J. Solid State Chem.* **32**, 329 (1980).

8. I. E. Grey, *Acta Cryst. B* **31**, 45 (1975).

9. J. T. Hoggins and H. Steinfink, *Acta Cryst. B* **33**, 673 (1977).

10. I. D. Brown, B. D. Cutforth, C. G. Davies, R. J. Gillespie, P. R. Ireland, and J. E. Vekris, *Can. J. Chem.* **52**, 791 (1975).

11. N. D. Miro et al., *J. Inorg. Nucl. Chem.* **40**, 1351 (1978); *Inorg. Chem.* **16**, 646 (1978).

12. R. Brouwer and F. Jellinek, *J. Phys. Colloque C7* **38**, C7 (1977).

13. R. Brouwer, thesis, University of Groningen, 1978.

14. E. Makovicky and B. G. Hyde, *Struct. Bonding* **46**, 101 (1981).

15. D. J. M. Bevan, in *Solid State Chemistry*, R. S. Roth and S. J. Schneider, Eds. NBS Spec. Publ. 364. U.S. Government Printing Office, Washington, DC, 1972, p. 749.

16. A. W. Mann and D. J. M. Bevan, *J. Solid State Chem.* **5**, 410 (1972).

17. D. J. M. Bevan and A. W. Mann, *Acta Cryst. B* **31**, 1406 (1975).

18. J. Mohyla, thesis, The Flinders University of South Australia, 1979.

19. R. J. Myers, thesis, University of Melbourne, 1976.

20. W. Jung and R. Juza, *Z. Anorg. Chem.* **399**, 129 (1973).

21. E. Matzat, *Acta Cryst. B* **35**, 133 (1979).

22. J. Takagi and Y. Takeuchi, *Acta Cryst. B* **28**, 649 (1972).

23. Y. Takeuchi and J. Takagi, *Proc. Jap. Acad. Sci.* **50**, 843 (1974).

24. E. Makovicky, *Neues Jahrb. Mineral. Abh.* **126**, 304 (1974).

25. T. B. Williams, thesis, Australian National University, 1986; T. B. Williams and B. G. Hyde, *Phys. Chem. Miner.,* in press (1988).

26. K. Kato, I. Kawada, and T. Takahashi, *Acta Cryst. B* **33,** 3437 (1977).

27a. T. Takahashi, T. Oka, O. Yamada, and K. Ametani, *Mater. Res. Bull.* **6,** 173 (1971).

27b. T. Takahashi, S. Osaka, and O. Yamada, *J. Phys. Chem. Solids* **34,** 1131 (1973).

27c. T. Murugesan, S. Ramesh, J. Gopalakrishnan, and C. N. R. Rao, *J. Solid State Chem.* **38,** 165 (1981).

28. C. Otero Diaz, J. FitzGerald, T. B. Williams, and B. G. Hyde, *Acta Cryst. B* **41,** 405 (1985); T. B. Williams and B. G. Hyde, *Acta Cryst.,* in press (1988).

29a. R. Papiernik and B. Frit, *Acta Cryst. B* **42,** 342 (1982).

29b. R. Papiernik, D. Mercurio, and B. Frit, *Acta Cryst. B* **36,** 1769 (1980).

30. H. Bärnighausen, in *Proceedings of the 12th Rare Earth Research Conference,* Vol. 1, C. E. Lundin, Ed. University of Denver, Denver, 1976, p. 404.

31. H. A. Eick, private communication, 1980.

32. J. Galy and R. S. Roth, *J. Solid State Chem.* **7,** 277 (1973).

33. J. Galy, private communication, 1978.

34. B. G. Hyde, L. A. Bursill, M. O'Keeffe, and S. Andersson, *Nature Phys. Sci.* **237,** 35 (1972).

35. E. Makovicky, *Fortschr. Mineral.* **59,** 137 (1981).

36. C. Otero Diaz and B. G. Hyde, unpublished work, 1981.

37. T. Srikrishnan and W. Nowacki, *Z. Kristallogr.* **140,** 114 (1974).

38. G. Dolino, J. P. Bachheimer, B. Berge, and C. M. E. Zeyen, *J. Phys.* **45,** 361 (1984).

39. J. M. Cowley et al., Eds., *Modulated Structures* (AIP Conf. Proc. No. 53). American Institute of Physics, New York, 1979.

40. A. Yamamoto and H. Nakazawa, *Acta Cryst. A* **38,** 79 (1982).

41a. C. Otero Diaz and B. G. Hyde, *Acta Cryst. B* **39,** 569 (1983).

41b. R. L. Withers, A. Prodan, and B. G. Hyde, submitted to *Acta Cryst.* (1988).

42a. B. G. Hyde, J. R. Sellar, and L. Stenberg, *Acta Cryst. B* **42,** 423 (1986).

42b. R. L. Withers, B. G. Hyde, and J. G. Thompson, *J. Phys. C: Solid State Phys.* **20,** 1653 (1987).

43. H. Böhm and H.-G. Von Schnering, *Z. Kristallogr.* **171,** 41 (1985).

CHAPTER XIII

Structures That Can Be Related to (Derived from) bcc or pc Packing

The structures of the following are considered in this chapter: the A15 type of "β-tungsten" (W_3O) and Cr_3Si, UCo, $Th_3P_4 = \gamma\text{-}La_3S_4 =$ anti-La_4Ge_3, langbeinite [$K_2Mg_2(SO_4)_3$] = eulytite [$Bi_4(SiO_4)_3$], $Au_3Sb_4Y_3$, Pu_2C_3 = anti-Rb_2O_3, $ThCl_4$, zircon ($ZrSiO_4$) (and rutile), K_2ZrF_6, scheelite ($CaWO_4$) (and fluorite), SiF_4, W_3Fe_3C and related subcarbides and suboxides, pyrochlore and cubic Sb_2O_3, etc., and $\alpha\text{-}Cr_4Al_{13}Si_4$.

Many metals have bcc structures—for example, the alkali and alkaline earth elements, the high-temperature forms of Ti, Zr, Hf, V, Nb, Ta, Cr, Mo, and W, and, of course, Fe. The structure type is that of tungsten, which has $a = 3.155$ Å, space group $Im3m$, with one W atom at 0,0,0 and one at $\frac{1}{2},\frac{1}{2},\frac{1}{2}$. Many other structures can be related to this one; those of CsCl (which, with due respect to many textbook authors, is *not* bcc) and Fe_3Si have been described earlier, and that of NaTl is another example.

However, there is no structure with a bcc *anion* array. It seems likely that, for geometrical reasons, such an arrangement is unsuitable for the accommodation of interstitial cations: the octahedral and tetrahedral interstices are far from regular.* However, using the concepts of Chapter XI, a slight distortion of such an arrangement of anions can be traced in several structures such as those of rutile, zircon ($ZrSiO_4$), $ThCl_4$, and Th_3P_4. Some

---

* The structure of "β-tungsten" = W_3O (= Cr_3Si, the A15 type; cf. Chapter XIV) may be an exception. If the atoms are ordered, as would be expected, then it seems likely that W will be on the Cr sites and O on the Si sites of the A15 type. The latter are at the corners and body center of the cubic unit cell—a bcc array. The W atoms then occupy half the (nonregular) tetrahedral interstices in the bcc array of O. The high-temperature solid electrolyte forms of AgI and Ag_2S are also described as having bcc anions, but in these cases the cation array is molten (1).

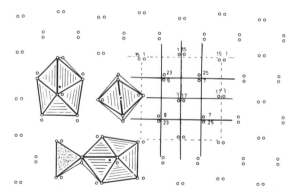

Figure 1. Deformed bcc packing (see text). Bisdisphenoids in three orientations and a tetrahedron are shown.

related structures have a similarly slightly distorted pc anion array; examples are scheelite ($CaWO_4$) and SiF_4. Many such ionic structures are common and important—for example, those of zircon and scheelite—and so we shall consider them in some detail. They resemble each other in having their larger cations (Zr, Ca) in eightfold (bisdisphenoid) coordination, while their smaller ones (Si, W) are tetrahedrally coordinated by oxygen.*

Among the alloy structures derived from bcc or pc arrays (albeit sometimes incomplete), we will discuss the structures of UCo and of (pyrochlore-related) W_3Fe_3C, pyrochlore itself, $Cr_4Al_{13}Si_4$, and cubic Sb_2O_3.

Distorted bcc Arrays

We start with a unit cell that is eight times the volume of the bcc unit cell (i.e., with a doubled cell edge) and shift each atom along the appropriate $\langle 111 \rangle$ direction by a distance $(a/32)$ $\langle 111 \rangle$. In this way the atoms can be relocated into positions $16(c)$ of the space group $I\bar{4}3d$ (No. 220): x,x,x, etc. The result, with $x = 1/32 = 0.03125$, is shown in Figure 1, in which the thin lines intersect at the atomic positions of the (parent) bcc array (and atomic heights are in units of $c/32$ of the $2 \times 2 \times 2$ cell). The figure also shows some of the polyhedral voids thus created: bisdisphenoids (in three of their six different orientations in the structure, all having their $\bar{4}$ axes parallel to a cube axis) and tetrahedra (ditto), only one of which is shown.

This atom arrangement defines a space-filling array of the two types of polyhedra—bisdisphenoids and tetrahedra—and is a useful starting point for several structures. Figure 2 shows the bisdisphenoid in plan and elevation (cf. Figure 16 of Chapter XI); in this case it is regular in the sense that all its

* Some compounds, which have these structures at higher temperatures, are ferroelastic and transform to related monoclinic polymorphs at lower temperatures (e.g., $DyVO_4$ and $BiVO_4$), a phenomenon of some technical, as well as scientific, importance.

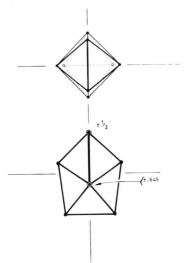

Figure 2. Plan (above) and elevation (below) of a regular bisdisphenoid. The heights of the atoms shown by double circles are above/below the midplane of the figure (in which the filled single circles lie), in units of the edge length of the figure.

edges are equal in length. An array of such *regular* bisdisphenoids and *regular* tetrahedra, similar to that in Figure 1, is not quite possible: the sum of the dihedral angles around an edge is 359.1° instead of the required 360°—hence the small distortions apparent in Figure 1.

The exact shape of a bisdisphenoid (two interpenetrating tetrahedra; see Figure 17 of Chapter XI) is variable. In particular, the distances from center to vertex (bond lengths for an atom at its center) fall into two groups of four each. Those to the vertices of one tetrahedron are all equal but are different from those to the vertices of the other tetrahedron. So the coordination number is usually 4 + 4 rather than 8. (In exceptional cases the two values are equal or nearly so; cf. Th_3P_4, below.)

The atomic shifts from bcc, $(a/n)\langle 111 \rangle$, give regular *tetrahedra* for $n = 32$ but *bisdisphenoids* for $n = 12$, which are regular in the (new) sense that all the center-to-vertex distances are the same. [In bcc, of course, the tetrahedra are not regular: two edges $= a_{bcc}$ and four edges $= a(\sqrt{3}/2)$.]

UCo

This differs slightly from the structure described in Figure 1. Structural data are given in Data Table 1. The $16(c)$ positions of $I\bar{4}3d$ have been split into two subsets $8(a)$ of $I2_1/3$, so that (1) the two types of atoms are ordered (i.e., it is a distortion of CsCl rather than bcc); (2) the symmetry is thereby reduced ($I\bar{4}3d$ to $I2_1/3$); and (3) the shift parameters differ slightly from 0.03125 and from each other—they are $0.035 \approx \frac{1}{29}$ for U and $0.294 - \frac{1}{4} = 0.044 \approx \frac{1}{23}$ for Co—but Figure 1 is still quite a good representation of the structure. (The atom positions deviate from the ideal by only 0.04 and 0.14 Å, respectively.) Both the bisdisphenoids and tetrahedra are empty in this structure.

Data Table 1   UCo (2)

Cubic, body-centered; space group $I2_1/3$, No. 199; $a = 6.3577(4)$ Å; $Z = 8$, $V = 256.98$ Å^3

Atomic Positions

All atoms in $8(a)$: $(0,0,0; \frac{1}{2},\frac{1}{2},\frac{1}{2}) +$

$(x,x,x; \frac{1}{2}+x,\frac{1}{2}-x,\bar{x}; \bar{x},\frac{1}{2}+x,\frac{1}{2}-x;$

$\frac{1}{2}-x,\bar{x},\frac{1}{2}+x)$

U: $x = 0.035$

Co: $x = 0.294$

Th$_3$P$_4$

This is generally regarded as a very difficult and complicated structure; certainly it is not easy to visualize. However, in terms of Figure 1 it is somewhat easier to appreciate. Structural data are given in Data Table 2. We note that its unit cell is again cubic and that the P atoms occupy positions $16(c)$ of $I\bar{4}3d$, as do the atoms in Figure 1, corresponding to both U and Co in the structure of UCo. The x parameter is now larger, $0.083 \approx \frac{1}{12}$ instead of $0.03125 = \frac{1}{32}$; which has the effect already mentioned of reducing the difference between the two sets of (bisdisphenoid) bond lengths to virtually zero (0.006 Å)—i.e., it makes the bisdisphenoids regular in the second sense (at the price of making the tetrahedra irregular).

The structure therefore consists of the same space-filling array of bisdisphenoids and tetrahedra of P atoms. The Th atoms occupy all the former; all the latter remain empty. The complete structure is still difficult to visualize, and Figures 3a—c are perhaps best used for building a model of the structure. In each of them all the P atoms are shown, but only one-third of the Th atoms (a different subset in each figure). It is not too difficult to see, in each of these figures, chains of edge-sharing ThP$_8$ bisdisphenoids that share additional edges with adjacent chains. The three three-dimensional structures, each of stoichiometry ThP$_4$, are superimposed (interpenetrate, with a common P$_4$ array) to give the overall stoichiometry Th$_3$P$_4$. The ThP$_8$ bisdisphenoids in the three different ThP$_4$ arrays share faces. (Remember that the structure is cubic, so that each drawing can also be regarded as a different view of the same ThP$_4$ portion of the structure.)

Data Table 2   Th$_3$P$_4$ (3)

Cubic, space group $I\bar{4}3d$, No. 220; $a = 8.600$ Å; $Z = 4$, $V = 636.1$ Å^3

Atomic Positions: $(0,0,0; \frac{1}{2},\frac{1}{2},\frac{1}{2})+$

Th in $12(a)$: $\frac{3}{8},0,\frac{1}{4}; \frac{1}{4},\frac{3}{8},0; 0,\frac{1}{4},\frac{3}{8}; \frac{1}{8},0,\frac{3}{4}; \frac{3}{4},\frac{1}{8},0; 0,\frac{3}{4},\frac{1}{8}$

P in $16(c)$: $(0,0,0$ and $\frac{1}{4},\frac{1}{4},\frac{1}{4}) + (x,x,x; \frac{1}{2}+x,\frac{1}{2}-x,\bar{x}; \bar{x},\frac{1}{2}+x,\frac{1}{2}-x;$

$\frac{1}{2}-x,\bar{x},\frac{1}{2}+x); x = 0.083$

Atomic Distances

Th–P = 2.98 Å (4×), 2.97 Å (4×)

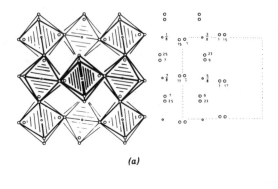

(a)

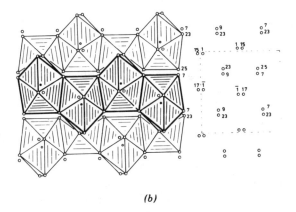

(b)

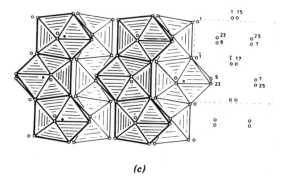

(c)

Figure 3.   The structure of Th$_3$P$_4$ [x idealized to $(a/32)\langle 111 \rangle$] projected on (001). *(a)*, *(b)*, and *(c)* each show the same array of P atoms but only one-third of the Th atoms in each case. ThP$_8$ bisdisphenoids are drawn.

This is also the so-called γ form of several sulfides of the lighter (larger) rare earth cations such as La_3S_4. A number of arsenides, antimonides, bismuthides, selenides, and tellurides of rare earths and actinides are also isostructural. In these cases the unit cell size is always ~9 Å, and $x \approx 0.068$–0.083 ($\sim\frac{1}{15}$–$\frac{1}{12}$). (Note that in early structure analyses $x = \frac{1}{12}$ was often assumed.) La_4Ge_3 has the antistructure.

Langbeinite (4), $K_2Mg_2(SO_4)_3$, eulytite (5), $Bi_4(SiO_4)_3$, and many isostructural compounds are oxygen-stuffed analogues of a Th_3P_4-like cation array, with S/Si in place of Th and (K,Mg)/Bi in place of P. Like Th_3P_4, eulytite is $I\bar{4}3d$ with $x(Bi) = 0.0857 \approx \frac{1}{12}$. The ordering in the K_2Mg_2 array in langbeinite does not correspond to that in the atomic array in UCo. Its symmetry is $P2_1/3$ compared with $I2_1/3$ for UCo. Each atom set 16(c) in Th_3P_4, split into two sets of 8(a) in UCo, is further split into four sets of 4(a) in langbeinite. The split 8(a) $\rightarrow$ 2 × 4(a) is due to U_8 (or Co_8) being replaced by K_4Mg_4. The four 4(a) atomic parameters corresponding to $x[16(c)]$ are 0.0667, 0.0482, 0.0851, and 0.0990—an average of $0.0748 \approx \frac{1}{13}$.

A different description of these structures, in terms of packed $\langle 111 \rangle$ rods (rows) of atoms, has been given elsewhere (6). In the same paper there is considerable discussion of eulytite (eulytine), langbeinite and the other structures related to the Th_3P_4 type.

$Au_3Sb_4Y_3$

The structure of $Au_3Sb_4Y_3$ is also related to that of Th_3P_4. (Data are given in Data Table 3.) The Sb_4Y_3 part is quite analogous, with Sb in the P positions and Y in the Th positions, i.e, in Sb_{4+4} bisdisphenoids. The Au atoms occupy the tetrahedra (which are empty in Th_3P_4).

Data Table 3 $Au_3Sb_4Y_3$ (7)

Cubic, space group $I\bar{4}3d$, No. 220; $a = 9.818$ Å; $Z = 4$, $V = 946.4$ Å³

Atomic Positions: $(0,0,0; \frac{1}{2},\frac{1}{2},\frac{1}{2})+$
Y in 12(a): $\frac{3}{8},0,\frac{1}{4}$ etc. (cf. Th_3P_4)
Sb in 16(c): x,x,x, etc. (cf. Th_3P_4), $x = 0.088$ (= 1/11.4)
Au in 12(b): $\frac{7}{8},0,\frac{1}{4}; \frac{1}{4},\frac{7}{8},0; 0,\frac{1}{4},\frac{7}{8}; \frac{5}{8},0,\frac{3}{4}; \frac{3}{4},\frac{5}{8},0; 0,\frac{3}{4},\frac{5}{8}$

Pu_2C_3

The Pu_2C_3 structure type, representing a rather large group of compounds, is closely related to the Th_3P_4 type. The Pu atoms of the former occupy the P sites of the latter, and C_2 groups [l(C–C) = 1.295 Å] take the Th positions. Consequently the stoichiometry is $(C_2)_3Pu_4 = Pu_4C_6 = 2Pu_2C_3$. It is, up to a point, an antiform of Th_3P_4. Rb_2O_3 and Cs_2O_3 are isostructural, with O_2 in place of C_2. (The axes of the X_2 groups are parallel to the $\bar{4}$ axes of the bisdisphenoids, i.e., to $\langle 100 \rangle$ of the cubic unit cell.) In the last two cases, $x(\text{Rb}) = x(\text{Cs}) = 0.054 \ (= 1/18.5)$. Three interesting points may be noted.

1. The alkali metal atom array is (as described earlier) distorted bcc, whereas in the metal itself it is bcc.
2. For Rb_2O_3 the equivalent bcc unit cell edge $= a(\text{Rb}_2\text{O}_3)/2 = 9.32/2 = 4.66$ Å is to be compared with 5.605 Å (at 78 K) for the bcc metal itself; i.e., the metal atoms are much closer together in the oxide.
3. There appear to be valency problems. According to the structure determination (8), Rb_2O_3 is to be regarded as $\text{Rb}_4^+(\text{O}_2^-)_2(\text{O}_2^{2-})$ —"as shown by magnetic measurements." (The exact parameters of the O atoms were not determined.)

By analogy with this description of Pu_2C_3, etc., we can describe eulytite and langbeinite as Th_3P_4 types with SiO_4 and SO_4 tetrahedra, respectively, in place of Th.

ThCl_4

This structure is readily derived from that of Th_3P_4 by substituting Cl for P and omitting two-thirds of the cations. The result is a structure close to *one* of Figures 3*a*, *b*, and *c*. The resulting symmetry is, of course, tetragonal rather than cubic, in agreement with the data in Data Table 4 for ThCl_4. It is for this reason that the bisdisphenoids are no longer tilted and twisted about their $\bar{4}$ axes; compare this structure in Figures 4*a* and *b* with those of Th_3P_4 in Figure 3.

Data Table 4 ThCl_4 (9)

Tetragonal, space group $I4_1/amd$, No. 141; $a = 8.48(1)$, $c = 7.46(1)$ Å;*
$c/a = 0.880$; $Z = 4$, $V = 536$ Å^3

Atomic Positions: $(0,0,0; \frac{1}{2},\frac{1}{2},\frac{1}{2})+$
Th in 4(a): $\pm(0,\frac{3}{4},\frac{1}{8})$
Cl in 16(h): $\pm(0,y,z; 0,\frac{1}{2}-y,z; \frac{1}{4}+y,\frac{1}{4},\frac{3}{4}+z; \frac{3}{4}-y,\frac{1}{4},\frac{3}{4}+z)$;
                $y = 0.0633$ and $z = 0.2008$
Bond Lengths
Th–Cl $= 2.72$ Å ($4\times$), 2.90 Å ($4\times$)

* According to ref. 10, these are $a = 8.490$, $c = 7.483$ Å.

According to Mason et al. (10), this structure is the high-temperature β form. The lower-temperature α-ThCl_4, while also tetragonal with ThCl_8 bisdisphenoids, is topologically distinct. It has a structure related to that of scheelite (see below) in the same way as the β form is related to that of

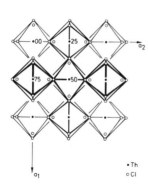 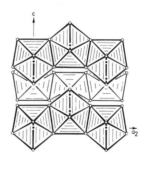

Figure 4. The structure of $ThCl_4$ projected on (a) (001) and (b) (100) of its tetragonal unit cell. (There is a shift of origin compared with the data in Data Table 4.)

zircon. Many tetrahalides of the actinide metals are isostructural with β-$ThCl_4$.

Zircon, $ZrSiO_4$

The ZrO_4 part of the structure is similar to that of $ThCl_4$ (in particular, the anion parameters are very close, see Data Table 5), but the appropriate subset of tetrahedra in the structure (empty in $ThCl_4$) is now occupied by Si. It is related to the structure of $Y_3Au_3Sb_4$ as that of $ThCl_4$ is related to that of Th_3P_4: subsets of one-third of both the bisdisphenoids and tetrahedra are occupied. The zircon structure is shown (as two bounded projections) in Figures 5a and b, and in Figure 5c.

Data Table 5 Zircon, $ZrSiO_4$ (11)

Tetragonal, space group $I4_1/amd$, No. 141; $a = 6.6164(5)$, $c = 6.0150(5)$ Å; $c/a = 0.909$; $Z = 4$, $V = 263.32$ Å^3

Atomic Positions: (2nd choice of origin, at $2/m$; $0,\frac{3}{4},\frac{1}{8}$ from 1st choice, $\overline{4}m2$):

$(0,0,0; \frac{1}{2},\frac{1}{2},\frac{1}{2})+$

Zr in 4(a): $\pm(0,\frac{3}{4},\frac{1}{8})$
Si in 4(b): $\pm(0,\frac{1}{4},\frac{3}{8})$
O in 16(h): $\pm(0,y,z; 0,\frac{1}{2}-y,z; \frac{1}{4}+y,\frac{1}{4},\frac{3}{4}+z; \frac{3}{4}-y,\frac{1}{4},\frac{3}{4}+z)$;
$\quad\quad\quad y = 0.067$ and $z = 0.198$

Bond Lengths
Zr–O = 2.15 Å (4×), 2.29 Å (4×)
Si–O = 1.61 Å (4×)

Elsewhere (ref. 12 and Chapter XI) we have shown that ABX_4 structures with AX_8 and BX_4 coordination, like that of zircon, are very closely related

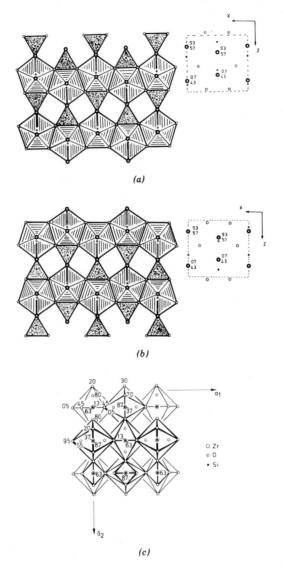

Figure 5. Bounded projections of the $ZrSiO_4$ structure projected on (010) [$\equiv$ (100)]: (a) $0.43 \le y/b \le 1.07$; (b) $0.93 \le y/b \le 1.57$; cations are at heights $\frac{3}{4}$ and $1\frac{1}{4}$, respectively. Note that bisdisphenoids share edges in both the x and y directions. Parallel to z there are infinite rods of alternating ZrO_8 bisdisphenoids and SiO_4 tetrahedra that also share edges. In (c) we show the (001) projection.

to structures MX_2, with MX_6 octahedra. In the case of zircon itself, the corresponding octahedral structure is that of rutile (with which that of zircon was originally confused). The former is derived from the latter by substituting alternately large (Zr) and small (Si) cations for Ti in the chains of edge-sharing TiO_6 octahedra in the rutile type. In this context, Figure 6 shows zircon projected on $(1\bar{1}0)$: the rods of edge-shared octahedra in rutile are

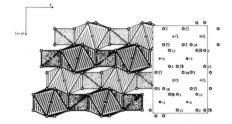

Figure 6. The structure of zircon projected on (1$\bar{1}$0).

replaced by rods of alternating edge-shared tetrahedra and bisdisphenoids in zircon, as explained in Chapter XI, where the emphasis was on the cation arrays.

The anion arrays in both rutile and zircon can be derived from bcc by small systematic shifts, reminiscent of the method used above for UCo, Th_3P_4, etc. Figure 7 shows (001) projections of both tetragonal structures. (Note that a doubled cell has been used for rutile, with a c axis twice the normal c but equivalent to the c axis of zircon. This facilitates the comparison.) The thin lines represent the projections of bcc unit cells of anions from which, in each case, the actual anion array in the structure is readily derived by small vector shifts. (Note also that a small change in axial ratio is needed; for these "bcc" cells, $c/a = 0.911$ for rutile and 0.909 for zircon.)

For rutile, the required shifts are from $\frac{1}{4},\frac{1}{4},0$ to $x,x,0$ [positions 4(f) in space group $P4_2/mnm$], $\Delta \approx 0.055a\langle110\rangle$ [$\equiv \sim(a/18)\langle100\rangle$ for the zircon-type cell with four times the volume of the normal rutile cell]—a distance of ~0.36 Å. For zircon they are $0.066,0,0.052 \approx \langle a/15,0,c/19\rangle$, or roughly

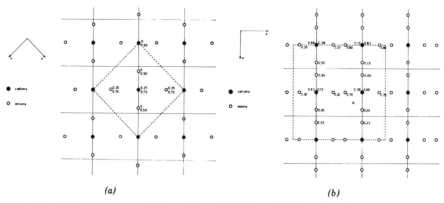

(a)   (b)

Figure 7. The structures of (a) rutile and (b) zircon, both projected on (001) of their tetragonal unit cells (which are outlined by broken lines). Atom heights are given as a proportion of $2c$ (rutile) $\equiv c$ (zircon). The light lines are projections of approximately bcc unit cells from which the actual anion arrays may be derived by vector sets that are different for the two structures (see text). The unit-cell origin for zircon has been shifted to aid the comparison with rutile.

⟨101⟩/17—a distance of ~0.5 Å. Hence, the vector forms are in directions ±[100] and ±[010] for rutile, but roughly ±[101], ±[$\bar{1}$01], ±[011], and ±[0$\bar{1}$1] for zircon—a more complex set for the latter, which is therefore not quite as simply related to a bcc anion array as is the former (and Th_3P_4).

Very many arsenates, vanadates, phosphates, etc. are isostructural with zircon. The smaller cation is usually B^{4+} or B^{5+}; compare scheelite, below.

The structure of K_2ZrF_6 is also related: K atoms formally substitute for anions to form a distorted bcc (K_2F_6) array similar to that in $ThCl_4$. But only half the bisdisphenoids are occupied by Zr, those consisting of only F [i.e., $Zr(K_2F_6)_{\Sigma=8}$]. These ZrF_8 bisdisphenoids form single layers (unconnected strings of edge-sharing bisdisphenoids) as in $ThCl_4$; but these are not connected (by similar edge-sharing) in the orthogonal direction (where the "bisdisphenoids" are now K_2F_6).

Distorted pc Arrays

Scheelite, CaWO₄

Systematic shifts of the anions, of about the same magnitude as those in the previous section for Th_3P_4 and so on, transform a pc array as shown in Figure 8, which relates the structure of scheelite, $CaWO_4$, to the fluorite type. (The relation between these two is rather similar to that between zircon and rutile.) On the right is shown the fluorite type, together with the vector scheme for distorting the anion array. The remaining part of the figure (after shifting the anions and substituting alternate Ca and W for the cations) is the structure of $CaWO_4$ projected along [100] = [010]. Crystallographic data are given in Data Table 6.

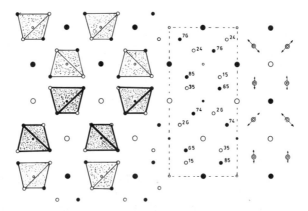

Figure 8. Derivation of the scheelite, $CaWO_4$, structure (on the left) from the fluorite type (on the right).

Data Table 6   Scheelite, CaWO$_4$ (13a, 13b)

Tetragonal, space group $I4_1/a$, No. 88; $a = 5.243$, $c = 11.376$ Å; $c/a = 2.170$; $Z = 4$, $V = 312.7$ Å^3

Atomic Positions: $(0,0,0; \frac{1}{2},\frac{1}{2},\frac{1}{2})+$
W in 4(a): $0,0,0; 0,\frac{1}{2},\frac{1}{4}$
Ca in 4(b): $0,0,\frac{1}{2}; 0,\frac{1}{2},\frac{3}{4}$
O in 16(f): $x,y,z; \bar{x},\bar{y},z; x,\frac{1}{2}+y,\frac{1}{4}-z; \bar{x},\frac{1}{2}-y,\frac{1}{4}-z; \bar{y},x,\bar{z};$
           $y,\bar{x},\bar{z}; \bar{y},\frac{1}{2}+x,\frac{1}{4}+z; y,\frac{1}{2}-x,\frac{1}{4}+z;$
           $x = 0.2413, y = 0.1511, z = 0.0861*$

Bond Lengths
W–O = 1.79 Å (4×)
Ca–O = 2.44 Å (4×), 2.48 Å (4×)

* From the neutron diffraction data (13a); the X-ray data (13b) are in excellent agreement.

In Figure 8 the dotted O$_4$ tetrahedra have W atoms at their centers (small circles; open at $x = 0$, filled at $x = \frac{1}{2}$). The medium-sized circles are oxygens (with their x parameters given inside the unit cell, which is outlined by broken lines). Large circles, open at $x = 0$ and filled at $x = \frac{1}{2}$, represent the Ca atoms. The WO$_4$ tetrahedron is rather regular; the coordination polyhedron around Ca is best described as a bisdisphenoid, CaO$_{4+4}$. Clearly, the topology of these two types of polyhedra is quite different from that in zircon.

In the right-hand (fluorite-like) part of the figure, double circles (with arrows) represent fluorines at $\pm\frac{1}{4}$; the filled and open circles represent cations, as in CaWO$_4$. The latter are in the same positions in both parts of the drawing. (However, there is a small tetragonal distortion of ~8.5%; cf. the c/a value.)

There are many compounds isostructural with scheelite. A large proportion of them are molybdates or tungstates; almost all of them have B^{6+} or B^{7+} as the smaller cation (cf. zircon). TlCl$_2$ is also isostructural (14); it is really Tl$^+$Tl^{3+}Cl$_4$, with Tl$^+$ in the bisdisphenoids and Tl^{3+} in the tetrahedra. (The isostructural pair GaCl$_2$ and TlBr$_2$ are also M$^+$M^{3+}X$_4$ types, with M$^+$ in X$_{4+4}$ bisdisphenoids and M^{3+} in tetrahedra; but their topology is different from that of TlCl$_2$.)

SiF$_4$

The structure of SiF$_4$ is shown in Figure 9, where the right-hand part again shows a pc anion array and the vector scheme to produce the actual SiF$_4$ anion array (which is shown on the left). Again (cf. Th$_3$P$_4$, above) the shifts are in $\langle 111 \rangle$ directions: for the ideal pc array the F parameter would be $x = \frac{1}{4}$; so the shifts are $\delta x = 0.085 = 0.46$ Å. Data are given in Data Table 7.

It can be regarded as a straightforward bcc packing of SiF$_4$ molecules. The large empty cavities between them are best described as bisdisphenoids.

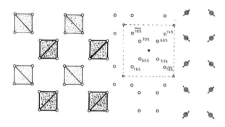

Figure 9. The structure of SiF_4 (left) derived from a pc anion array (right).

Data Table 7 SiF_4 (15)

Cubic, space group $I\bar{4}3m$, No. 217; $a = 5.41$ Å; $Z = 2$, $V = 158$ Å^3

Atomic Positions: $(0,0,0; \frac{1}{2},\frac{1}{2},\frac{1}{2})+$
Si in $2(a)$: $0,0,0$
F in $8(c)$: $x,x,x; x,\bar{x},\bar{x}; \bar{x},x,\bar{x}; \bar{x},\bar{x},x; x = 0.165$
Bond Lengths
Si–F = 1.56 Å (4×)

Stellae Quadrangulae

Figure 10 shows how a so-called *stella quadrangula* can be exactly derived from a cube of atoms by a "small topological distortion." A (regular) stella quadrangula consists of four regular tetrahedra, each sharing one face with a central, fifth tetrahedron, which is also regular (or four regular tetrahedra each sharing three edges, one with each of its three neighbors). Using the unit of this figure we can calculate the resulting (cubic) unit cell to have $a = 4e\sqrt{2}/3$, where e is the tetrahedron edge. The atomic parameter is $x = e/(2\sqrt{2}a) = 3/16 = 0.1875$.

This gives another way of describing the structure of SiF_4: as stellae quadrangulae, SiF_{4+4}, sharing corners, as shown in Figure 11. This description of SiF_4 is rather accurate, the only distortion being the shrinking of the central (Si-containing) tetrahedron by ~0.2 Å (for an obvious reason). The ideal anion shift (giving regular tetrahedra in the stellae quadrangulae) is clearly $(a/16)\langle 111 \rangle = 0.0625a\langle 111 \rangle$; the actual shift in SiF_4 is $0.085a\langle 111 \rangle$, a discrepancy of 0.2 Å. (Note that the SiF_4 "parent"—with a pc array of anions—is the $NiHg_4$ type, described in Chapter VIII, in which $NiHg_8$ *cubes* share corners. Here SiF_{4+4} stellae quadrangulae share corners.)

Similarly, the scheelite structure can also be described in terms of WO_{4+4} stellae quadrangulae instead of WO_4 tetrahedra, but now they share edges (see Figure 12).

Figure 10. Derivation of a stella quadrangula from a cube.

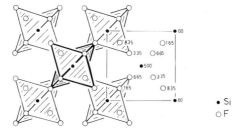

Figure 11. The structure of SiF$_4$ depicted as a corner-connected array of SiF$_{4+4}$ stellae quadrangulae. Compare Figures 9 and 10.

AX$_2$ and ABX$_4$ Structure Relationships

It may be noted that zircon types (often, at least) transform to scheelite types under high pressure, a transformation of some interest (16) because there is no change in cation coordination numbers; cf. olivine → spinel.

In view of this and the structural relations between rutile and zircon and between fluorite and scheelite, and also the fact that the rutile type transforms to the fluorite type at high pressure, it is worth comparing all four structure types. This is done in Figure 13, in which the unit cells defined by the dotted lines should be compared. Rutile → zircon (Figure 13a → b) involves small anion shifts (as discussed earlier), as does fluorite → scheelite (Figure 13c → d). [Of course, the patterns of anion shifts are not identical in the two cases—cf. the anion heights for (b) and (d).] The unit-cell shapes are similar within each pair. The high-pressure transformations (a → c and b → d) involve most obviously a change in unit-cell shape but also, again, some small anion shifts (shuffles). The topological similarity of all four structure types is clear.

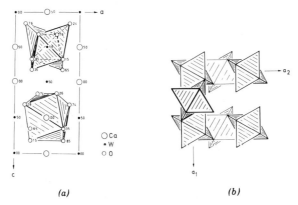

(a)   (b)

Figure 12. (a) The scheelite (CaWO$_4$) structure projected on (010). One CaO$_{4+4}$ bisdisphenoid and one WO$_{4+4}$ stella quadrangula are shown. Compare Figure 8. (b) The topology of (edge-sharing) WO$_{4+4}$ stellae quadrangulae in the scheelite type [projected on (001)].

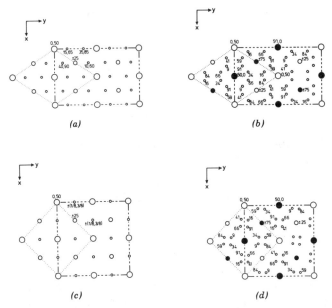

Figure 13. Corresponding projections of (*a*) rutile, TiO$_2$, (*b*) zircon, ZrSiO$_4$, (*c*) fluorite, CaF$_2$, and (*d*) scheelite, CaWO$_4$, all on the same scale. In each case two different types of unit cells are outlined (by dotted and broken lines). The relations of the larger ones (broken lines) to the conventional unit cells are (*a*) 002/200/0$\bar{2}$0, (*b*) 001/110/$\bar{1}$10, (*c*) 1$\bar{1}$0/110/001, and (*d*) also 1$\bar{1}$0/110/001. The two larger circles represent cations; the smallest circles represent anions in each case; and atomic heights are in units of *c*/100 for either outlined cell. In (*b*) and (*d*) the largest open circles indicate Zr/Ca at *z* = 0 and Si/W at *z* = $\frac{1}{2}$; the largest filled circles, Si/W at *z* = 0 and Zr/Ca at *z* = $\frac{1}{2}$; the medium-sized circles are Zr/Ca at *z* = $\frac{1}{4}$, Si/W at *z* = $\frac{3}{4}$ (open), and Si/W at *z* = $\frac{1}{4}$, Zr/Ca at *z* = $\frac{3}{4}$ (filled). Compare the shape and contents of the *dotted* cells in each case.

W$_3$Fe$_3$C, Pyrochlore (A$_2$B$_2$X$_6$Y), and Related Structures (17)

We now turn to some face-centered cubic structures, at first sight complex and with unit cells large enough (*a* ≈ 10–11 Å) to make them difficult to draw clearly in projection. They become more straightforward when regarded as two more or less independent, interpenetrating, three-dimensional structural frameworks.

In the structure of W$_3$Fe$_3$C (the "zeta carbide," E9$_3$ type) one of these frameworks is the array of corner-connected stellae quadrangulae (SQs) shown in Figure 14*a*. Note that its topology is distinct from that in the anion array of SiF$_4$ (above). This is because it can be generated from an *incomplete* (75%) pc array of Fe atoms, as shown in Figure 14*b*, by shifting two-thirds of the Fe atoms (those in the central tetrahedra) by ideally *a*⟨111⟩/20, where *a* is

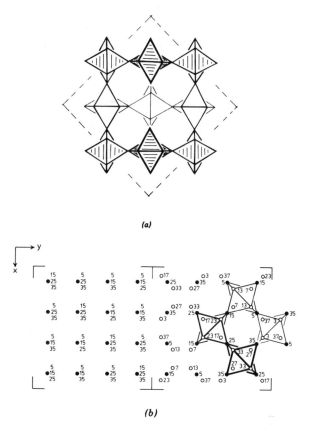

Figure 14. (*a*) The array of corner-connected stellae quadrangulae of Fe atoms in W₃Fe₃C, projected on (001) of the cubic unit cell. (*b*) The generation of an array of corner-connected stellae quadrangulae, on the right, by small, concerted shifts of some of the atoms in three-fourths of a pc array on the left. The filled circles on the right are unmoved atoms, the open circles have been moved by the ideal $a\langle 111\rangle/20$ to give regular tetrahedra (and SQs): they are, respectively, at positions 16(*d*) and 32(*e*) (with $x = \frac{17}{40}$) of $Fd3m$. Heights are in units of $c/40$.

the edge of the final cubic unit cell. (Compare Figure 14*b* with the derivation of the anion array of SiF₄, in Figures 10 and 11, from a complete pc array.)

The other framework in W₃Fe₃C (the W₃ part) is the array of corner-connected octahedra shown in Figure 15*a*. This is called (for a reason that will become obvious) the "pyrochlore framework." Note that its topology is distinct from that in the ReO₃ or perovskite-type array of corner-connected octahedra. It too may be generated from an *incomplete* (75%) pc array of (W) atoms, as shown in Figure 15*b*, by shifting the atoms by (ideally) $a\langle 100\rangle/16$.

The building units of the two frameworks can be taken to be those shown in Figure 16*a*, a stella quadrangula, and 16*b*, a "pyrochlore unit," respec-

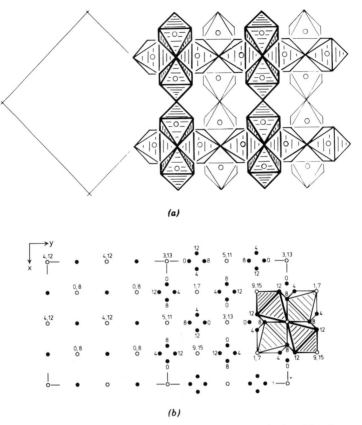

Figure 15. (*a*) The array of corner-connected CW_6 octahedra (the "pyrochlore framework") in W_3Fe_3C, projected on (001) of the cubic unit cell. (*b*) The generation of an array of corner-connected octahedra (the "pyrochlore framework"), on the right, from three-fourths of a pc array, on the left, by small, concerted shifts of the atoms, ideally $a\langle 100\rangle/16$ to give regular octahedra. One "pyrochlore unit" is emphasized. On the left, the open circles are half-filled [001] rows of atoms, and the filled circles are complete rows (with atoms at 0, 4, 8, 12; all heights in units of $c/16$). The former are shifted by $a[001]/16$, the latter by $a[100]/16$ or $a[010]/16$. Note that there is an additional octahedron (not emphasized) at the center of each pyrochlore unit.

tively; but note that there is another octahedron (unshaded) at the center of the latter unit. The two units are related; one is a central tetrahedron capped on all four faces by additional tetrahedra, the other a central octahedron capped on four (of its eight) faces by additional octahedra. Because, in W_3Fe_3C, the two frameworks interpenetrate—each occupying the interstices in the other—their unit cell edges are identical in size. If we denote the cell edge by a and the edges of the (regular) tetrahedra and octahedra by e_t and e_o, respectively, then, by simple geometry, $a = e_t(10\sqrt{2}/3) = e_o(8\sqrt{2}/3)$.

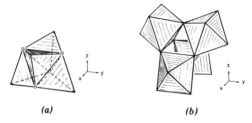

Figure 16. Clinographic projections of (*a*) a stella quadrangula and (*b*) a pyrochlore unit. The first is a central tetrahedron (vertices, ○), capped, on each face, by another tetrahedron. The second is a central octahedron (unshaded), capped, on alternate faces, by four other octahedra (shaded).

Hence the ratio of the edge lengths of the octahedra and tetrahedra is $e_o/e_t = 10/8 = 1.25$.

In W_3Fe_3C the carbon atoms occupy the shaded octahedra in Figure 16*b*, and the central octahedron is empty. But in W_6Fe_6C the situation is reversed: the central (unshaded) octahedron is occupied by C, and the shaded octahedra are now empty.

The mineral pyrochlore is $(Na,Ca)_2(Nb,Ti)_2(O,F)_6(O,F)$ [$= A_2B_2X_6Y$], but there are many isostructural compounds, such as $Y_2Ti_2O_7$. Like W_3Fe_3C [$= \frac{1}{2} \times Fe_2C_2W_6(Fe_4) = \frac{1}{2} \times A_2B_2X_6Y$], it has $Fd3m$ symmetry, with the anions X occupying the octahedron corners (in place of W), the cations B the centers of the shaded (Figure 16*b*) octahedra (in place of C), the cations A the *shared* corners of the stellae quadrangulae [in place of one-third of the Fe atoms, Fe(1)], and the single anions Y replacing the central tetrahedron of each stella quadrangula [in place of the other two-thirds of the Fe atoms,

TABLE 1   Some Atomic Positions in Space Group $Fd3m$ (No. 227), Origin at $\bar{4}3m$

Positions	Corresponding Atoms in Formula $A_2B_2X_6Y$	Wyckoff Symbol	Coordinates
Octahedra (Figure 16*b*)			
Corners	X	48(*f*)	$x_1,0,0$, etc.
Centers of outer octahedra	B	16(*c*)	$\frac{1}{8},\frac{1}{8},\frac{1}{8}$, etc.
Centers of inner octahedra		8(*a*)	$0,0,0$, etc.
Stellae quadrangulae (Figure 16*a*)			
Corners of outer tetrahedra	A	16(*d*)	$\frac{5}{8},\frac{5}{8},\frac{5}{8}$, etc.
Corners of inner tetrahedra		32(*e*)	x_2,x_2,x_2, etc.
Centers[a]	Y	8(*b*)	$\frac{1}{2},\frac{1}{2},\frac{1}{2}$, etc.

[a] A special case of 32(*e*) = 8(*b*).

TABLE 2 Lattice Parameters (a), Site Occupancies, and Atomic Parameters of W_3Fe_3C, Pyrochlore, and Some Compounds with Similar Structures

Compound	a (Å)	8(a)	8(b)	16(c)	16(d)	32(e)/x_2	48(f)/x_1	Ref.
Ideal (regular polyhedra)						$\frac{3}{8} + \frac{1}{20} = 17/40$ $= 0.425^a$	$\frac{3}{16} = 0.1875$	
W_3Fe_3C	11.087	—	—	C	Fe(1)	Fe(2)/0.420	W/0.198	19
W_6Fe_6C	10.934	C	—	—	Fe(1)	Fe(2)/0.421	W/0.197	19
Co_2W_4C	11.21	—	—	C	W(1)	Co/0.425	W(2)/0.195	20
$M_3Ti_3O^b$	~11.3	—	—	O	M	M/0.415	Ti/0.185	21
$NiTi_2$	11.319	—	—	—	Ti	Ni/0.410	Ti/0.189	22
Pyrochlores (various)	~10.4	—	(O,F)	(Nb,Ti)	(Na,Ca)	—	(O,F)/0.170–0.187	23
Senarmontite = cubic Sb_2O_3	11.152	—	—	—	—	Sb/0.3647	O/0.1863	24
$AgSbO_3$	10.32	—	—	Ag	Sb	—	O/0.165	25
$RbNbTeO_6$	10.25	—	Rb	(Nb,Te)	—	—	O/?	26
$KTaWO_6 \cdot H_2O$	10.48	—	H_2O	(Ta,W)	$K_{1/2}{}^c$	—	O/?	26
α-$Cr_4Al_{13}Si_4{}^d$	10.917	—	Al(1)$_{1/2}{}^e$	—	Si	$Cr_{1/2}$/0.3421^f	Al(2)$_{1/2}$/0.1897^g and Al(3)$_{1/2}$/0.1900	27

[a] There is an error in ref. 16: for W_3Fe_3C the ideal x parameter for 32(e) is given as 0.050 = 1/20. This is, of course, the shift of Fe(2) from ideal pc positions. The ideal x is this shift + $\frac{3}{8} = \frac{1}{20} + \frac{3}{8} = 0.425$, as given here.
[b] M = Co, Cu, Fe, Mn, or Ni.
[c] Occupancy = 0.5.
[d] Space group is $F\bar{4}3m$ (No. 216), but the parameters listed are appropriate to $Fd3m$.
[e] $\frac{1}{2} \times 8(b)$ of $Fd3m \to 4(a)$ of $F\bar{4}3m$.
[f] $\frac{1}{2} \times 32(e) \to 16(e)$.
[g] 48(f) $\to$ 24(f) + 24(g).

348

Fe(2)$_4$]. One array is of corner-connected BX$_6$ octahedra, the other is of corner-connected YA$_4$ tetrahedra (the anti-form of the idealized β-cristobalite structure, C9 type), instead of corner-sharing stellae quadrangulae. [The anion coordination about each A cation is then a deformed cube, or puckered hexagonal bipyramid, of eight anions, X$_6$Y$_2$.]

Pyrochlore was formerly described as M$_4$N$_7$ [= (A$_2$B$_2$)(X$_6$Y)], a "defect fluorite with one-eighth of its anion sites unoccupied." In the light of the above, that is clearly an inadequate, although not inaccurate, description [cf. Wadsley (18) and the above discussion, especially of Figure 15].

The space group positions involved in these and related structures are shown in Table 1; while Table 2 shows the occupancies of the various sites in the (related) structures of several compounds, and the appropriate parameters. (Both tables are based on a unit cell with its origin at $\overline{4}3m$—*not* the alternative at $\overline{3}m$.) These are quite straightforward and emphasize the versatility of the structure, which is also apparent in the above description. The only serious cause for comment is the last item in Table 2: α-Cr$_4$Al$_{13}$Si$_4$. This has space group $F\overline{4}3m$ instead of $Fd3m$, due to a simple and interesting modification of the first framework: stellae quadrangulae in W$_3$Fe$_3$C but tetrahedra in pyrochlore. Now it is a hybrid of the two—stellae quadrangulae alternating with tetrahedra. The common corners [A sites, 16(d)] are occupied by Si, half their centers by Cr$_4$ [in half of 32(e)], and the other half by Al(1) [in half of 8(b)]. The pyrochlore-like framework of (empty) corner-connected Al$_6$ octahedra is unchanged. The stoichiometry is therefore A$_2$B$_2$X$_6$Y = Si$_2\square_2$Al$_6$[(Cr$_4$)$_{1/2}$ Al$_{1/2}$] = $\frac{1}{2}$ $\times$ Cr$_4$Al$_{13}$Si$_4$.

The cubic Sb$_2$O$_3$ structure is also a little less "obvious" than the others. The oxygen atoms in 48(f) of $Fd3m$ form the pyrochlore-like framework, but all the octahedra are empty. The Sb atoms [in 32(e)] occupy the positions of the vertices of the *inner* tetrahedron of each stella quadrangula, thereby capping (pyramidally) the unshared faces of the *central* octahedron in each pyrochlore unit (unshaded in Figure 16b), and forming SbO$_3$ pyramids. The lone pair on each Sb^{3+} then points toward the center of the inner tetrahedron, also in position 32(e) but with $x = 0.4196$, if we assume that the center of the lone pair is 1.06 Å from the Sb atom (28) (cf. Table 2 of Chapter X).

References

1. M. O'Keeffe and B. G. Hyde, *Phil. Mag.* **33**, 219 (1976).
2. N. C. Baenziger, R. E. Rundle, A. I. Snow, and A. S. Wilson, *Acta Cryst.* **3**, 34 (1950).
3. K. Meisel, *Z. Anorg. Chem.* **240**, 300 (1939).
4. K. Mereiter, *Neues Jahrb. Mineral. Monatsh.* **1979**, 182.
5. D. J. Segal, R. P. Santoro, and R. E. Newnham, *Z. Kristallogr.* **123**, 73 (1966).

6. M. O'Keeffe and S. Andersson, *Acta Cryst. A* **33**, 914 (1977).

7. A. E. Dwight, *Acta Cryst. B* **33**, 1579 (1977).

8. A. Helms and W. Klemm, *Z. Anorg. Chem.* **242**, 201 (1939).

9. K. Mucker, G. S. Smith, Q. Johnson, and R. E. Elson, *Acta Cryst. B* **25**, 2362 (1969).

10. J. T. Mason, M. C. Jha, and P. Chiotti, *J. Less-Common Metals* **34**, 143 (1974).

11. I. R. Krstanovic, *Acta Cryst.* **11**, 896 (1958).

12. H. Nyman, B. G. Hyde, and S. Andersson, *Acta Cryst. B* **40**, 441 (1984).

13a. M. I. Kay, B. C. Frazer, and I. Almodovar, *J. Chem. Phys.* **40**, 504 (1964).

13b. A. Zalkin and D. H. Templeton, *J. Chem. Phys.* **40**, 501 (1964).

14. G. Thiele and W. Rink, *Z. Anorg. Chem.* **414**, 231 (1975).

15. M. Atoji and W. N. Lipscomb, *Acta Cryst.* **7**, 597 (1954).

16. M. O'Keeffe and B. G. Hyde, *J. Solid State Chem.* **44**, 24 (1982); *Struct. Bonding* **61**, 77 (1985).

17. H. Nyman, S. Andersson, and B. G. Hyde, *J. Solid State Chem.* **26**, 123 (1978).

18. A. D. Wadsley, in *Non-Stoichiometric Compounds* (Adv. Chem. Ser., Vol. 39), R. F. Gould, Ed. American Chemical Society, Washington, DC, 1963, p. 29.

19. Z. Bojarski and J. Leciejewicz, *Struct. Rep.* **32A**, 45 (1967); *Arch. Hutn. Polska* **12**, 255 (1967).

20. R. Kiessling, *Proc. International Symposium on the Reactivity of Solids.* Elanders Boktryckeri, Gothenburg, 1952, p. 1065; *Struct. Rep.* **22**, 889 (1958).

21. N. Karlsson, *Nature (Lond.)* **168**, 558 (1951); *Struct. Rep.* **22**, 889 (1958).

22. G. A. Yurko, J. W. Barton, and J. G. Parr, *Acta Cryst.* **12**, 909 (1959); *Struct. Rep.* **22**, 889 (1958).

23. G. Perrault, *Can. Mineral.* **9**, 383 (1968).

24. C. Svensson, *Acta Cryst. B* **31**, 2016 (1975).

25. N. Schrewelius, *Strukturbericht* **6**, 120 (1938); *Z. Anorg. Allgem. Chem.* **238**, 241 (1938).

26. B. Darriet, M. Rat, J. Galy, and P. Hagenmuller, *Mater. Res. Bull.* **6**, 1305 (1971).

27. K. Robinson, *Acta Cryst.* **6**, 854 (1953).

28. J. Galy, G. Meunier, S. Andersson, and A. Åström, *J. Solid State Chem.* **9**, 92 (1975).

CHAPTER XIV

Alloy Structures with Edge-Capped Tetrahedra, Octahedra, Square Antiprisms, Stellae Quadrangulae, and/or Tetraedersterns

The structures dealt with in this chapter are rather complex, and some are quite difficult. Nevertheless they are important. And the structural principles used are simple and lead to useful, unifying descriptions and hence are worthwhile. The structures considered in this chapter include the following: γ-brass (Cu_5Zn_8 etc.), Mn_5Si_3, $TlFe_3Te_3$, W_5Si_3, Cr_3Si (A15, "β-W"), Zr_4Al_3, $CuAl_2$ (Fe_2B), "Mo_3CoSi," Fe_3P (Ni_3P), $Ni_{12}P_5$, α-V_3S, β-V_3S, Ti_3P, Ta_3As, $CaZn_5$, $CoSn$, CrO_3, the σ phases (e.g., σ-FeCr), the Friauf-Laves phases $MgCu_2$, $MgZn_2$, $MgNi_2$, etc., the μ phase (e.g., μ-Co_7Mo_6), the M phases, W_2FeSi, melilite [$(Ca,Na)_2(Mg,Al,Si)_3O_7$], the χ phase.

γ-Brass (Cu_5Zn_8), Mn_5Si_3, and W_5Si_3

The structure of γ-brass, Cu_5Zn_8 (and many isostructural compounds, not all of 5:8 stoichiometry) was originally described as slightly distorted bcc, with two atoms missing from its large (~9 Å) unit cell. Bradley and Jones (2) were the first (in 1933) to describe it in terms of atomic clusters, each containing 26 atoms: starting from its center and working outwards, an inner tetrahedron [IT = Zn(1)], followed by a larger (outer) tetrahedron [OT = Cu(1)], then a larger octahedron [OH = Cu(2)], and finally a large, slightly irregular cuboctahedron [CO = Zn(2)]. Our description is different and uses regular tetrahe-

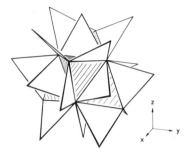

Figure 1. A stella quadrangula capped on each edge, forming the 26-atom cluster of the γ-brass structure.

dra and regular (equilateral) triangles to generate the cluster. In this way we arrive at not only an accurate model, but also a structure that can be related to several others (3).

We start with a stella quadrangula, which consists of the inner plus outer tetrahedra of the previous description (Figures 10 and 16a of Chapter XIII), each of the four faces of the former [Zn(1)$_4$] being capped by a Cu(1) atom. Each of the six edges of the inner tetrahedron is then capped by an additional atom [Cu(2)] to form an approximately equilateral triangle. This Cu(2)$_6$ array forms the octahedron [as in Th$_6$Mn$_{23}$ (3), which we will not discuss here]. The cluster is completed by putting a similar "triangular cap" [Zn(2)] on each of the 12 remaining (outer) edges of the stella quadrangula, yielding the outer, slightly irregular cuboctahedron. A clinographic projection of this structure unit is shown in Figure 1. The γ-brass structure is a bcc array of such clusters. It is shown in Figure 2a (from which the body-centering translation is omitted in the right-hand part of the figure). A "polyhedral" model is shown in Figure 2b. (see color plates). Structural data are given in Data Table 1.

Data Table 1   Cu$_5$Zn$_8$ (1)

Cubic, space group $I\bar{4}3m$, No. 217; a = 8.878 Å; Z = 4, V = 699.8 Å^3

Atomic Positions: $(0,0,0; \frac{1}{2},\frac{1}{2},\frac{1}{2})+$

Zn(1) in 8(c):	$x,x,x; x,\bar{x},\bar{x}; \bar{x},x,\bar{x}; \bar{x},\bar{x},x; x$ = 0.1089
Cu(1) in 8(c):	x = 0.8280
Cu(2) in 12(e):	$\pm(x,0,0; 0,x,0; 0,0,x); x$ = 0.3558
Zn(2) in 24(g):	$x,x,z; z,x,x; x,z,x; \bar{x},x,\bar{z}; \bar{z},x,\bar{x}; \bar{x},z,\bar{x}; x,\bar{x},\bar{z}; z,\bar{x},\bar{x}; x,\bar{z},\bar{x}; \bar{x},\bar{x},z;$
	$\bar{z},\bar{x},x; \bar{x},\bar{z},x; x$ = 0.3128, z = 0.0366

Within experimental error, this model agrees with the experimentally determined structure. The calculated (ideal) unit cell edge is a = $e(1/\sqrt{2} + \sqrt{3} + 1)$, where e is the edge length of tetrahedra and triangles. If θ is the dihedral angle of the regular tetrahedron (70°32′) then, using sin(3θ/2) = 5/(3$\sqrt{3}$) and cos(3θ/2) = $\sqrt{2}$/(3$\sqrt{3}$), it can be readily proved that the various

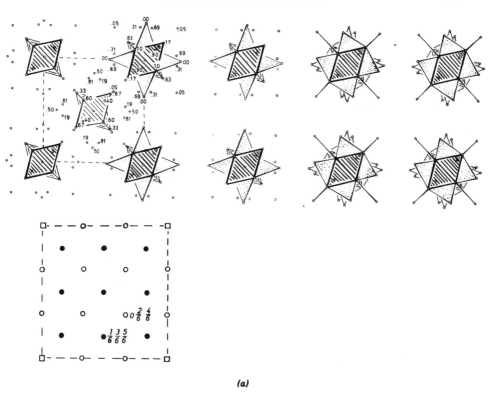

(a)

Figure 2. (*a*) The real structure of γ-brass, Cu₅Zn₈; the stellae quadrangulae on the left are gradually capped toward the right to form the 26-atom clusters of Figure 1. A 3 × 3 × 3 supercell of bcc is also shown; the atomic shifts to form the cluster are obvious. (The positions corresponding to the atoms missing from bcc are at the centers of the stellae quadrangulae, at 0,0,0 and $\frac{1}{2},\frac{1}{2},\frac{1}{2}$.) For (*b*), a polyhedral model of the γ-brass structure, see color plate XIV.2*b*.

TABLE 1   Atomic Parameters for γ-Brass Structures

Wyckoff positions in $I\bar{4}3m^a$	Calculated Coordinates (n_1, n_2)	Observed Coordinates (3)			
		Cu₅Zn₈	Cu₅Cd₈	Fe₃Zn₁₀	Avg
IT, 8(c)	x = 0.103 (3,0)	0.109	0.094	0.103	0.102
OT, 8(c)	x = −0.171 (−5,0)	−0.172	−0.161	−0.167	−0.167
OH, 12(e)	x = 0.355 (3,6)	0.356	0.351	0.354	0.354
CO, 24(g)	x = 0.305 (4,4)	0.313	0.297	0.305	0.305
	z = 0.053 (4,−2)	0.037	0.057	0.049	0.048

ᵃ IT = inner tetrahedron, OT = outer tetrahedron, OH = octahedron, CO = cuboctahedron.

atomic parameters of the structure are given by

$$(x,z) = \frac{n_1\sqrt{2} + n_2\sqrt{3}}{6(\sqrt{2} + 2\sqrt{3} + 2)}$$

Calculated and observed values are given in Table 1, the average differences being (reading down from the top of the table) 0.001, 0.004, 0.001, 0, and 0.005, corresponding to no more than 0.05 Å.

Data Table 2   Mn$_5$Si$_3$ (4)

Hexagonal, space group $P6_3/mcm$, No. 193; $a = 6.910$, $c = 4.814$ Å, $c/a = 0.6967$; $Z = 2$, $V = 199.1$ Å^3

Atomic Positions
Mn(1) in 4(d): $\pm(\frac{1}{3},\frac{2}{3},0; \frac{1}{3},\frac{2}{3},\frac{1}{2})$
Mn(2) in 6(g): $\pm(x,0,\frac{1}{4}; 0,x,\frac{1}{4}; x,x,\frac{3}{4})$, $x = 0.2358(6)$
Si in 6(g): $x = 0.5992(15)$

As the triangular (edge) capping of tetrahedra is used to generate the γ-brass structure, so the edge capping of octahedra can be used to generate the hexagonal structure of Mn$_5$Si$_3$, shown in projection in Figure 3a. (Structural data are given in Data Table 2.) Columns of face-sharing Mn(2)$_6$ octahedra (as in hcp) occur at [00z]. All their "sloping" edges [those *not* parallel to (0001)] are capped by Mn(1) atoms at $\pm[\frac{1}{3}, \frac{2}{3}, z]$, each of which is common to three columns, and all those edges parallel to (0001) are capped by Si atoms (which are not shared between columns). If all the octahedra and capping triangles are perfectly regular, then the atomic parameters and the value of c/a are readily calculated. The observed and calculated values are listed in Table 2. The biggest discrepancy is for Si. This clearly arises from the fact that Si is smaller than Mn, so that the Mn(2)$_2$Si caps are more likely to be isosceles than equilateral; cf. Figure 3a. Figures 3b,c (see color plates) show a "polyhedral" model of the structure.

It is possible that this structure is stabilized by trace amounts of carbon [presumably in the Mn(2)$_6$ octahedra].

If Mn(2) is replaced by Fe, Si by Te, and the Mn(1) at 0 and $\frac{1}{2}$ by single

TABLE 2   Calculated and Observed (4) Parameters for Mn$_5$Si$_3$

	Observed	Calculated
Mn(2)	$x_1 = 0.236$	$2/[3(1 + \sqrt{3})] = 0.244$
Si	$x_2 = 0.599$	$1 - 4/[3(1 + \sqrt{3})] = 0.512$
	$c/a = 0.697$	$4\sqrt{2}/[3(1 + \sqrt{3})] = 0.690$

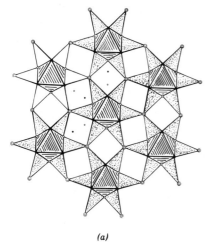

Figure 3. (*a*) The structure of Mn_5Si_3 projected on (0001). Columns of face-sharing $Mn(2)_6$ octahedra at the origin have their basal plane edges capped with Si atoms (some of which are shown as small circles in the empty areas) and their other edges capped by Mn(1) atoms (double circles, the caps being the dotted triangles). Single circles are at $\pm\frac{1}{4}$ (open and filled); double circles are at 0 and $\frac{1}{2}$. For (*b,c*), a polyhedral model of part of the Mn_5Si_3 structure, see color plates XIV.3*b,c*.

(a)

atoms of Tl at $\pm(\frac{1}{3}, \frac{2}{3}, \frac{3}{4})$, we have the structure of $TlFe_3Te_3$ and space group $P6_3/m$.

Data Table 3   W₅Si₃ (5)

Tetragonal, space group $I4/mcm$, No. 140; $a = 9.645$, $c = 4.969$ Å, $c/a = 0.515$; $Z = 4$, $V = 462.3$ Å

Atomic Positions: $(0,0,0; \frac{1}{2},\frac{1}{2},\frac{1}{2})+$
Si(1) in 4(*a*): $\pm(0,0,\frac{1}{4})$
W(1) in 4(*b*): $(0,\frac{1}{2},\frac{1}{4}; \frac{1}{2},0,\frac{1}{4})$
Si(2) in 8(*h*): $\pm(x,\frac{1}{2}+x,0; \frac{1}{2}+x,\bar{x},0)$; $x = 0.17$
W(2) in 16(*k*): $\pm(x,y,0; \bar{y},x,0; x,\bar{y},\frac{1}{2}; y,x,\frac{1}{2})$; $x = 0.074$, $y = 0.223$

The structure of W_5Si_3 (Figure 4 and Data Table 3) is constructed in a manner similar to that of Mn_5Si_3, but the [00z] columns are of square anti-

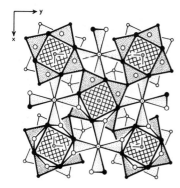

Figure 4. The W_5Si_3 structure projected on (001) of its tetragonal unit cell. Large circles are W; small circles are Si; open at $z = 0$, filled at $z = \frac{1}{2}$, dotted at $z = \pm\frac{1}{4}$. Si-centered W_8 square antiprisms are cross-hatched; triangular caps (on the edges of the square antiprisms) are dotted. (In the lower half the latter are cut away to expose the antiprisms.)

prisms [W(2)$_8$] instead of triangular antiprisms (i.e., octahedra). There are similar columns at [$\frac{1}{2}$, $\frac{1}{2}$, z] also. Again "sloping" edges are capped with the same sort of metal atom [W(1)], each cap now being common to four columns, and the edges parallel to (001) are capped by Si(2) atoms, which are now common to two adjacent columns. In Mn$_5$Si$_3$ the octahedra are empty (but see the note above); in W$_5$Si$_3$ the square antiprisms are all occupied by Si(1) atoms. (Note that the square antiprisms are not quite regular, in the sense that their top and bottom faces are rotated with respect to each other by ~35° instead of 45°; cf. Chapter XI.)

The Cr$_3$Si (A15) Type and Some Related "Tetrahedrally Close Packed" Structures

The Cr$_3$Si structure, for which data are given in Data Table 4, is an important structure type, sometimes known as the β-W structure. It is the structure of some superconductors with relatively high critical temperatures,* for example, Nb$_3$Ge. Shown in Figure 5, it too is readily derived from bcc (cf. the left-hand side of this figure). The derivation involves a contraction of 30% in the direction of the projection axis. (Another precise, though formal, description of this structure is as bcc Si, with half its tetrahedral sites occupied by Cr. The Cr atoms are at alternate vertices of a perfectly regular truncated octahedron and define a slightly irregular icosahedron, SiCr$_{12}$).

Data Table 4   Cr$_3$Si (6)

Cubic, space group $Pm3n$, No. 223; $a = 4.555(3)$ Å; $Z = 2$, $V = 94.5$ Å^3

Atomic Positions
Si in 2(a):   0,0,0; $\frac{1}{2},\frac{1}{2},\frac{1}{2}$
Cr in 6(c): $\pm(\frac{1}{4},0,\frac{1}{2}; \frac{1}{2},\frac{1}{4},0; 0,\frac{1}{2},\frac{1}{4})$

A characteristic and important feature of the structure is the short Cr–Cr distance in the rows parallel to the directions of the three axes of the cubic unit cell. This is emphasized when the structure is described as a rod packing (7), as in Figure 6. The conversion to bcc is simple: Two of the three rods simply translate by $a\langle100\rangle/4$.

The line-shaded figures in Figure 5 are tetraedersterns (8). They are topologically identical to stellae quadrangulae; elevations of the two are compared in Figure 7. To obtain a tetraederstern, the central tetrahedron of a stella quadrangula is elongated in the direction of its $\bar{4}$ axis (the projection axis of Figure 5) so that the dihedral angle between the faces of the capping

---

* But compare the addendum on the "new" high-temperature superconductors in Chapter II, p. 31.

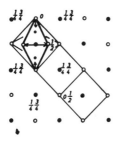 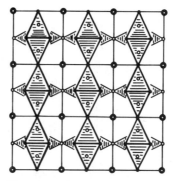

Figure 5. The "β-W" structure of Cr_3Si projected on (001) with, at the left, the $\sqrt{2} \times \sqrt{2} \times 2$ bcc array from which it may be formally derived. On the right the structure is shown as tetraedersterns. Inner tetrahedra of Cr are capped by Si atoms (common to four tetraedersterns)—all open at $z = 0$, filled at $z = \frac{1}{2}$. The tunnels between them are occupied by additional Cr atoms (double circles) at $z = \pm\frac{1}{4}$.

tetrahedra (at the top and bottom), 148°24′ in the stella quadrangula, becomes 180° in the tetraederstern; that is, the pairs of faces concerned become coplanar [and parallel to (001) in Figure 5]. We shall see that the tetraederstern is a very useful polyhedron for describing a considerable number of complex (especially alloy) structures. It also helps in relating these to more straightforward structures, such as the $CuAl_2 = Fe_2B$ type (see below). It is not too difficult to see that the Cr atoms in the rods parallel to the projection axis cap four of the edges of the central (Cr_4) tetrahedron of each tetraederstern (cf. Figure 8).

The Zr_4Al_3 structure is shown in Figure 9, projected on ($11\bar{2}0$). The resemblance to Figure 8 is obvious; the difference is that in the horizontal direction, [$\bar{1}100$], the tetraedersterns share edges instead of (in Cr_3Si) corners. Clearly then, Zr_4Al_3 is a (one-dimensional) CS derivative of Cr_3Si. (Note

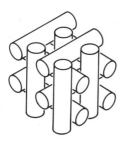

Figure 6. Rod packing as in Cr_3Si. The rods are of strongly bonded Cr atoms. Si atoms are in the interstices between the rods.

Figure 7. Elevations of stella quadrangula (left) and tetraederstern (right).

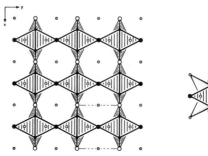

Figure 8. The Cr_3Si structure projected on (001) as in Figure 5, showing, on the right, how the Cr atoms in the [001] rods cap four of the edges of the central tetrahedron in each tetraederstern.

that, again, the "rod" atoms cap the edges of the inner tetrahedron of each tetraederstern.)

Structural data for Zr_4Al_3 are given in Data Table 5.

Data Table 5 Zr_4Al_3 (9)

Hexagonal, space group $P\bar{6}$, No. 174; $a = 5.433$, $c = 5.390$ Å; $c/a = 0.992$; $Z = 1$, $V = 159.1$ Å^3

Atomic Positions
Zr(1) in 1(b): $0,0,\frac{1}{2}$
Zr(2) in 1(f): $\frac{2}{3},\frac{1}{3},\frac{1}{2}$
Zr(3) in 2(h): $(\frac{1}{3},\frac{2}{3},z; \frac{1}{3},\frac{2}{3},\bar{z})$; $z = \frac{1}{4}$
Al in 3(j): $x,y,0; \bar{y},x - y,0; y - x,\bar{x},0$; $x = \frac{1}{3}$, $y = \frac{1}{6}$

Figure 10 shows how Zr_4Al_3 can be derived from an incomplete bcc array (with 14/16 of the bcc sites occupied).

Already considered from a different point of view earlier (Chapter XI Figure 26b), Figure 11 depicts $CuAl_2$ appropriately for the present context. Above is shown the Al array of $CuAl_2$ (or the Fe array of the isostructural Fe_2B, etc.) as tetraedersterns. Below is shown a corresponding element of

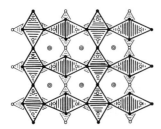

Figure 9. The Zr_4Al_3 structure projected on ($11\bar{2}0$) of its hexagonal unit cell. (The **c** axis is vertical, and [$\bar{1}100$] is horizontal.) Tetraederstern atoms are on 0 or $\frac{1}{2}$, and the rods parallel to the projection axis consist of atoms at $\pm\frac{1}{4}$. The latter and the common *corners* between tetraedersterns are Al; the remaining atoms (forming all the inner tetrahedra of the tetraederstern) are Zr.

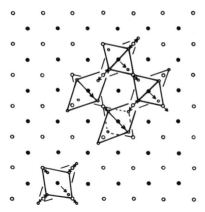

Figure 10. The derivation of the Zr_4Al_3 type from a deficit bcc array (with one-eighth of its atoms missing).

Zr_4Al_3. Clearly the former derives from the latter by another CS operation (with CS planes orthogonal to those in $Cr_3Si \rightarrow Zr_4Al_3$), so that the $CuAl_2$ type can be described as a double (two-dimensional) CS derivative of Cr_3Si—a column structure. Both Zr_4Al_3 and $CuAl_2$ are the end members of, respectively, one- and two-dimensional CS on Cr_3Si; with the CS planes as close together as possible. (The Cu atoms in $CuAl_2$, or B atoms in Fe_2B, lie at $\pm\frac{1}{4}$, thus forming [001] rods in the tunnels between the tetraedersterns, exactly as seen in the previous two structures. Again, these atoms cap the

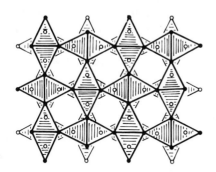

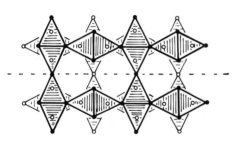

Figure 11. Above, the Al array of $CuAl_2$ (= the Fe array of Fe_2B) is projected along [001] of the tetragonal unit cell, and the atoms are connected to emphasize Al_8 tetraedersterns. (The Cu or B atoms form rods parallel to the projection axis.) Below, an element of the tetraederstern array of Zr_4Al_3 is shown. The broken line indicates the position of the CS plane for the transformation $Zr_4Al_3 \rightarrow CuAl_2$.

inner tetrahedra of the tetraedersterns.) Structural data for CuAl$_2$ are given in Data Table 6.

Data Table 6   CuAl$_2$ (10)

Tetragonal, space group $I4/mcm$, No. 140; $a = 6.063$, $c = 4.872$ Å, $c/a = 0.804$; $Z = 4$, $V = 179.1$ Å^3

Atomic Positions: $(0,0,0; \frac{1}{2},\frac{1}{2},\frac{1}{2})$ +
Cu in 4(a): $\pm(0,0,\frac{1}{4})$
Al in 8(h): $\pm(x,\frac{1}{2} + x,0; \frac{1}{2} + x,\bar{x},0)$; $x = 0.1541$

Reference 10 lists 47 isostructural compounds.

Figure 12 shows how the **CuAl$_2$** structure can be derived from an incomplete bcc (strictly, **CsCl**) array of atoms, this time with 12/16 of the bcc sites occupied. (Compare Figures 10 and 5.) This deficit bcc array is exactly the PtHg$_2$ structure type (Chapter VIII, Figure 6) from which the CuAl$_2$ type is derived by collapsing its empty cubes to tetraedersterns—the PtHg$_8$ cubes simultaneously (and necessarily) transforming to square antiprisms.

The crystallographic shear relationships described above suggest the possibility of families of both one- and two-dimensional derivatives of **Cr$_3$Si**. We consider now a 2 × 2 column (double CS) derivative analogous to the *octahedral* example of AlNbO$_4$ in Chapter II—that of "Mo$_3$CoSi," in which both sets of CS planes are twice as far apart as in CuAl$_2$. This structure is shown in Figure 13. Crystallographic data are in Data Table 7.

In Cr$_3$Si the atoms in the [001] rods lie in (bicapped) hexagonal antiprisms (CN = 12 + 2 = 14); in Zr$_4$Al$_3$ the corresponding atoms, in [11$\bar{2}$0] rods, are in pentagonal antiprisms (CN = 10 + 2 = 12); and in CuAl$_2$ they are in square antiprisms (CN = 8 + 2 = 10). In "Mo$_3$CoSi" they lie in all three types of antiprism—hexagonal, pentagonal, and square (and in all cases they cap the edges of the inner tetrahedra of the tetraedersterns).

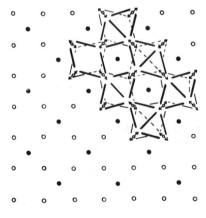

Figure 12.   The derivation of the CuAl$_2$ type from a deficit bcc array (with one-fourth of its atoms missing).

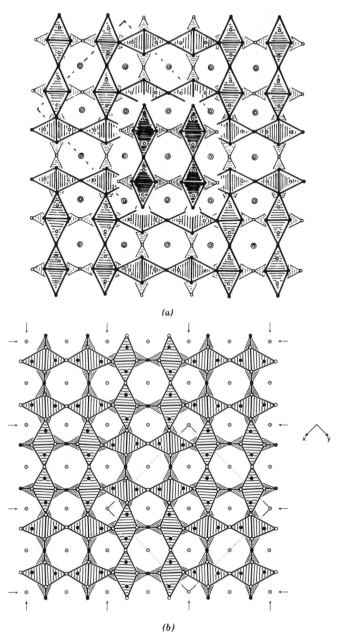

(a)

(b)

Figure 13. (*a*) Formal derivation of Mo₃CoSi from Cr₃Si by double CS. Note the 2×2 columns of Cr₃Si joined at the CS planes as in the CuAl₂ type (Figure 11). (*b*) The exact structure of Mo₃CoSi projected on (001). There is mixed occupancy of most sites, but this is not shown. Open circles are atoms at $z \approx \frac{1}{2}$; filled circles are atoms at $z \approx 0$; dotted circles are atoms at $z = \pm\frac{1}{4}$. Light arrows at the periphery indicate the CS planes by which **Cr₃Si** is transformed to **Mo₃CoSi**.

Data Table 7 "Mo₃CoSi" (11)

Tetragonal, space group $I\bar{4}c2$, No. 120; $a = 12.649$, $c = 4.889$ Å, $c/a = 0.3865$; $Z = 56$ atoms in the unit cell, $V = 782.2$ Å³

Atomic Positions: $(0,0,0; \frac{1}{2},\frac{1}{2},\frac{1}{2})+$

(Si$_{2.9}$Co$_{1.1}$) in 4(a):	$\pm(0,0,\frac{1}{4})$
(Mo$_{2.8}$Co$_{1.2}$) in 4(c):	$\pm(0,\frac{1}{2},\frac{1}{4})$
(Si$_{5.2}$Co$_{2.8}$) in 8(e):	$x,x,\frac{1}{4}; \bar{x},\bar{x},\frac{1}{4}; x,\bar{x},\frac{3}{4}; \bar{x},x,\frac{3}{4}; x = \frac{1}{4}$
(Co$_{4.8}$Si$_{3.2}$) in 8(h):	$\pm(x,\frac{1}{2} + x,0; x,\frac{1}{2} - x,\frac{1}{2}); x = 0.138$
(Mo$_{12.8}$Co$_{3.2}$) in 16(i):	$(x,y,z; \bar{x},\bar{y},z; \bar{x},y,\frac{1}{2} + z; x,\bar{y},\frac{1}{2} + z; \bar{y},x,\bar{z}; y,\bar{x},\bar{z}; y,x,\frac{1}{2} - z;$
	$\bar{y},\bar{x},\frac{1}{2} - z); x = 0.067, y = 0.294, z = 0.013$
Mo in 16(i):	$x = 0.155, y = 0.085, z = 0.013$

Stoichiometric atomic sums $= $ Mo$_{31.6}$Co$_{13.1}$Si$_{11.3}$ $\equiv$ 13.1 Mo$_{2.41}$Co$_{1.00}$Si$_{0.86}$ or 11.3 Mo$_{2.80}$Co$_{1.16}$ Si$_{1.00}$.

The last three structures—Zr₄Al₃, CuAl₂, and "Mo₃CoSi"—were derived from the Cr₃Si type by CS, which is a translation operation. The application of other operations, such as cyclic twinning and rotation, to Cr₃Si leads to other structure types.

For example, Figure 14 shows how the W₅Si₃ structure (on the right) is derived by rotating columns in Cr₃Si (on the left) by ±45° and vice versa. The operation doubles the a and b axes of the unit cell so that, ideally, c/a should change from 1.000 for the cubic cell of Cr₃Si to 0.500 for the tetrago-

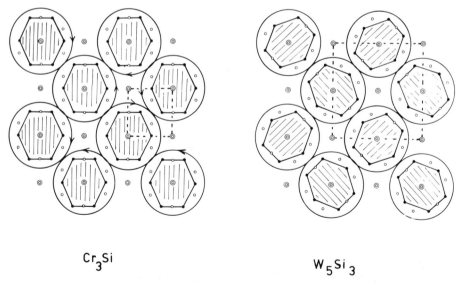

Cr₃Si   W₅Si₃

Figure 14. The transformation of Cr₃Si (left) to W₅Si₃ (right) by rotation of columns of structure.

nal cell of W_5Si_3. In fact, the latter has $c/a = 0.515$—a small distortion. (Crystallographic data for W_5Si_3 were given earlier, in Data Table 3.)

The shaded hexagons on the left of Figure 14 are (Cr_4Si_2) faces of half the hexagonal antiprisms in Cr_3Si, readily recognizable in Figures 5 and 8; those on the right can be delineated in the W_5Si_3 structure in Figure 4 (although perhaps less readily; they are centered on the midpoints of the unit-cell edges). *Within* the rotated cylinders the same atomic sites are occupied by metal (Cr or W) and Si in both structures; their stoichiometries are therefore unchanged. On the other hand, the atomic rods *between* these cylinders are metal (Cr) in Cr_3Si but Si in W_5Si_3. Therefore, in terms of the contents of the larger unit cell (that of $W_5Si_3 \equiv 2 \times 2 \times 1$ times that of Cr_3Si), the composition changes from $(Cr_{10}Si_4)_2[Cr_2]_2$ to $(W_{10}Si_4)_2[Si_2]_2$, that is, 8 $Cr_3Si \rightarrow 4$ W_5Si_3.

The rotations result in a twin relation between adjacent columns. Figure 15 shows the formation of a W_5Si_3 element by cyclic twinning. The latter is, of course, the same operation as in Figure 14; but there it has been regularly repeated. In Figure 16 the same cyclic twinning operation with a parallel *but different* composition plane generates a structure close to that of the Fe_3P or Ni_3P type, shown in Figure 17. (Compare Figures 16 and 18.)

Data Table 8   Ni₃P (12)

Tetragonal, space group $I\bar{4}$, No. 82; $a = 8.954$, $c = 4.386$ Å, $c/a = 0.4898$; $Z = 8$, $V = 351.6$ Å^3

Atomic Positions
All atoms in 8(*g*): $(0,0,0; \frac{1}{2},\frac{1}{2},\frac{1}{2}) + (x,y,z; \bar{x},\bar{y},z; y,\bar{x},\bar{z}; \bar{y},x,\bar{z})$, with

	x	y	z
Ni(1)	0.0775(3)	0.1117(3)	0.2391(15)
Ni(2)	0.3649(3)	0.0321(3)	0.9765(15)
Ni(3)	0.1689(3)	0.2200(3)	0.7524(15)
P	0.2862(5)	0.0487(5)	0.4807(28)

The Ni_3P or Fe_3P structure (Data Table 8) is attained by a number of compounds, especially transition metal pnictides (phosphides, arsenides, and antimonides). The best determined structure appears to be that of the mineral rhabdite, Fe_2NiP, from a meteoritic source (13) (with $R = 4.2\%$).

The main difference between the real Ni_3P structure (Figure 17) and the one derived from the Cr_3Si type in Figure 16 lies in the regions between the Ni_4P_4 tetraederstern columns. In Figure 16 these are occupied by square columns of atoms, approximately cubes sharing faces in the projection direction, whereas in Figure 17 these Ni_8 "cubes" are seen to be collapsed into tetraedersterns, centered at 0,0,0, and $\frac{1}{2},\frac{1}{2},\frac{1}{2}$.

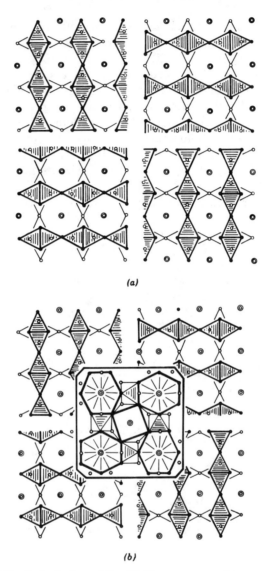

(a)

(b)

Figure 15. (*a*) Cyclic twinning of the Cr$_3$Si structure to form a fourling. (*b*) The atomic positions are identical to those in (*a*), but the central region has been redrawn to emphasize that it is a column of W$_5$Si$_3$ type. (Compare with Figure 4.) Note particularly the formation of a square antiprism at the center of the drawing, as well as the "octahedra."

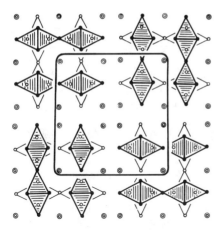

Figure 16. Cyclic twinning of Cr_3Si as in Figure 15, but with (parallel) composition planes at slightly different positions, so that they have a different structure, close to that of Ni_3P, etc. (see below).

It is therefore interesting that the structure of $Ni_{12}P_5$, shown in Figure 18, is closely related to that of Ni_3P ($= \frac{1}{4} \times Ni_{12}P_4$) shown in Figure 17 and to the twinned element at the center of Figure 16. The extra P atom is inserted into the centers of the Ni_8 tetraedersterns in the former, restoring the tetraedersterns in Ni_3P to cubes, as is clear from a comparison of Figures 17 and 18, $(Ni_8)_{te}Ni_4P_4 \rightarrow (PNi_8)_{cube}Ni_4P_4 = Ni_{12}P_5$. The result is a small change in symmetry and a slight change in the unit cell parameters a, c, and V, and in c/a (cf. Data Tables 8 and 9).

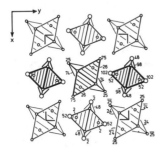

Figure 17. The structure of Ni_3P projected on (001). Small circles are Ni; large circles are P atoms; heights are in units of $c/100$. Structure is drawn as Ni_{4+4} and Ni_4P_4 stellae quadrangulae. (Note how the atoms in one tetraederstern cap the edges of adjacent tetraedersterns.)

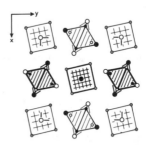

Figure 18. The structure of $Ni_{12}P_5$ projected on (001). Small circles are Ni; large circles are P atoms: open at $z = 0$, filled at $z = \frac{1}{2}$, dotted at $z = \pm\frac{1}{4}$. The structure is drawn as Ni_4P_4 tetraedersterns and P-centered Ni_8 "cubes." (Note how the atoms in the TEs cap the faces and edges of the cubes and the cube atoms cap the edges of the TEs.) Compare Figures 16 and 17.

Data Table 9   Ni$_{12}$P$_5$ (14)

Tetragonal, space group $I4/m$, No. 87; a = 8.646, c = 5.070 Å, c/a = 0.5864; Z = 2, V = 379.0 Å^3

Atomic Positions: $(0,0,0; \frac{1}{2},\frac{1}{2},\frac{1}{2})+$

P(1) in 2(a):   0,0,0
P(2) in 8(h): $\pm(x,y,0; \bar{y},x,0)$; x = 0.1939, y = 0.4132
Ni(1) in 8(h): x = 0.3655, y = 0.0609
Ni(2) in 16(i): $\pm(x,y,z; x,y,\bar{z}; \bar{y},x,z; y,\bar{x},z)$; x = 0.1166, y = 0.1812, z = 0.2490

Figure 19a shows how the structure of Ni$_{12}$P$_5$ can be derived more accurately and more directly from that of Cr$_3$Si by another form of cyclic twinning: n-glide rather than the simple c-glide of Figure 16. Figure 19b shows another possibility, in which the Ni$_{12}$P$_5$ element is generated by a cyclic intergrowth of **Cr$_3$Si** and **CsCl**.

We now return to the Ni$_3$P structure (Figure 17). In Figure 20a it is shown slightly idealized so that the {110} planes are mirrors, and the symmetry is increased from $I\bar{4}$ to $I\bar{4}2m$ (by slightly rotating all the tetraederstern columns about c). This higher symmetry structure is that reported for α-V$_3$S (a high-temperature polymorph, stable above 950°C). In this figure only the M$_8$ tetraedersterns are emphasized. The M$_4$X$_4$ tetraedersterns that lie between them are not drawn in; only their atomic positions are shown. Note, however, that the latter tetraedersterns point in opposite directions in alternate columns (as do those of M$_8$).

If half of these M$_4$X$_4$ tetraederstern columns (those within the circles) are rotated by 90°, so that they all point in the same direction (at the same level), the I-centering is destroyed, and the resulting structure is that reported for β-V$_3$S (the low-temperature polymorph, stable below 825°C). This structure is shown in Figure 20b. (A translation of these columns by $c/2$ is equivalent to their rotation by 90°.) The space group is now $P4_2/nbc$, and a slight rotation of all the columns about c (analogous to that which transforms α-V$_3$S to Ni$_3$P) yields the lower-symmetry ($P4_2/n$) structure of Ti$_3$P (not drawn).

The higher-symmetry α-V$_3$S and β-V$_3$S and the lower-symmetry Ni$_3$P and Ti$_3$P are therefore all rather similar.

Figure 21 again shows the β-V$_3$S (or idealized Ti$_3$P) structure. If the M$_4$X$_4$ tetraederstern columns lying in alternate sheets parallel to {110} (as indicated in Figure 21 by some circles) instead of sheets parallel to {100} (Figure 20a) are rotated by 90° (or translated by $c/2$), one gets a new and different structure. It has a doubled unit cell volume ($\sqrt{2} \times \sqrt{2} \times 1$ times that of the previous group of structures), and, if of the lower-symmetry type without mirror planes, it is the monoclinic structure of Ta$_3$As. (The monoclinic distortion is slight, γ = 90.57°.)

Clearly, in principle, a whole range of structures is possible, depending on the relative heights of similarly oriented M$_4$X$_4$ tetraedersterns in their

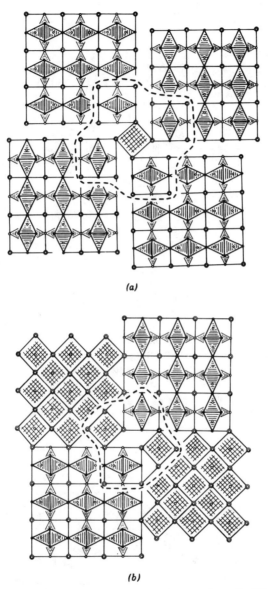

(a)

(b)

Figure 19. (a) The derivation of an $Ni_{12}P_5$-type element from Cr_3Si by n-glide reflection twinning. (b) The derivation of an $Ni_{12}P_5$ type element from cyclic intergrowth of the **Cr₃Si** and **CsCl** types.

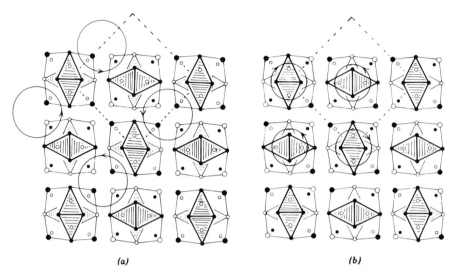

Figure 20. (*a*) The idealized Ni_3P structure (= α-V_3S) projected on (001) but emphasizing the M_8 tetraedersterns (shaded) (cf. Figure 17). These TE atoms (medium-sized circles) are at heights $\frac{1}{4}$ (open) and $\frac{3}{4}$ (filled). The other atoms are M (large) and X (small) at 0, $\frac{1}{2}$ (open, filled). Rotation of alternate columns (circled) by 90° or (equivalently) translation by $c/2$ interchanges these open and filled circles and gives (*b*) the idealized Ti_3P (= β-V_3S) structure. [Atom key as for (*a*).] Rotation of the *tetrahedral* columns in (*b*) by 45° regenerates the Cr_3Si type (if the M_8 capping atoms—the atoms at the centers of the edges of the outlined "blocks"—coalesce in projection to give rectilinear rows).

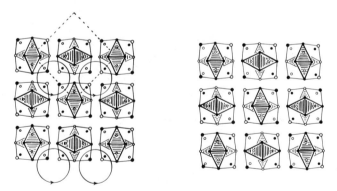

Figure 21. The β-V_3S (= idealized Ti_3P) type as in Figure 20*b*. Rotation of *all* the (undrawn) TE columns in *alternate* horizontal rows (see circles) produce the Ta_3As type, with a unit cell $\sqrt{2} \times \sqrt{2} \times 1$ times that of β-V_3S type (outlined). As in Figure 20, the large and small circles are M and X, respectively, open at 0, filled at $\frac{1}{2}$, and the medium circles (defining the TEs) are M at $\pm\frac{1}{4}$.

columns (or relative orientations at the same height). If the two possibilities are denoted by $+$ and $-$, then, for the doubled Ta_3As-like unit cell we have

$$Ti_3P/\beta\text{-}V_3S = \begin{smallmatrix} + & + \\ + & + \end{smallmatrix} = \begin{smallmatrix} - & - \\ - & - \end{smallmatrix}, \qquad Ni_3P/\alpha\text{-}V_3S = \begin{smallmatrix} + & - \\ - & + \end{smallmatrix} = \begin{smallmatrix} - & + \\ + & - \end{smallmatrix},$$

$$Ta_3As = \begin{smallmatrix} + & - \\ + & - \end{smallmatrix} = \begin{smallmatrix} + & + \\ - & - \end{smallmatrix}, \text{ etc.}$$

The crystallographic data for these structures are given in Data Tables 10–13.

Data Table 10 α-V_3S (15)

Tetragonal, space group $I\bar{4}2m$, No. 121; $a = 9.470$, $c = 4.589$ Å, $c/a = 0.4846$; $Z = 8$, $V = 411.5$ Å^3

Atomic Positions (origin at $\bar{4}2m$): $(0,0,0; \frac{1}{2},\frac{1}{2},\frac{1}{2})+$
V(1) in 8(i): $x,x,z; \bar{x},\bar{x},z; x,\bar{x},\bar{z}; \bar{x},x,\bar{z}; x = 0.0932, z = 0.750$
V(2) in 8(i): $x = 0.2000, z = 0.250$
V(3) in 8(f): $\pm(x,0,0; 0,x,0); x = 0.3550$
S in 8(g): $\pm(x,0,\frac{1}{2}; 0,x,\frac{1}{2}); x = 0.2851$
Atomic Distances
V–V = 2.50–3.00 Å
V–S = 2.32–2.46 Å

Data Table 11 β-V_3S* (15)

Tetragonal, space group $P4_2/nbc$, No. 133; $a = 9.381$, $c = 4.663$ Å, $c/a = 0.4971$; $Z = 8$, $V = 410.4$ Å^3

Atomic Positions (origin at $\bar{4}$, first choice):
V(1) in 8(j): $x,\frac{1}{2} + x,0; \bar{x},\frac{1}{2} - x,0; \bar{x},\frac{1}{2} + x,\frac{1}{2}; x,\frac{1}{2} - x,\frac{1}{2}; \frac{1}{2} - x,\bar{x},\frac{1}{2}; \frac{1}{2} + x,x,\frac{1}{2};$
                   $\frac{1}{2} + x,\bar{x},0; \frac{1}{2} - x,x,0; x = 0.4080$
V(2) in 8(j): $x = 0.2028$
V(3) in 8(i): $x,0,\frac{3}{4}; \bar{x},0,\frac{3}{4}; \frac{1}{2},\frac{1}{2} + x,\frac{1}{4}; \frac{1}{2},\frac{1}{2} - x,\frac{1}{4}; \frac{1}{2} - x,\frac{1}{2},\frac{3}{4}; \frac{1}{2} + x,\frac{1}{2},\frac{3}{4}; 0,\bar{x},\frac{1}{4}; 0,x,\frac{1}{4};$
                   $x = 0.1486$
S in 8(h): $x,0,\frac{1}{4}; \bar{x},0,\frac{1}{4}; \frac{1}{2},\frac{1}{2} + x,\frac{3}{4}; \frac{1}{2},\frac{1}{2} - x,\frac{3}{4}; \frac{1}{2} - x,\frac{1}{2},\frac{1}{4}; \frac{1}{2} + x,\frac{1}{2},\frac{1}{4}; 0,\bar{x},\frac{3}{4}; 0,x,\frac{3}{4};$
                   $x = 0.2171$
Atomic Distances
V–V = 2.44–3.05 Å
V–S = 2.31–2.47 Å

* No other compounds have been reported to be isostructural with α- and β-V_3S.

Finally, a simple, direct relation between the β-V_3S and Cr_3Si structures can be pointed out. If the *centers* of the M_8 tetraederstern columns of the former (some of which are circled in Figure 20b) are rotated by 45° and simultaneously translated by $\mathbf{c}/4$ so that they are capped by X atoms (shared

Data Table 12　Ti₃P* (16)

Tetragonal, space group $P4_2/n$, No. 86; a = 9.9592, c = 4.9869 Å, c/a = 0.5007; Z = 8, V = 494.6 Å³

Atomic Positions
All atoms in 8(g) (origin at $\bar{1}$; $\frac{1}{4},\frac{1}{4},\frac{1}{4}$ from $\bar{4}$):　$\pm(x,y,z; \frac{1}{2} + x,\frac{1}{2} + y,\bar{z};$
$$\bar{y},\frac{1}{2} + x,\frac{1}{2} + z; \frac{1}{2} + y,\bar{x},\frac{1}{2} + z)$$

	x	y	z
Ti(1)	0.1661	0.6428	0.715
Ti(2)	0.1101	0.2785	0.530
Ti(3)	0.0696	0.5334	0.241
P	0.0440	0.2919	0.035

Atomic Distances
Ti–Ti = 2.712(5)–3.786(4) Å
Ti–P = 2.490(8)–2.627(6) Å

* There are many isostructural transition metal phosphides, arsenides, silicides, and germanides (of Ti, Nb, Ta, Zr, Hf) as well as Y₃Sb and some borides and boride phosphides.

Data Table 13　Ta₃As* (17)

Monoclinic, space group $B2/b$ (nonstandard setting of $C2/c$), No. 15; a = 14.6773(6), b = 14.5505(4), c = 5.0954(2) Å, γ = 90.572(3)°, $\sqrt{2}\, c/a$ = 0.4910, $\sqrt{2}\, c/b$ = 0.4952; Z = 16, V = 2 × 544.1 Å³

Atomic Positions
All atoms in 8(f):　at $(0,0,0; \frac{1}{2},0,\frac{1}{2})$ + $(x,y,z; x,\frac{1}{2} + y,\bar{z})$

	x	y	z
Ta(1)†	0.4087	0.7469	0.7717
Ta(11)	−0.0041	0.8416	0.7459
Ta(2)	0.1715	−0.0667	0.4998
Ta(22)	0.3215	−0.0745	0.0115
Ta(3)	0.2025	0.7577	0.7467
Ta(33)	−0.0060	−0.0484	0.2491
As(1)	0.1553	0.6138	0.0146
As(11)	0.1402	−0.0999	0.0012

Atomic Distances
Average Ta–Ta = 3.10–3.14 Å
Average As–Ta = 2.77 Å

* This compound was previously thought to have the Ti₃P-type structure. Hf₃As is isostructural.
† Atom numbering to correspond with that in Ti₃P (Data Table 12).

with the original M_4X_4 tetraedersterns; the small circles in Figure 20b), then the Cr_3Si framework is recovered. If the old M_4 caps shift slightly (so that zigzag rows in the **c** direction become rectilinear), the whole structure is Cr_3Si type (Figure 8).

An Alternative Description of Zr₄Al₃, Some Related Structures, and the σ Phase

Figure 22 shows the structure of Zr_4Al_3 projected on the basal plane of its hexagonal unit cell and utilizes a description quite different from that used earlier (Figure 9). The various atoms have coordination numbers of 12 for Al, 14 for Zr(3), and 15 for Zr(1) and Zr(2)—respectively, the not-quite-regular icosahedron characteristic of the A15 structure (Al_4Zr_8) and Al_6Zr_8 and Al_6Zr_9, which are two of the well-known "Frank-Kasper polyhedra" (18).

Data Table 14   CaZn₅ (19)

Hexagonal, space group $P6/mmm$, No. 191; $a = 5.416$, $c = 4.191$, $c/a = 0.774$; $Z = 1$, $V = 106.5$ Å³

Atomic Positions
Ca in 1(a): $0,0,0$
Zn(1) in 2(c): $\pm(\frac{1}{3},\frac{2}{3},0)$
Zn(2) in 3(g): $\frac{1}{2},0,\frac{1}{2}; 0,\frac{1}{2},\frac{1}{2}; \frac{1}{2},\frac{1}{2},\frac{1}{2}$

The (0001) projection of the CaZn₅ structure (Data Table 14) looks exactly like Figure 22. The difference is simply that the double circles in that figure,

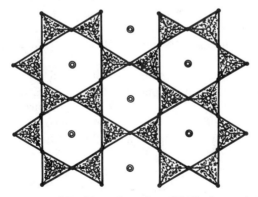

Figure 22.   The structure of Zr_4Al_3 projected on (0001). Open circles are Zr atoms, single at $z = \frac{1}{2}$, double at $z = \pm\frac{1}{4}$; filled circles are Al at $z = 0$. (The shaded triangles are trigonal bipyramids.)

which represent Zr(3) atoms at $z = \pm\frac{1}{4}$ are now Ca atoms at $z = \frac{1}{2}$. The remaining atoms [Zr(1), Zr(2), and Al in Zr_4Al_3] are now all Zn atoms. [Schubert (8) lists almost 60 isostructural AB_5 compounds.]

The (0001) projection of CoSn also looks the same as Figure 22. It has the same space group as $CaZn_5$ but with Co in 3(g) and Sn in 2(c) (both Zn in $CaZn_5$) and Sn in 1(b) at $0,0,\frac{1}{2}$ [instead of Ca in 1(a) at 0,0,0].

There are several ways by which the structures of Zr_4Al_3, $CaZn_5$, and CoSn can be derived from simpler structures. One example is shown in Figure 23, in which the $CaZn_5$ type in (b) is formed from an ordered hcp array in (a) by topological transformation of $3^6 \rightarrow 3 \cdot 6 \cdot 3 \cdot 6$ nets discussed in Chapter XI (and shown in Figure 6 of that chapter)—the "umbrella" distortion again.

Data Table 15 CrO_3 (20)

Orthorhombic, space group $C2cm$, No. 40; $a = 4.789(5)$, $b = 8.557(5)$, $c = 5.743(4)$ Å; $Z = 4$, $V = 235.4$ Å^3

Atomic Positions: $(0,0,0; \frac{1}{2},\frac{1}{2},0)+$
O(1) in 4(a): $x,0,0; x,0,\frac{1}{2}; x = 0.3841(2)$
Cr in 4(b): $x,y,\frac{1}{4}; x,\bar{y},\frac{3}{4}; x = 0.0, y = 0.4023(5)$
O(2) in 4(b): $x = 0.8755(14), y = 0.2323(5)$
O(3) in 4(b): $x = 0.3284(9), y = 0.3922(5)$

The CrO_3 structure, shown in Figure 24, consists of unconnected strings of corner-shared CrO_4 tetrahedra. At first sight it seems an unlikely relative of Zr_4Al_3, but the tetrahedral chains are stacked in the x direction so that the CrO_4 tetrahedra share faces with empty tetrahedra below (and corners with empty tetrahedra above); the Cr atoms are not far from the centers of trigonal bipyramids. Obvious CS on (010) makes these chains of bipyramids corner-connected and (except for the unoccupied sites in the rods of atoms parallel to the projection axis) transforms the CrO_3 structure to the framework of bipyramids in Zr_4Al_3 shown in Figure 22. Structural data are given in Data Table 15.

The anion array in CrO_3 is similar to the metal array in Ni_2In (or, for example, NiTiSi or the high-temperature form of MnCoGe), which was described earlier (Chapter VI, Figure 14). This, in turn, is related to the ω-phase structure, as discussed in Chapter V. These relations again emphasize the absence of any real gap between the geometry of alloy structures and "inorganic" structures.

The "tetrahedrally close-packed" structure of the "sigma phases," such as σ-FeCr, has been discussed by Frank and Kasper (18) in terms of planar nets (see also ref. 22) and Frank–Kasper polyhedra. It is shown in Figure 25 and clearly consists of elements of **Cr_3Si** (Figure 5) and **Zr_4Al_3** (Figure 22). Figure 26 shows a cyclic intergrowth of these two component structure

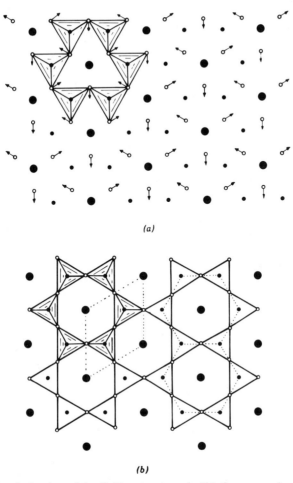

(a)

(b)

Figure 23. The derivation of the CaZn$_5$ structure, in (b), from an ordered hcp array, in (a). Filled circles are at height 0, and open circles at $\frac{1}{2}$. Large circles are Ca, and small circles are Zn. In (b), dotted lines form a honeycomb (6^3) net, and full lines form a kagome (3·6·3·6) net. Some trigonal bipyramids are emphasized in the top left of (a) and (b).

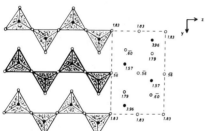

Figure 24. The structure of CrO$_3$ projected on (100). Atomic heights are given in ångströms.

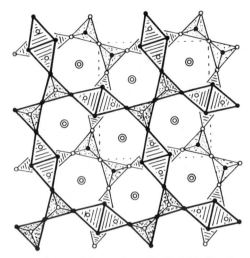

Figure 25. The real structure of the σ-phase FeCr. Filled and open circles represent atoms at 0 and $\frac{1}{2}$, respectively; double circles are atoms at $\pm\frac{1}{4}$.

types, which has an element of the σ-phase structure at its center. Clearly the complete structure of the σ phase can be generated from the two component structures by regular repetition of such an intergrowth. This description is accurate, and we recall that (1) both the component structures can be derived from incomplete bcc arrays (Figures 5 and 10) and (2) both constituents (Fe and Cr) have bcc structures at room temperature. One can imagine

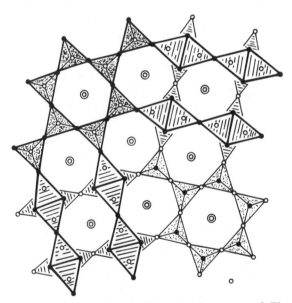

Figure 26. An element of the σ-phase structure (at the center, cf. Figure 25) formed by cyclic intergrowth of **Cr$_3$Si** and **Zr$_4$Al$_3$**.

that, at high temperature, the alloy is a disordered, incomplete bcc packing that orders on cooling and transforms topologically to give the tetrahedra and tetraedersterns, etc. of Figure 25.

Structural data for σ-FeCr are given in Data Table 16.

Data Table 16 σ-FeCr (21)

Tetragonal, space group $P4_2/mnm$, No. 136; $a = 8.7995(4)$, $c = 4.544(2)$ Å, $c/a = 0.5164$; $Z = 30$ atoms, $V = 351.9$ Å^3

Atomic Positions*
$2(a)$: $0,0,0; \frac{1}{2},\frac{1}{2},\frac{1}{2}$
$4(f)$: $\pm(x,x,0; \frac{1}{2} + x,\frac{1}{2} - x,\frac{1}{2}); x = 0.3981(6)$
$8(i)$: $\pm(x,y,0; y,x,0; \frac{1}{2} + x,\frac{1}{2} - y,\frac{1}{2}; \frac{1}{2} + y,\frac{1}{2} - x,\frac{1}{2})$;
        $x = 0.4632(7), y = 0.1316(6)$
$8(i)$: $x = 0.7376(6), y = 0.0653(6)$
$8(j)$: $\pm(x,x,z; x,x,\bar{z}; \frac{1}{2} + x,\frac{1}{2} - x,\frac{1}{2} + z; \frac{1}{2} + x,\frac{1}{2} - x,\frac{1}{2} - z)$;
        $x = 0.1823(6), z = 0.2524(6)$

* Ordering of Fe and Cr (atomic numbers 26 and 24, respectively) is not known.

Other important (tetrahedrally close-packed) alloy structures are related to, and may be readily derived from, that of the σ phase. Twinning and shearing (reflection and translation operations on) the σ-phase structure generates the P phase (23). The same operations on a fourling of the σ phase generates the complex ν-phase structure (24).

Structures of the Friauf-Laves Phases, and More Tetrahedrally Close-Packed Structures

The MgCu$_2$ structure (Data Table 17) is related to ccp in the following way. Ccp copper metal can be regarded as two interpenetrating, C9-like (i.e., "ideal high-cristobalite"-like, see Chapter XV) arrays of corner-connected Cu$_4$ tetrahedra. If one of these arrays is replaced by half as many Mg atoms by substituting Mg at its center for each Cu$_4$ tetrahedron, then a diamondlike Mg array interpenetrates the remaining array of Cu atoms (also a diamond-like array, but of corner-connected Cu$_4$ tetrahedra); and this is the MgCu$_2$ structure. This derivation is shown in Figure 27.*

Figure 28 picks out various polyhedra, including *large* polyhedra, from a ccp array of atoms including, in order of increasing size, a tetrahedron, octahedron, cuboctahedron, and truncated tetrahedra and truncated octahedra. It also shows how these may share faces. [In a complete ccp array,

* Alternatively, it may be related to bcc. The latter can be regarded as two independent inter-penetrating diamond-like arrays, with $a_{\text{diamond}} = 2a_{\text{bcc}}$. If one of these is replaced by Cu$_4$ tetrahedra (with Cu at the midpoints of the bonds so that the tetrahedra share corners) and the other by Mg atoms, we again get the MgCu$_2$ structure.

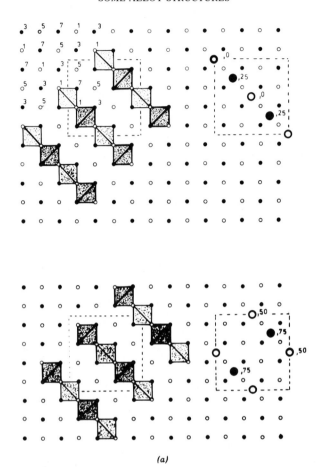

(a)

Figure 27. (*a*) The derivation of the MgCu$_2$ structure from a 2 × 2 × 2 supercell of ccp, shown as two bounded (001) projections. The digits at the top left are Cu atomic heights in units of $c/8$: alternate atoms of ccp are missing. The large circles on the right are Mg atoms, with heights in units of $c/100$ (i.e., 0, $\frac{1}{4}$, $\frac{1}{2}$, $\frac{3}{4}$). (*b*) The same structure represented as MgCu$_{12}$ truncated tetrahedra.

Data Table 17   MgCu$_2$ (25)

Cubic, space group $Fd\bar{3}m$, No. 227; $a = 7.02$ Å; $Z = 8$, $V = 346$ Å^3

Atomic Positions: $(0,0,0;\ 0,\frac{1}{2},\frac{1}{2};\ \frac{1}{2},0,\frac{1}{2};\ \frac{1}{2},\frac{1}{2},0)+$
Mg in 8(*a*): $0,0,0;\ \frac{1}{4},\frac{1}{4},\frac{1}{4}$
Cu in 16(*d*): $\frac{5}{8},\frac{5}{8},\frac{5}{8};\ \frac{5}{8},\frac{7}{8},\frac{7}{8};\ \frac{7}{8},\frac{5}{8},\frac{7}{8};\ \frac{7}{8},\frac{7}{8},\frac{5}{8}$
Atomic Distances
Cu–Cu = $\sqrt{2}\ a/4 = 2.48$ Å (6×)
Mg–Mg = $\sqrt{3}\ a/4 = 3.04$ Å (4×)
Mg–Cu = $\sqrt{11}\ a/8 = 2.91$ Å (12×)

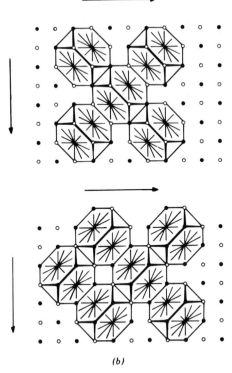

(b)

Figure 27.   (*Continued*)

there is, of course, no atom at the center of a tetrahedron or an octahedron, but there is one atom at the center of a cuboctahedron, a tetrahedron of four atoms at the center of a truncated tetrahedron (cf. the description of $MgCu_2$ above), and a face-centered cube of 14 atoms at the center of a truncated octahedron.]

In the $MgCu_2$ structure, $MgCu_{12}$ truncated tetrahedra share each of their four hexagonal faces with neighboring truncated tetrahedra. This means that each $MgCu_{12}$ coordination polyhedron is capped by four (additional) Mg atoms, so that the CN (Mg) = 12 + 4. The ratio of the distances Mg–Mg to Mg–Cu is $\sqrt{12/11} = 1.044$. The space remaining between the truncated tetrahedra is the corner-connected array of tetrahedra, defined by the Cu atom positions. (The tetra-capped truncated tetrahedron is called a Friauf polyhedron.)

The $MgZn_2$ structure, shown in Figure 29, is derived from an **hc** array of Zn as the $MgCu_2$ type is derived from ccp Cu: half the Zn_4 tetrahedra are replaced by Mg atoms at their centers. This is emphasized in Figure 29*a*, which shows the Mg atoms and the array of corner-connected Zn_5 trigonal bipyramids (or face- and corner-shared tetrahedra). Figure 29*b* shows the

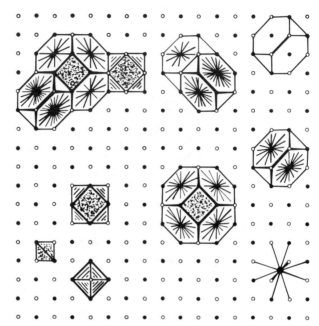

Figure 28. A ccp array, projected down the fcc cube edge, showing the various polyhedra present in the structure: tetrahedron, octahedron, and cuboctahedron (bottom left), truncated tetrahedra (top right), and truncated octahedron (center). At the top center, a pair of truncated tetrahedra (at different levels) share a hexagonal face. At the top left, a truncated octahedron shares a hexagonal face with a truncated tetrahedron and a square face with a cuboctahedron.

$MgZn_{12}$ truncated tetrahedra (also tetracapped by four Mg atoms) and the way they are connected. Structural data are given in Data Table 18.

Data Table 18   MgZn₂ (26, 27)

Hexagonal, space group $P6_3/mmc$, No. 194; $a = 5.221$, $c = 8.567$, $c/a = 1.641$; $Z = 4$, $V = 202.2$ Å³

Atomic Positions
Zn(1) in 2(a): $0,0,0; 0,0,\frac{1}{2}$
Mg in 4(f): $\pm(\frac{1}{3},\frac{2}{3},z; \frac{1}{3},\frac{2}{3},\frac{1}{2} - z); z = 0.0630$
Zn(2) in 6(h): $\pm(x,2x,\frac{1}{4}; 2\bar{x},\bar{x},\frac{1}{4}; x,\bar{x},\frac{1}{4}); x = 0.83053$
Atomic Distances
Zn(1)–Zn(2) = 2.63_4 Å (6×)
Mg–Mg = 3.20_4 Å, 3.20_2 Å (3×)
Mg–Zn(1) = 3.062 Å (3×)
Mg–Zn(2) = 3.064 Å (3×), 3.06_3 Å (6×)

MgZn₂ is a two-layer hexagonal stacking of truncated tetrahedra, and MgCu₂ can be described, in a hexagonal unit cell, as a related three-layer

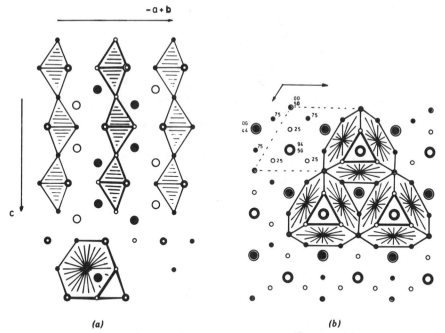

Figure 29. (*a*) The hexagonal structure of MgZn$_2$ projected on (11$\bar{2}$0) and depicted as chains of Zn$_5$ trigonal bipyramids parallel to **c** with intervening Mg atoms. At the bottom, one MgZn$_{12}$ truncated tetrahedron is drawn. Double circles are Zn at $\pm\frac{1}{4}$; open and filled small circles are Zn at 0, $\frac{1}{2}$, respectively. Large circles are Mg at 0 (open) and $\frac{1}{2}$ (filled). (*b*) MgZn$_2$ projected on (0001) showing the MgZn$_{12}$ truncated tetrahedron. Large double circles are Mg; small double circles are two superimposed Zn atoms. Atomic heights are given in units of $c/100$.

stacking. MgNi$_2$ is the next simplest stacking variant, a four-layer type based on **hc**^{3}Ni. Komura *et al.* have reported many, more complex variants (see, e.g., ref. 27). These correspond to, respectively, a hexagonal diamond (wurtzite)like array of Mg atoms (MgZn$_2$), a cubic diamond (zinc blende)like array (MgCu$_2$), an hc array, and various other polytypes, exactly as in the ZnS and SiC systems. The three simplest are shown as polyhedral models in Figure 30 (see color plates). These encompass many AB$_2$ intermetallic compounds; over 200 are reported with the MgCu$_2$-type structure. Here, however, we will concentrate only on the first two. (Note: there are also compounds Mg$_2$Cu and Mg$_2$Ni that have quite different structures from MgCu$_2$ and MgNi$_2$).

For our present purposes, MgCu$_2$ and MgZn$_2$ are best described and related by different projections: along [110]$_{cub}$ and [11$\bar{2}$0]$_{hex}$, respectively, as shown in Figures 31*a* and *b*, which emphasize Mg$_3$Cu$_2$ (Mg$_3$Zn$_2$) trigonal bipyramids composed of pairs of Mg$_2$Cu$_2$ (Mg$_2$Zn$_2$) tetrahedra sharing a face. [These A$_3$B$_2$ bipyramids (A = Mg, B = Cu, Zn) are, of course, not quite

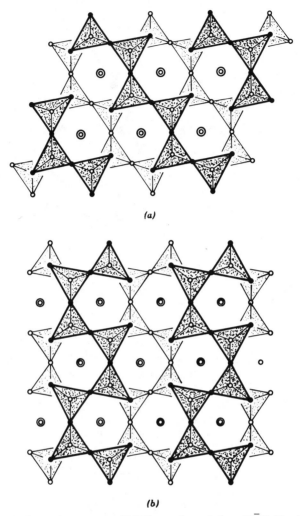

Figure 31. (*a*) The real structure of MgCu$_2$ projected along [1$\bar{1}$0]. Double circles are superimposed Cu atoms at $\pm\frac{1}{4}$; the remaining Cu atoms are at 0 or $\frac{1}{2}$ (open and filled circles), where trigonal bipyramids (shaded) share *corners*. The Mg atoms, also at 0, $\frac{1}{2}$, are where the trigonal bipyramids share *edges*. (*b*) The real structure of MgZn$_2$ projected on (11$\bar{2}$0). The atomic ordering and heights are as in (*a*).

regular: the equatorial plane is AB$_2$, and the apical atoms are both A, so that they each have one edge B–B, six edges A–B, and two edges A–A. In MgCu$_2$ their lengths are 2.48, 2.91 and 3.04 Å, respectively (in a ratio of $\sqrt{2/3}:1:\sqrt{12/11} = 0.853:1:1.044$). In MgZn$_2$ they are 2.63, 3.06, and 3.20 Å, respectively (in a ratio of $0.860:1:1.046$), and not fixed by symmetry. Nevertheless, they are not far from regular.]

If Figures 31*a* and *b* are compared with Figure 22, it is clear that, described in this way, both **MgCu₂** and **MgZn₂** are CS derivatives of **Zr₄Al₃**. For the former the CS plane is parallel to ($1\bar{1}00$) of **Zr₄Al₃**, the CS vector being [$1\bar{1}03$]/6. For the latter the CS plane is parallel to ($11\bar{2}0$) of **Zr₄Al₃**, the CS vector being [$10\bar{1}3$]/6 and [$01\bar{1}3$]/6 (there is some distortion), an average of [$11\bar{2}3$]/6.

The μ-Co₇Mo₆ structure is shown in Figure 32. It is clearly a regular intergrowth of narrow (0001) slabs of Laves phase (**MgCu₂** or **MgZn₂**) and of the crystallographically sheared **Cr₃Si** shown in Figure 11, i.e., **Zr₄Al₃** or **CuAl₂**. Data are given in Data Table 19.

Data Table 19 μ-Co₇Mo₆ (28)

Rhombohedral, space group $R\bar{3}m$, No. 166; $a_{rh} = 8.970$ Å, $\alpha = 30°47'$; $Z = 1$; $a_{hex} = 4.762(1)$, $c_{hex} = 25.615$ Å, $c/a = 5.379$; $Z = 3$, $V = 503.0$ Å³

Atomic Positions (in the hexagonal unit cell): $(0,0,0; \frac{1}{3},\frac{2}{3},\frac{2}{3}; \frac{2}{3},\frac{1}{3},\frac{1}{3})+$
Co(1) in 3(a):   0,0,0
Mo(1) in 6(c): $\pm(0,0,z)$; $z = 0.1655(3)$
Mo(2) in 6(c): $z = 0.3483(3)$
Mo(3) in 6(c): $z = 0.4518(3)$
Co(2) in 18(h): $\pm(x,\bar{x},z; x,2x,z; 2x,x,\bar{z})$; $x = 0.8333(2)$, $z = 0.2562(2)$
Atomic Distances
Co–Co = 2.38₁–2.40₈ Å
Mo–Mo = 2.46₉–3.02₆ Å
Co–Mo = 2.60₅–2.86₅ Å

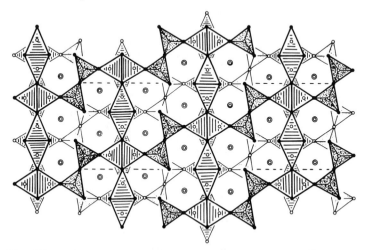

Figure 32.   The μ-phase structure of, e.g., W₆Fe₇ projected on ($11\bar{2}0$). (The **c** axis is horizontal.) Filled and open circles are on 0 and $\frac{1}{2}$; Fe are where trigonal bipyramids share only corners with each other or with tetraedersterns; W are where trigonal bipyramids share edges and at the corners of the central tetrahedron in each tetraederstern. This drawing, composed of ideal **Zr₄Al₃** and **MgCu₂** elements, is very close to the real μ-phase structure. (Double circles are Fe at $\pm\frac{1}{4}$.)

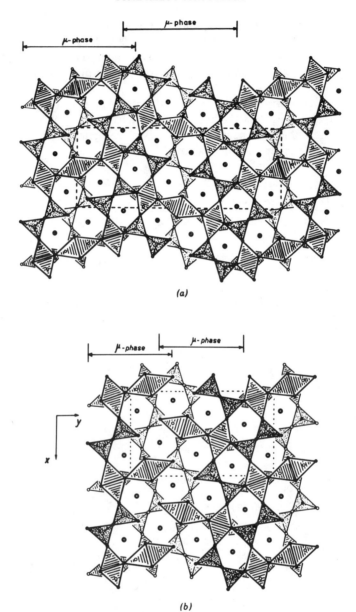

(a)

(b)

Figure 33. (a) A hypothetical structure produced from that of the μ phase (Figure 32) by twinning. The width of the μ-phase lamellae is three $\mathbf{Zr_4Al_3}$ elements. (b) The M-phase structure derived by twinning the μ phase by the same operation as in (a)

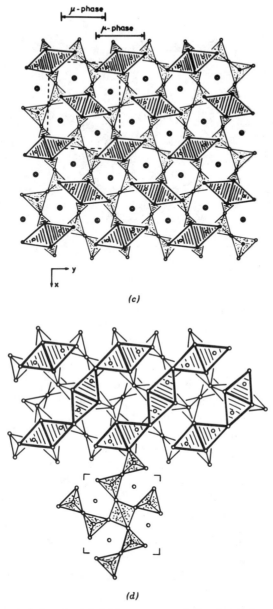

(c)

(d)

but with narrower twin bands. (*c*) Still narrower twin bands produce the structure of W_2FeSi from that of the μ phase. (*d*) The melilite structure on (001); SiO_4 tetrahedra and the unit cell are shown. The alkali metal atoms lie in the pentagonal channels.

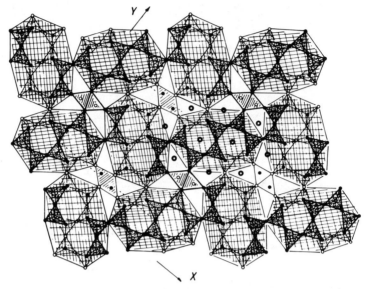

Figure 34. The χ-phase structure of (Mn, Co, Si) projected on (001) and constructed by cyclic twinning of **MgZn₂**, the columns of which are bounded by thin lines and cross-hatched. This drawing is close to the real structure. Atoms are on 0 and ½, except for the double circles, which are atoms at ±¼ (and not all of which are drawn).

Figure 33*a* shows a hypothetical structure derived from that of the μ phase by regularly repeated twinning. At the composition planes the trigonal bipyramids and tetraedersterns now share edges (instead of just corners, as in the μ phase). Reducing the width of the twin bands tends to obscure the relation to the μ phase but generates, by exactly the same twinning operation, the complex structures of the orthorhombic M phase of Al₁₃Nb₄₈Ni₃₉ (29), shown in Figure 33*b*, and of W₂FeSi (30), shown in Figure 33*c*.

The structure of melilite [(Ca,Na)₂(Mg,Al,Si)₃O₇] (31) can be similarly derived, with even smaller μ-phase columns (see Figure 33*d*).

Another tetrahedrally close-packed structure, that of the orthorhombic χ phase* (32), is shown in Figure 34 as fourlings of **MgZn₂**, which simplifies its description. The elliptical columns of the Laves phase (cross-hatched) are connected so that, with the insertion of extra pairs of atoms, columns of tetraedersterns are created in the interstices.

Other tetrahedrally close-packed structures have been described in similar terms elsewhere (22, 23, 33).

---

* This is a Co + Mn + Si alloy of uncertain composition. The unit cell (~15.4 × 12.4 × 4.75 Å³) contains 74 atoms on 16 crystallographically distinct types of sites, at least some of which are apparently of mixed occupancy.

References

1. J. K. Brandon, R. Y. Brizard, P. C. Chieh, R. K. McMillan, and W. B. Pearson, *Acta Cryst. B* **30**, 1412 (1974).
2. A. J. Bradley and P. Jones, *J. Inst. Metals* **51**, 131 (1933).
3. H. Nyman and S. Andersson, *Acta Cryst. A* **35**, 580 (1979).
4. B. Aronsson, *Acta Chem. Scand.* **14**, 1414 (1960).
5. B. Aronsson, *Acta Chem. Scand.* **9**, 1107 (1955).
6. B. Boren, *Ark. Kem. Mineral. Geol.* **11a**, 1 (1933).
7. M. O'Keeffe and S. Andersson, *Acta Cryst. A* **33**, 914 (1977).
8. K. Schubert, *Kristallstrukturen zweikomponentiger Phasen.* Springer-Verlag, Berlin, 1964, p. 150.
9. C. G. Wilson, D. K. Thomas, and F. J. Spooner, *Acta Cryst.* **13**, 56 (1960).
10. E. E. Havinga et al., *J. Less-Common Metals* **27**, 169, 187, 269, 281 (1972); summarized in *Struct. Rep.* **38A**, 5 (1972).
11. E. I. Gladyshevskii, P. I. Kripyakevich, and R. V. Skolozdra, *Sov. Phys.— Dokl.* **12**, 755 (1968).
12. S. Rundqvist, E. Hassler, and L. Lundvik, *Acta Chem. Scand.* **16**, 242 (1962).
13. F.-D. Doenitz, *Z. Kristallogr.* **131**, 222 (1970).
14. E. Larsson, *Arkiv Kemi* **23**, 335 (1965).
15. B. Pedersen and F. Grønvold, *Acta Cryst.* **12**, 1022 (1959).
16. T. Lundström and P.-O. Snell, *Acta Chem. Scand.* **21**, 1343 (1967).
17. Yu Wang, L. D. Calvert, E. J. Gabe, and J. B. Taylor, *Acta Cryst. B* **35**, 1447 (1979).
18. F. C. Frank and J. S. Kasper, *Acta Cryst.* **11**, 184 (1958); **12**, 483 (1959).
19. W. Haucke, *Z. Anorg. Allgem. Chem.* **244**, 17 (1940).
20. J. S. Stephens and D. W. J. Cruickshank, *Acta Cryst. B* **26**, 222 (1970).
21. G. Bergman and D. P. Shoemaker, *Acta Cryst.* **7**, 857 (1954).
22. M. O'Keeffe and B. G. Hyde, *Phil. Trans. Roy. Soc. (Lond.) A* **295**, 553 (1980).
23. S. Andersson, *J. Solid State Chem.* **23**, 191 (1978).
24. L. Stenberg and S. Andersson, *Z. Krist.* **158**, 205 (1982).
25. J. B. Friauf, *J. Am. Chem. Soc.* **49**, 3107 (1927).
26. J. B. Friauf, *Phys. Rev.* **29**, 35 (1927).
27. Y. Komura and K. Tokunaga, *Acta Cryst. B* **36**, 1548 (1980).
28. J. B. Forsyth and L. M. d'Alte da Veiga, *Acta Cryst.* **15**, 543 (1962).
29. C. B. Shoemaker and D. P. Shoemaker, *Acta Cryst.* **23**, 231 (1967).
30. P. I. Kripyakevich and Ya. P. Yarmolyuk, see *Struct. Rep.* **43A**, 73 (1977).
31. J. V. Smith, *Am. Mineral.* **38**, 643 (1953).
32. P. C. Manor, C. B. Shoemaker, and D. P. Shoemaker, *Acta Cryst. B* **28**, 1211 (1972).
33. S. Andersson, in *Structure and Bonding in Crystals,* Vol. 2, M. O'Keeffe and A. Navrotsky, Eds. Academic Press, New York, 1981, p. 233.

CHAPTER XV

Silicate Structures

The following structures are described in this chapter: garnet, benitoite BaTiSi$_3$O$_9$, kaolinite, muscovite, chlorite, cristobalite, quartz, gismondine, phillipsite, merlinoite, paulingite, sanidine (feldspar), gmelinite, chabazite, cancrinite, sodalite, offretite, erionite, Linde A, faujasite, and mordenite.

Except at very high pressure—for example, stishovite, SiO$_2$ of the rutile type and KAlSi$_3$O$_8$ of the hollandite type, silicon atoms are usually in tetrahedral coordination in oxides.* The silicon–oxygen distance is about 1.6 Å, and the oxygen–oxygen distance 2.6–2.7 Å. Very often there is a partial replacement of silicon by aluminum.

Isolated SiO$_4^{4-}$ tetrahedra exist in olivine, (Mg,Fe)$_2$SiO$_4$ (described in Chapter III); in phenacite, Be$_2$SiO$_4$ (Chapter VI); and in zircon, ZrSiO$_4$ (Chapters XI and XIII). These are called orthosilicate structures. A very important structure type, garnet, also belongs here and will be discussed below. (Si$_2$O$_7$)$^{6-}$ groups (= cornerconnected pairs of SiO$_4$ tetrahedra) occur in thortveitite, Sc$_2$Si$_2$O$_7$, and also in åkermanite, Ca$_2$MgSi$_2$O$_7$ (which belongs to the melilite group, Chapter XIV). (Si$_3$O$_9$)$^{6-}$ rings (of three corner-connected SiO$_4$ tetrahedra) exist in the mineral benitoite, BaTiSi$_3$O$_9$ (described below) and also in the related wadeite, K$_2$ZrSi$_3$O$_9$. "4-Rings" (Si$_4$O$_{12}$)$^{8-}$ occur in axinite and "6-rings" (Si$_6$O$_{18}$)$^{12-}$ in beryl and in tourmaline.

Chains of tetrahedra sharing corners, (SiO$_3^{2-}$)$_\infty$, similar to those in CrO$_3$ exist in pyroxenes, such as CaMg(SiO$_3$)$_2$. If two pyroxene chains join up to form a double chain as shown in Figure 1, its composition and valence are (Si$_4$O$_{11}^{6-}$)$_\infty$. Such double chains exist in amphiboles, a typical example being tremolite of composition Ca$_2$Mg$_5$(Si$_4$O$_{11}$)$_2$(OH)$_2$.

---

* But not always, e.g., ^{VI}Si^{IV}P$_2^{IV}$O$_7$.

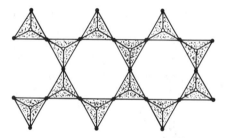

Figure 1. Double CrO_3 or pyroxene chain as it occurs in the amphiboles. (The dotted triangles are tetrahedra.)

Pyroxene and amphibole chains can condense to form wider chains (triple chains in jimthomsonite, double and triple in chesterite) and eventually complete sheets, which are the basis for the structures of talc, mica, chlorite, and kaolinite. Some of these structures will also be discussed here.

Continuous (three-dimensional) frameworks of tetrahedra linked by sharing corners occur in structures such as the various modifications of SiO_2, the feldspars, and zeolites and similar structures. These will also be discussed.

Garnet, an Orthosilicate Structure Type

The garnet structure, typified by grossularite, $Ca_3Al_2Si_3O_{12}$, is of great interest in mineralogy and solid state physics. Many hundreds of synthetic compounds and minerals have this structure, and the crystal data given here (Data Table 1) are for the ferrimagnetic $Y_3Fe_5O_{12}$ ($= Y_3Fe_2Fe_3O_{12}$).

Data Table 1 Yttrium Iron Garnet, $Y_3Fe_5O_{12}$ (1a, 1b)

Cubic, space group $Ia3d$, No. 230; $a = 12.376$ Å; $Z = 8$, $V = 1895.6$ Å^3

Atomic Positions: $(0,0,0; \frac{1}{2},\frac{1}{2},\frac{1}{2})+$

Y in 24(c): $\pm(\frac{1}{8},0,\frac{1}{4}; \frac{1}{4},\frac{1}{8},0; 0,\frac{1}{4},\frac{1}{8}; \frac{3}{8},0,\frac{1}{4}; \frac{3}{4},\frac{3}{8},0; 0,\frac{3}{4},\frac{3}{8})$

Fe(1) in 16(a): $0,0,0; 0,\frac{1}{2},\frac{1}{2}; \frac{1}{2},0,\frac{1}{2}; \frac{1}{2},\frac{1}{2},0; \frac{1}{4},\frac{1}{4},\frac{1}{4}; \frac{1}{4},\frac{3}{4},\frac{3}{4}; \frac{3}{4},\frac{1}{4},\frac{3}{4}; \frac{3}{4},\frac{3}{4},\frac{1}{4}$

Fe(2) in 24(d): $\pm(\frac{3}{8},0,\frac{1}{4}; \frac{1}{4},\frac{3}{8},0; 0,\frac{1}{4},\frac{3}{8}; \frac{1}{8},0,\frac{3}{4}; \frac{3}{4},\frac{1}{8},0; 0,\frac{3}{4},\frac{1}{8})$

O in 96(h): $\pm(x,y,z; \frac{1}{2}+x,\frac{1}{2}-y,\bar{z}; \bar{x},\frac{1}{2}+y,\frac{1}{2}-z; \frac{1}{2}-x,\bar{y},\frac{1}{2}+z; z,x,y;$
$\frac{1}{2}+z,\frac{1}{2}-x,\bar{y}; \bar{z},\frac{1}{2}+x,\frac{1}{2}-y; \frac{1}{2}-z,\bar{x},\frac{1}{2}+y; y,z,x;$
$\frac{1}{2}+y,\frac{1}{2}-z,\bar{x}; \bar{y},\frac{1}{2}+z,\frac{1}{2}-x; \frac{1}{2}-y,\bar{z},\frac{1}{2}+x; \frac{1}{4}+x,\frac{1}{4}+z,\frac{1}{4}+y;$
$\frac{1}{4}+y,\frac{1}{4}+x,\frac{1}{4}+z; \frac{1}{4}+z,\frac{1}{4}+y,\frac{1}{4}+x; \frac{3}{4}+x,\frac{1}{4}-z,\frac{3}{4}-y;$
$\frac{3}{4}+y,\frac{1}{4}-x,\frac{3}{4}-z; \frac{3}{4}+z,\frac{1}{4}-y,\frac{3}{4}-x; \frac{3}{4}-x,\frac{3}{4}+z,\frac{1}{4}-y;$
$\frac{3}{4}-y,\frac{3}{4}+x,\frac{1}{4}-z; \frac{3}{4}-z,\frac{3}{4}+y,\frac{1}{4}-x; \frac{1}{4}-x,\frac{3}{4}-z,\frac{3}{4}+y;$
$\frac{1}{4}-y,\frac{3}{4}-x,\frac{3}{4}+z; \frac{1}{4}-z,\frac{3}{4}-y,\frac{3}{4}+x),$
$x = -0.0269(1), y = 0.0581(3), z = 0.1495(1)$

Atomic Distances

Y–O = 2.419 Å (4×), 2.368 Å (4×)

Fe(1)–O = 2.012 Å (6×)

Fe(2)–O = 1.880 Å (4×)

O–O = 2.675–3.163 Å

The structure is best described in terms of a bcc rod packing. [We have used rod packing earlier in order to describe the Cr_3Si structure. A general description of rods in space is given elsewhere (2).] In garnet, each rod consists of alternating octahedra and trigonal prisms sharing faces along the nonintersecting trigonal axes of $Ia3d$. The trigonal prisms are empty; the octahedral cations, Al in grossularite [Fe(1) in YIG] are in special positions $16(a)$ at the centers of O_6 octahedra. Between these polyhedral rods, there are tetrahedral interstices occupied by Si [Fe(2)] and twisted cubes occupied by Ca [Y]. Figure 2 shows the bcc rod packing: four sets of identical, interpenetrating rod systems. (The rods in each set are parallel to one of the four $\langle 111 \rangle$ or body diagonal directions of the cubic cell.) Figure 3 shows a polyhedral model of garnet, and Figures 4 and 5 show the structure projected along a body diagonal. Only the empty trigonal prisms are shown in Figure 4, and those of each of the four different rod systems are shown in different colors. Figure 5 is the same projection, but now the octahedra and tetrahedra are shown. (See color plates for Figures 3–5.)

If the octahedra and trigonal prisms were regular, the unit cell edge would be $a = 4e(\sqrt{3} + \sqrt{2})/3$, where e is the polyhedron edge. With $e = 2.80$ Å, $a = 11.7$ Å. The oxygen parameters for this ideal structure are

$$
\begin{aligned}
x &= (2\sqrt{3} - 3\sqrt{2} + \sqrt{6} - 2)/8 = -0.0411 \\
y &= (\sqrt{6} - 2)/8 \qquad\qquad\quad = 0.0562 \\
z &= (3\sqrt{2} - 2\sqrt{3} + \sqrt{6} - 2)/8 = 0.1535
\end{aligned}
$$

Deviations of the real garnet structures from this ideal structure are described as being of three kinds: (1) Compression or expansion of the octahedra along the trigonal axis; (2) umbrella distortions in which O atoms move in the plane perpendicular to the trigonal axis; and (3) rotation of the octahedra, clockwise or anticlockwise. The significance, with examples, of these distortions has been described (2).

For grossularite, the observed parameters are $a = 11.855$ Å, $x = -0.0382$,

Figure 2. Four identical interpenetrating rod systems: bcc rod packing as it occurs in the garnet structure.

$y = 0.0457$, $z = 0.1512$. They differ from the ideal values because of (1) a 3.0% expansion, (2) a -3.8% umbrella distortion, and (3) a 3.6° rotation.

An interesting defect in grossularite (and in other garnets and other silicates) is the replacement of a Si atom by four H atoms, so that the SiO_4 tetrahedron becomes $(H_4)O_4$. The degree of replacement can range from 0 to 100% without a change in structure but with a substantial change in the edge length of the unit cell: from 11.855 Å for $Ca_3Al_2Si_3O_{12}$ to 12.576 Å for $Ca_3Al_2(D_4)_3O_{12} = Ca_3Al_2(OD)_{12}$ (3). The general formula of such "hydrogrossulars" is $Ca_3Al_2Si_{3-x}(H_4)_xO_{12}$; the mineral names are hibschite for $0 < x \le \frac{3}{2}$ and katoite for $\frac{3}{2} < x \le 3$ (4).

The trigonal prism plus octahedron rod is a common unit in many other structures also. Examples are cataplete, $Na_2ZrSi_3O_9$, wadeite, $K_2ZrSi_3O_9$, and benitoite, $BaTiSi_3O_9$, the latter shown in Figure 6 (color plate XV.6). Such rods can also be traced in metal alloys like TiP.

Some Sheet Silicates

The structures of the sheet silicates are based on octahedra and hexagonal networks of tetrahedra. We divide them into the following groups:

I. Pyrophyllite, $Al_2(Si_4O_{10})(OH)_2$; talc, $Mg_3(Si_4O_{10})(OH)_2$; bentonite, $Na_{0.67}(Al_{3.33}Mg_{0.67})[Si_8O_{20}(OH)_4]$; and the mica group, for example muscovite, $KAl_2(AlSi_3O_{10})(OH)_2$, and biotite, $K(Mg,Fe)_3AlSi_3O_{10}(OH)_2$.

II. The chlorite group with the general formula $(Mg,Al,Fe)_6(Si,Al)_4O_{10}(OH)_8$, and vermiculites, $(Mg,Al,Fe)_3(Si,Al)_4O_{10}(OH)_2 \cdot 4H_2O$.

III. The kaolinite group with kaolinite $= Al_2(Si_2O_5)(OH)_4$. In this group we also include the serpentine minerals, $Mg_3Si_2O_5(OH)_4$, for example chrysotile (asbestos).

The grouping is mainly organized on a structural basis. Figure 7a shows one layer of kaolinite, in which a hexagonal layer of SiO_4 tetrahedra sharing corners also shares corners with a layer of AlO_6 octahedra. Tetrahedra share corners with each other and with the octahedra; octahedra share edges. Such layers yield the structure of kaolinite according to the scheme shown in Figure 7b. Kaolinite is triclinic, with

$$a = 5.14\ (5.3)\ \text{Å}, \qquad \alpha = 91.8°$$
$$b = 8.93\ (9.2)\ \text{Å}, \qquad \beta = 104.7°$$
$$c = 7.37\ (14.6 = 2 \times 7.3)\ \text{Å}, \qquad \gamma = 90°$$

Figures within parentheses are unit cell edges of the structurally related chrysotile (asbestos).

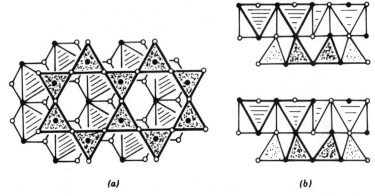

(a)   *(b)*

Figure 7. (*a*) A single layer of kaolinite. (*b*) Layers of kaolinite projected along the **a** axis.

In chrysotile, Mg_3 replaces the Al_2 in kaolinite. (Note the empty octahedra in Figure 7*a*.) The two structures are otherwise identical, but the larger dimensions of the brucitelike octahedral layer in chrysotile* and the special appearance of diffraction photographs led Pauling (already in 1930) to suggest that in chrysotile, stacked kaolinite sheets curl into a cylindrical spiral, with the silicon–oxygen sheets on the inner side (smaller perimeter), giving the characteristic fiber structure of asbestos. This has been beautifully confirmed by electron microscopy (5).

The structure of muscovite is shown in Figure 8, projected along **a** and slightly idealized. The unit layers now consist of sheets of tetrahedra on *both* sides of the octahedral layer, and these composite layers are held together by alkali metal atoms in 12-coordination (hexagonal prism). Bentonite has the same structure but with fewer alkali metal atoms between the layers. Finally, in talc and pyrophyllite, the layers have no charge (and hence no interleaving alkali metal atoms), but otherwise the structures are the same.

Muscovite has the cell dimensions shown in Data Table 2 (talc in brackets).

Data Table 2   Muscovite and Talc*

Space group $C2/c$, No. 15, a = 5.19 Å (5.26); b = 9.00 Å (9.10), c = 20.10 Å (18.81); β = 95.5° (100°); Z = 4, V = 924.6 Å^3

* Data for talc are in parentheses.

In the chlorite group, the structural scheme is given in Figure 9. Here the mica layers are interleaved with brucite layers.

* $r(Mg^{2+})$ = 0.86 Å, $r(Al^{3+})$ = 0.67 Å.

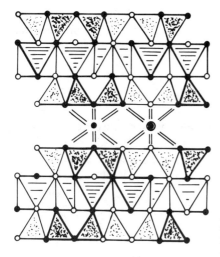

Figure 8. Slightly idealized structure of muscovite.

From these structures it is not surprising that disorder and polytypism are common phenomena among the sheet silicates. There are also many more related minerals in the above-described groups. Minerals normally called clays include some of these. They are soft and hydrated and exhibit cation exchange and intercalation of organic molecules. Their structures are found in the kaolinite group (montmorillonite is a typical clay) and also in some chlorite-related minerals. The clays are important as constituents of soil and also have many industrial applications. Because of the swelling property of

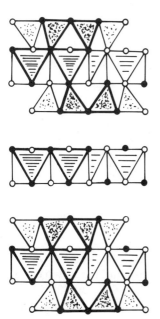

Figure 9. The structural scheme of the chlorite group.

bentonite (montmorillonite), it is used in oil-well drilling as a thickener. Vermiculite is used as a filler in plastics, and many clays are important as catalysts.

Framework Silicates

Here we shall describe some common and important structures: cristobalite, quartz, feldspar, and many zeolites. Among the latter we discuss the phillip-site group, the natrolite group, the gmelinite group, and some commercially important zeolites such as Linde A, faujasite, and mordenite.

High- and Low-Temperature Cristobalite

The structure given for the high-temperature (β) form of cristobalite in most texts is the C9 type shown in Figure 10 in two projections. Each corner of each SiO_4 tetrahedron is common to two tetrahedra. (There is, of course, no edge-sharing.) This is cubic, space group $Fd3m$, but it is now known to be incorrect. The real β-cristobalite structure is topologically identical to C9, but the corner-connected array of tetrahedra has partly collapsed, reducing the SiÔSi bond angles from 180° to ~147°, the symmetry to tetragonal (space group $F\bar{4}d2$ for the same unit cell, $I\bar{4}2d$ for the equivalent cell of half the volume), and the structure to that shown in Figure 11. The collapse involves a tilt/rotation of each tetrahedron about parallel $\bar{4}$ axes, alternately clockwise and anticlockwise so that the connectivity is retained.

Data Table 3 β-Cristobalite, SiO_2 (6, and references therein)

Tetragonal, space group $I\bar{4}2d$, No. 122; $a = 5.07$ ($= 7.17/\sqrt{2}$), $c = 7.17$ Å; $c/a = \sqrt{2}$; $Z = 4$, $V = 184._3$ Å^3 ($= 4 \times 46.1$ Å^3)

Atomic Positions: $(0,0,0; \frac{1}{2},\frac{1}{2},\frac{1}{2})+$
Si in 4(a): $0,0,0; 0,\frac{1}{2},\frac{1}{4}$
O in 8(d): $x,\frac{1}{4},\frac{1}{8}; \bar{x},\frac{3}{4},\frac{1}{8}; \frac{3}{4},x,\frac{1}{8}; \frac{1}{4},\bar{x},\frac{7}{8}; x = -0.09$
Atomic Distances
Si–O = 1.62 Å
O–O = 2.62 Å (2×), 2.69 Å (4×) (tetrahedra not quite regular)
Bond Angle
SiÔSi = 147°

The low-temperature, α-cristobalite structure (Data Table 4) is also a slightly collapsed version of C9. But now the tetrahedra tilt about two equiv-alent (in C9), mutually orthogonal $\bar{4}$ axes, to give a different tetragonal structure (space group $P4_12_12$), again topologically identical to C9 and there-

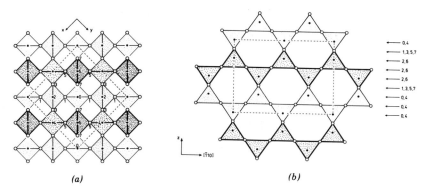

Figure 10. The C9 structure (idealized high-temperature cristobalite) shown in two projections. Atomic heights in units of (a) $c/8$, (b) $a[110]/8$.

fore to β-cristobalite. This structure is shown in Figure 12. (The $\alpha \rightleftharpoons \beta$ transition temperature is in the range $\sim$200–275°C.)

A number of compounds have structures similar to these two cristobalite types. The relationships have been discussed in some detail elsewhere (6). *Note Added in Proof*

The description of cristobalite given above represented the state of

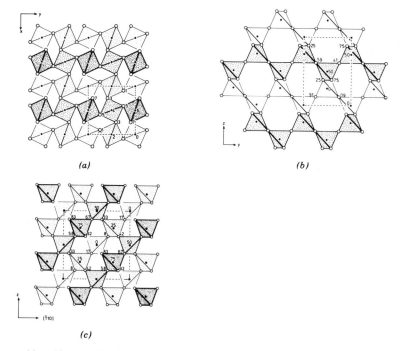

Figure 11. The β-cristobalite structure in three projections; compare Figure 10. Atomic heights in units of (a) $c/8$, (b) $a/100$, (c) $a[110]/100$.

Data Table 4   α-Cristobalite, SiO₂ (6, and references therein)

Tetragonal, space group $P4_12_12$, No. 92; $a = 4.978$, $c = 6.948$ Å, $c/a = 1.396$;
$Z = 4$, $V = 172.4$ Å³ ($= 4 \times 43.0$ Å³)

Atomic Positions
Si in 4(a): $x,x,0; \bar{x},\bar{x},\frac{1}{2}; \frac{1}{2} - x,\frac{1}{2} + x,\frac{1}{4}; \frac{1}{2} + x,\frac{1}{2} - x,\frac{3}{4}; x = 0.300$
O in 8(b): $x,y,z; \bar{x},\bar{y},\frac{1}{2} + z; \frac{1}{2} - y,\frac{1}{2} + x,\frac{1}{4} + z; \frac{1}{2} + y,\frac{1}{2} - x,\frac{3}{4} + z; y,x,\bar{z};$
              $\bar{y},\bar{x},\frac{1}{2} - z; \frac{1}{2} - x,\frac{1}{2} + y,\frac{1}{4} - z; \frac{1}{2} + x,\frac{1}{2} - y,\frac{3}{4} - z;$
              $x = 0.239, y = 0.105, z = 0.179$

Atomic Distances
Si–O = 1.61 Å
O–O = 2.60 Å (2×), 2.62 Å, 2.63 Å (2×), 2.66 Å
Bond Angles
SiÔSi = 146° at room temperature, 149° at 230°C

knowledge at the end of 1987. A subsequent electron diffraction study has radically altered the picture for β-cristobalite (25). A high-purity sample of cristobalite exhibited the $\alpha \rightleftarrows \beta$ transition at 265 ± 5°C on heating and 240 ± 10°C on cooling. In the temperature range 275–650°C the diffraction pattern of the β-form showed spectacular, intense sheets of diffuse scattering normal to 110$_{C9}^*$ directions (in addition to the Bragg reflections).

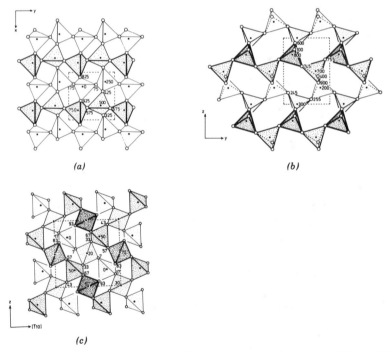

(a)

(b)

(c)

Figure 12.   The α-cristobalite structure in three projections; compare Figures 10 and 11. Atomic heights in units of (a) c/1000, (b) a/1000, (c) (a + b)/100.

Electron microscopy revealed no evidence of a domain texture: the diffuse scattering is undoubtedly dynamic, rather than static in origin. It could be reproduced in lattice dynamic calculations—but only with a C9-type model (and not with a domain model, such as would be essential for e.g. an $I\bar{4}2d$ structure). It is the result of "disorder" due to the substantial excitation of low frequency phonons ($\sim$1.4Thz $\equiv E/h = 67$ K). These appear to involve coupled rotations (librations) of the corner-connected SiO_4 tetrahedra in (quasi-independent) $\langle110\rangle_{C9}$ rods; cf. Figure 10, especially 10b. (In contrast, micrographs of the α form showed clear domain boundaries as expected for twinning of this lower-symmetry structure, the nature of which is not in doubt. The corresponding diffraction patterns exhibited diffuse scattering of the same form as in β; but its intensity was relatively very weak.)

Clearly, more work is needed for, as was pointed out 60 years ago, there are severe crystal chemical problems with a simple C9 structure (in which the average oxygen position is on an Si . . . Si axis). With a cubic unit cell parameter $a \approx 7.17$ Å (and $Fd3m$ symmetry) the bond length l(Si–O) = $\sqrt{3}a/8 = 1.55$ Å, which is much too short. It implies an apparent bond valence $v' = 1.21$ and therefore atomic valences $V'(\text{Si}) = 4v' = 4.85$ and $V'(\text{O}) = 2v' = 2.43$ (instead of the expected 4.0 and 2.0). The oxygens must therefore be displaced off the Si . . . Si axis, but in a correlated rather than random way (to preclude excessively short, high-energy O . . . O distances: these are already fairly short in a normal, regular SiO_4 tetrahedron—$\sim$2.65 Å for l(Si–O) = 1.624 Å).

High- and Low-Temperature Quartz

Under ambient conditions the cristobalite types of SiO_2 are metastable. (Cristobalite is the stable polymorph just below the melting point of SiO_2.) The stable polymorph at and immediately above room temperature is quartz, which is topologically distinct from C9 and cristobalite. Interconversion is difficult; it necessitates rupturing the very strong Si–O bonds. (It is therefore a reconstructive transformation.) For this reason cristobalite is readily quenched to room temperature. Whereas cristobalites consist of 6-rings and fourfold spirals of corner-connected SiO_4 tetrahedra, quartz contains three- and sixfold spirals. It also occurs in two forms; α at low temperature and β at high temperature. (The transition temperature is 573°C.) They are topologically identical and differ only in the tilt of the tetrahedra. Drawings of this structure are perhaps even more difficult to comprehend than those of cristobalite: they are given in Figure 13. Structural data are given in Data Tables 5 and 6.

In β-quartz (Figure 13b) the tetrahedra are in a symmetrical orientation, from which they are tilted in α-quartz (Figure 13a). Viewed along the *positive x,y* or [11$\bar{2}$0] directions (Figures 13c and d), the tilt may be clockwise or

Data Table 5 β-Quartz, SiO$_2$ (Right-Handed Enantiomorph)

Hexagonal, space group $P6_222$, No. 180; $a = 5.01$, $c = 5.47$ Å (at ~600°C), $c/a = 1.092$; $Z = 3$, $V = 118.9$ Å^3 (= 3 × 39.6 Å^3)

Atomic Positions
Si in 3(*c*): $\frac{1}{2},0,0; \frac{1}{2},\frac{1}{2},\frac{1}{3}; 0,\frac{1}{2},\frac{2}{3}$
O in 6(*j*): $x,\bar{x},\frac{5}{8}; \bar{x},x,\frac{5}{8}; x,2x,\frac{1}{2}; \bar{x},2\bar{x},\frac{1}{2}; 2x,x,\frac{1}{6}; 2\bar{x},\bar{x},\frac{1}{6}; x = 0.197$
Atomic Distances
Si–O = 1.62 Å
O–O = 2.50 Å (2×), 2.67 Å (4×)
Bond Angle
SiÔSi = 147°

Data Table 6 α-Quartz, SiO$_2$ (Right-Handed Enantiomorph)

Trigonal, space group $P3_221$, No. 154*; $a = 4.916$, $c = 5.4054$ Å, $c/a = 1.100$; $Z = 3$, $V = 113.13$ Å^3 (= 3 × 37.71 Å^3)

Atomic Positions
Si in 3(*a*): $x,0,0; 0,x,\frac{2}{3}; \bar{x},\bar{x},\frac{1}{3}; x = 0.4697$
O in 6(*c*): $x,y,z; \bar{y},x - y,\frac{2}{3} + z; y - x,\bar{x},\frac{1}{3} + z; y,x,\frac{2}{3} - z; \bar{x},y - x,\frac{1}{3} - z;$
     $x - y,\bar{y},\bar{z}; x = 0.4135, y = 0.2669, z = 0.1158$
Atomic Distances
Si–O = 1.614 Å (2×), 1.605 Å (2×)
O–O = 2.612 Å, 2.631 Å, 2.645 Å (2×), 2.617 Å (2×)
Bond Angles
SiÔSi = 143.7°

* It is customary to shift the unit cell origin by $[0,0,-\frac{1}{3}]$ from that in the *International Tables of X-Ray Crystallography*, as has been done. This has the advantage of making the units cells of α- and β-quartz correspond exactly.

(as in Figures 13*a* and *c*) anticlockwise, giving two equivalent twin forms of α—the so-called Dauphiné twins. There is also another twinning possibility: the threefold spirals of corner-connected tetrahedra about $\pm[\frac{1}{3},\frac{2}{3},z]$ may be left-handed (as in Figure 13*a*) or right-handed—two enantiomorphic forms. The former is dextrorotatory, the latter levorotatory. These are called Brazil twins. Whereas Dauphiné twins are readily interconverted (by tilting the tetrahedra without changing their topology), Brazil twins can only be inter-converted reconstructively: to invert the handedness of the spirals of tetra-hedra, Si–O bonds must be broken and remade. (Note: Changing from one Dauphiné twin to the other does not affect the handedness of quartz.)

Compare the volumes per molecule of these different SiO$_2$ polymorphs; and note that the Si–O bond lengths and SiÔSi bond angles are virtually constant.

GeO$_2$ has been reported with the α-cristobalite and quartz structures (and the rutile type, of course).

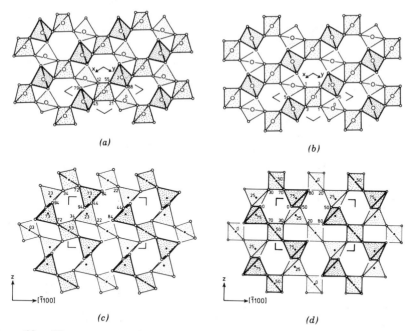

(a)

(b)

(c)

(d)

Figure 13. The structures of α- and β-quartz in projection: (a) α on (0001), (b) β on (0001), (c) α on (11$\bar{2}$0), (d) β on (11$\bar{2}$0). In (a) and (b) the small circles are oxygens, with heights in units of $c/100$ in (a) and $c/6$ in (b); large circles are silicons, with heights in units of $c/3$. In (c) and (d) the larger open circles are oxygens and the smaller filled circles are silicons, with all heights in units of $(a[11\bar{2}0]/3)/100 \equiv a/3$.

Feldspar and Zeolites

Ordinary feldspar and the zeolites of the phillipsite group are among the most common of all inorganic compounds in nature. Phillipsite and harmotome are found in many geological environments, such as in the huge layers of deep-sea sediments on the bottom of (particularly) the Pacific and Indian oceans, while feldspar is a rock-forming mineral that, with varying composition, constitutes 58% of the earth's crust and thus is encountered everywhere.

In Figures 14 and 15 we show idealized drawings of gismondine and phillipsite. The structures are continued vertically by mirror planes parallel to the plane of the paper. Rotation of half the 4-ring columns in the gismondine framework, as indicated in Figure 14, transforms it to the phillipsite framework, Figure 15. The unit-cell data are given in Data Tables 7 and 8.

Data Table 7   Gismondine, $Ca(Al_2Si_2)O_8 \cdot 4H_2O$ (7)

Monoclinic, space group $P2_1c$, No. 14; $a = 10.02$, $b = 10.62$, $c = 9.84$ Å, $\beta = 92°25'$; $V = 1109$ Å^3

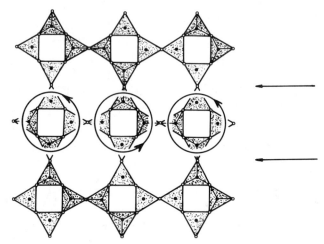

Figure 14. Idealized drawing of the gismondine structure. Rotation of the ringed columns yields the phillipsite structure.

The linking of tetrahedra in three dimensions readily creates large cavities where the alkali metal ions and water are situated. Phillipsite is obtained by unit-cell twinning of gismondine, and in a fourling construction a third zeolite, merlinoite (Data Table 9) may be created, as shown in the center of Figure 16.

Figure 15. Idealized drawing of the phillipsite structure.

Data Table 8   Phillipsite, $(K_xNa_{1-x})_5Si_{11}Al_5O_{32}\cdot10H_2O$ (8)

Orthorhombic, space group $B2mb$, No. 40; $a = 9.96$, $b = 14.25$, $c = 14.25$ Å; $V = 2024$ Å^3

Data Table 9   Merlinoite, $(K,Ca,Na,Ba)_7Si_{23}Al_9O_{64}\cdot23H_2O$ (9)

Orthorhombic, space group $Immm$, No. 71; with $a = 14.116$, $b = 14.229$, $c = 9.946$ Å; $V = 1998$ Å^3

Paulingite is one of the most complex zeolites, with a cubic unit cell of 35.1 Å edge, space group $Im3m$. Six fourling units of Figure 16 (almost merlinoite) are put together into a formidable sixling, as shown in the color picture in Figure 17 (see color plates). The structure was determined by Samson and co-workers (10), and the description given here is from ref. 11.

Thomsonite and edingtonite belong to the natrolite group of zeolites. They are easily related, as shown in Figure 18. Twinning of the natrolite structure produces thomsonite, and if this is continued to a fourling operation, the edingtonite structure is obtained.

The structure of the monoclinic potassium feldspar sanidine, $KAlSi_3O_8$, can be derived from a hypothetical zeolite related to the gismondine group by a two-dimensional translation operation (12). Figure 19 gives the detailed structure. The hypothetical zeolite structure is shown in Figure 20; its tetrag-

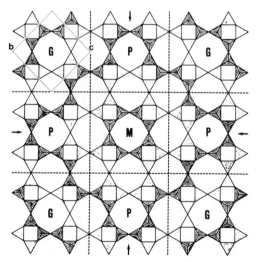

Figure 16. Fourling construction of gismondine to give the merlinoite structure. Framed area = blue unit in Figure 17. G = gismondine, P = phillipsite, M = merlinoite.

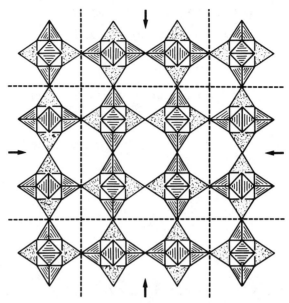

Figure 18.   Fourling of natrolite down [001] with thomsonite along the twin planes (arrowed) and edingtonite in the center. Different heights of chains are not indicated.

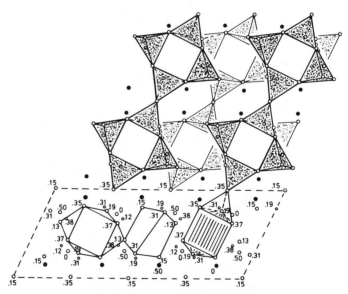

Figure 19.   The structure of sanidine, $KAlSi_3O_8$, projected along **b**. Only atoms in the lower part of the unit cell are shown; the rest are repeated by a mirror plane at $y = \frac{1}{2}$. Slightly distorted (bicapped) trigonal prisms are also shown in the lower part of the figure.

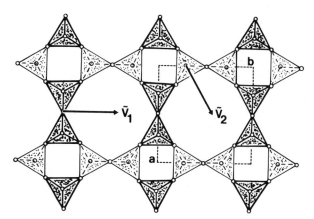

Figure 20. The hypothetical zeolitic parent to the feldspar framework, projected along **c**. Mirror planes as in Figure 21. Note that the shear vector $\mathbf{V}_2$ is not in the plane of the paper in Figures 20 and 21.

onal unit cell has $a = e(\sqrt{3} + 1)$ and $c = 4e\sqrt{2/3}$, e being the edge of an ideal, regular tetrahedron. The same hypothetical zeolite is projected along **b** in Figure 21. The shear vectors for the operation to produce the feldspar structure are indicated; note that V_2 is not in the plane of the paper in Figure 20. Figures 22 and 23 show the feldspar structure (with regular tetrahedra) obtained from the double translation operation. Finally a polyhedral model in color is shown in Figure 24 (see color plates).

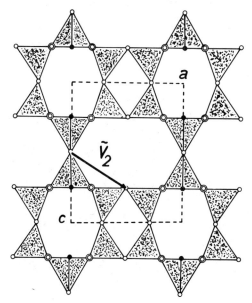

Figure 21. The hypothetical zeolite framework of Figure 20 projected along **b** (or V_1).

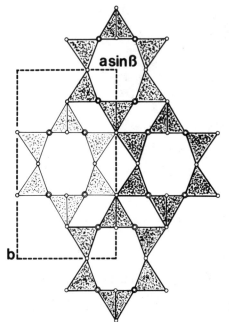

Figure 22. The resulting idealized feldspar framework projected along **b**.

This model of feldspar is very accurate: $a = 2e(1/\sqrt{3} + 1)$; $b = 2e\sqrt{6}$;
$c = e(\sqrt{3} + 1)$; $\sin(180° - \beta) = \sqrt{3}/2$; and with an average O–O distance
$e = 2.68$ Å, we obtain

$$a_{calc} = 8.45 \text{ Å}; \quad b_{calc} = 13.13 \text{ Å}; \quad c_{calc} = 7.32 \text{ Å}; \quad \beta_{calc} = 120°;$$

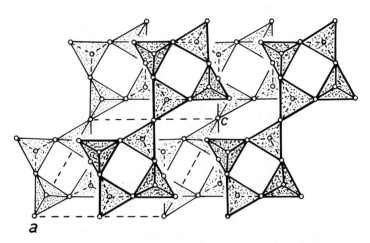

Figure 23. The idealized feldspar framework projected along **c**.

TABLE 1 Atomic Parameters for Feldspar

Atom[a]	Calculated			Observed		
	x	y	z	x	y	z
O(1) in 4(i)	0.71	0.00	0.33	0.63	0.00	0.29
O(2) in 4(g)	0.00	0.17	0.00	0.00	0.15	0.00
O(3) in 8(j)	0.82	0.17	0.21	0.83	0.15	0.23
O(4) in 8(j)	0.00	0.33	0.21	0.03	0.31	0.26
O(5) in 8(j)	0.18	0.17	0.42	0.18	0.13	0.40
T_1 in 8(j)	0.00	0.21	0.21	0.01	0.19	0.22
T_2 in 8(j)	0.71	0.13	0.33	0.71	0.12	0.34

[a] T_1 and T_2 are the tetrahedrally coordinated cations, Al and Si.

compared with

$$a_{obs} = 8.56 \text{ Å}; \quad b_{obs} = 13.03 \text{ Å}; \quad c_{obs} = 7.17 \text{ Å}; \quad \beta_{obs} = 116°.$$

The space group is $C2/m$, No. 12, $Z = 4$.

The observed atomic parameters, after Ribbe (13), compared with those calculated from the above model are given in Table 1.

We now describe six more zeolites that are structurally related to one another: gmelinite and chabazite, cancrinite and sodalite, offretite and erionite. Their structural data are given in Data Tables 10–15.

Data Table 10 Gmelinite, $(Na_2,Ca)(Al_2Si_4)O_{12} \cdot 6H_2O$ (14)

Hexagonal, space group $P6_3/mmc$, No. 194; $a = 13.75$, $c = 10.05$ Å; $Z = 4$, $V = 1645.5$ Å^3

Data Table 11 Chabazite, $(Ca,Na_2)Al_2Si_4O_{12} \cdot 6H_2O$ (15a, 15b)

Hexagonal (rhombohedral), space group $R\bar{3}m$, No. 166; $a = 13.7$, $c = 14.9$ Å; $Z = 6$, $V = 2422$ Å^3

Data Table 12 Cancrinite, $Na_6Al_6Si_6O_{24} \cdot CaCO_3 \cdot 2H_2O$ (16)

Hexagonal, space group $P6_3$, No. 173; $a = 12.75$, $c = 5.14$ Å; $Z = 1$, $V = 723.6$ Å^3

Data Table 13 Sodalite, $Na_4Al_3Si_3O_{12}Cl$ (17)

Cubic, space group $P\bar{4}3n$, No. 218; $a = 8.91$ Å; $Z = 2$, $V = 707.4$ Å^3

Data Table 14   Offretite, $(K_2,Ca,Mg)_{2.5}Al_5Si_{13}O_{36} \cdot 15H_2O$ (18)

Hexagonal, space group $P\bar{6}m2$, No. 187; $a = 13.29$, $c = 7.58$ Å; $Z = 1$, $V = 1160$ Å^3

Data Table 15   Erionite, $(Ca,Mg,Na_2,K_2)_{4.5}Al_9Si_{27}O_{72} \cdot 27H_2O$ (19)

Hexagonal, space group $P6_3/mmc$, No. 194; $a = 13.26$, $c = 15.12$ Å; $Z = 1$, $V = 2300$ Å^3

The structures of gmelinite and chabazite are shown in Figures 25*a* and *b*. A combined twin plus translation operation (the twin plane is the plane of the paper) transforms one structure into the other. Readers can find out for themselves simply by using transparent paper. The gmelinite structure is similar to the structures in the gismondine group; but instead of rings of four and eight tetrahedra, there are 6-rings, which build 4-rings as well as huge 12-rings of tetrahedra.

The orientations of the tetrahedra are somewhat different in the cancrinite and sodalite structures, but otherwise the relationship between these two structures is identical to that between gmelinite and chabazite. The repetition along the projection axes is also shown in the depiction of these structures in Figures 26*a* and *b*. [A detailed description of the sodalite structure and its relation to other, including alloy, structures has been given elsewhere (20).]

The two different orientations of tetrahedra in the 6-rings that compose the two pairs of structures shown in Figures 25 and 26 are combined or intergrown into a new pair—the offretite and erionite structures—shown in Figures 27*a* and *b*. Again a combined twin plus translation operation transforms one structure into the other.

More complex zeolites, produced by varying the size of the blocks to which these operations are applied, as well as related disorder in crystals have been observed.

Linde A, faujasite, and mordenite are all relatively easy to synthesize; they are the zeolites that are produced commercially.

The structure of Linde A (which does not exist in nature) is perhaps the simplest and most beautiful of all the zeolites. The building unit is an empty cuboctahedron capped by eight tetrahedra, as shown in Figure 28 (see color plates). Such units share corners to form a cubic structure, a model of which is shown in Figure 29 (see color plates). Another way to demonstrate the structure is shown in Figure 30 (see color plates). Here only the Si (or Al) framework atoms are shown, and the oxygens have to be imagined sitting approximately at the centers of the connecting rods. Truncated octahedra can be traced, and these are joined via the cubes. If such truncated octahedra are joined directly via their hexagonal faces, the structure of sodalite is

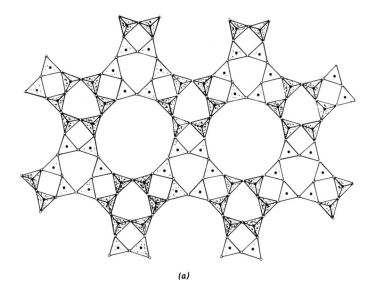

(a)

(b)

Figure 25. (a) Slightly idealized structure of gmelinite. (b) Idealized drawing of the structure of chabazite.

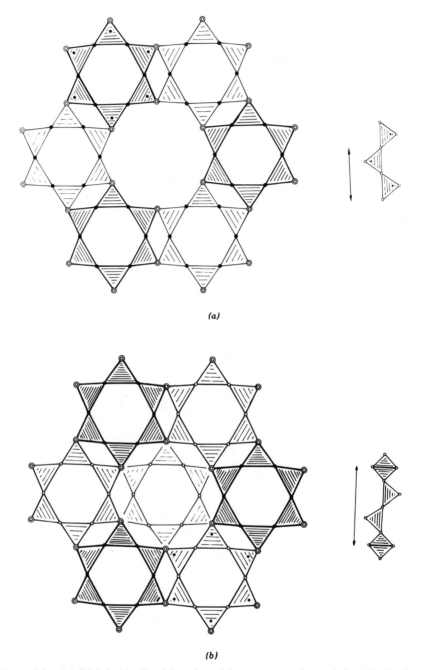

(a)

(b)

Figure 26. (*a*) Slightly idealized drawing of the structure of cancrinite. (*b*) Idealized drawing of sodalite.

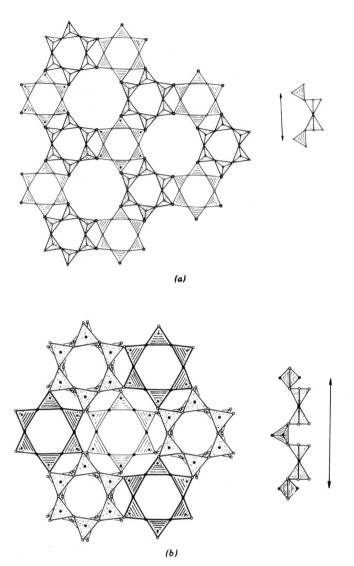

Figure 27. (*a*) Idealized drawing of the structure of offretite. (*b*) Idealized drawing of the structure of erionite.

obtained. The (Si, Al) cubes in Linde A are those units shown in Figure 28 (color plates).

If identical truncated octahedra are joined via intervening hexagonal prisms, the structure of faujasite is derived; it is shown in Figure 31 (see color plates). The hexagonal prisms created between the truncated octahedra are simply double hexagonal rings of tetrahedra. Commercial faujasite is

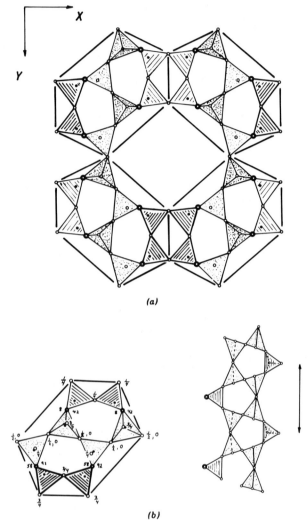

Figure 32. (*a*) The real structure of mordenite. (*b*) The building block unit of mordenite, with heights given. The same unit is used to build the structure of dachiderite.

called zeolite X. It has sodium as the caged cation, and a ratio of Si/Al = 1–1.5. When this ratio is 1.5–3, still with sodium as cation, it is called zeolite Y.

The structure of mordenite is shown in Figures 32*a* and *b*. Building units are framed in Figure 32*a* and, in Figure 32*b*, atomic heights are given as well as the mode of joining tetrahedra in the projection direction. 5-Rings are dominant in this mordenite structure (and in the related structure of dachiderite). Mordenite is more stable to acids than other zeolites, which has led to certain applications. During acid leaching, aluminum can be removed

TABLE 2 Characteristic Parameters of Some Zeolite Structures

Zeolite	Number of Si + Al per 1000 Å^{3a}	Free Apertures of Channels in Å	Number of Tetrahedra per Ring
Natrolite	17.8	2.6–3.9	8
Gmelinite	14.6	6.9	12
Chabazite	14.6	3.7–4.2	8
Erionite	15.7	3.6–4.8	8
Cancrinite	16.7	6.2	12
Sodalite	17.2	2.6	6
Gismondine	15.3	2.8–4.9	8
Phillipsite	15.8	2.8–4.8	8
Merlinoite	16.1	2.8–4.9	8
Mordenite	17.2	6.7–7.0	12
Faujasite	12.7	7.4	12
Linde A	12.9	4.1	8
Paulingite	15.5	3.9	8

^{a} Compare 21.7 and 26.6 per 1000 Å^3 for the cristobalites and quartzes.

without loss of crystallinity, and this is probably due to Al in O$_4$ tetrahedra (plus counter cations) being replaced by (H$_4$) groups, i.e., substitution of Al^{3+} plus, for example, Na$^+$ by 4H$^+$.* The resulting framework is empty (free of large cations), nonionic (uncharged), and hydrophobic.

Crystal data for these last three zeolites are given in Data Tables 16–18.

Data Table 16 Linde A, NaAlSiO$_4$* (22)

Cubic, space group $Pm3m$, No. 221; $a = 12.28$ Å; $Z = 12$, $V = 1851.8$ Å^3

* Ordering of Al and Si in their tetrahedra doubles the unit-cell axis.

Data Table 17 Faujasite, Na$_2$CaAl$_4$Si$_{10}$O$_{28}$·20H$_2$O (23)

Cubic, space group $Fd3m$, No. 227; $a = 24.74$ Å; $Z = 16$, $V = 15,142$ Å^3

Data Table 18 Mordenite, NaAlSi$_5$O$_{12}$·3H$_2$O (24)

Orthorhombic, space group $Cmcm$, No. 63; $a = 18.13$, $b = 20.49$, $c = 7.52$ Å; $Z = 8$, $V = 2794$ Å^3

* The substitution of 2H$^+$, 3H$^+$, or 4H$^+$ groups for tetrahedrally coordinated cations, such as Al^{3+}, Si^{4+}, and P^{5+}, is not uncommon (21a–21c). In the garnet grossularite, Ca$_3$Al$_2$(OH)$_{12}$, all the Si may be replaced by (H$_4$), giving the "isostructural" katoite (3), as discussed earlier.

Zeolites are, of course, very open framework structures and in Table 2 the framework density is expressed by the number of $(Si,Al)O_4$ tetrahedra per 1000 Å^3. "Window" dimensions and the number of tetrahedra in the corresponding rings are also given. The first measures the (reciprocal of the) absorption capacity of the zeolite; the second indicates the maximum size of molecule that can enter the structure (the molecular sieve effect), which is related to the third parameter.

References

Some general references for zeolites and silicates are:

L. Bragg and G. F. Claringbull (and W. H. Taylor), *The Crystalline State. Crystal Structures of Minerals*. G. Bell and Sons, London, 1965.

F. Liebau, *Structural Chemistry of Silicates*. Springer-Verlag, Berlin, 1985.

Mineralogical Society of America, *Reviews in Mineralogy,* Vols. 2, 4, 5, 7, 9a, 9b, and 13.

D. W. Breck, *Zeolite Molecular Sieves. Structure, Chemistry and Use*. Wiley, New York, 1974.

W. M. Meier and D. H. Olson, *Atlas of Zeolite Structure Types,* Juris, Zürich, 1978.

1a. A. Batt and B. Post, *Acta Cryst.* **15,** 1268 (1962).

1b. F. Bertaut and F. Forrat, *C.R. Acad. Sci. Paris* **243,** 38 (1956).

2. M. O'Keeffe and S. Andersson, *Acta Cryst. A* **33,** 914 (1977).

3. D. W. Foreman, *J. Chem. Phys.* **48,** 3037 (1968).

4. J. A. Zilczer, *Am. Mineral.* **70,** 873 (1985).

5. K. Yada, *Acta Cryst. A* **27,** 659 (1971); *Am. Mineral.* **62,** 958 (1977); *Can. Mineral.* **17,** 679 (1979).

6. M. O'Keeffe and B. G. Hyde, *Acta Cryst. B* **32,** 2923 (1976).

7. A. Alberti and G. Vezzalini, *Acta Cryst. B* **35,** 2866 (1979).

8. H. Steinfink, *Acta Cryst.* **15,** 644 (1962).

9. E. Pasaglia, D. Pongiluppi, and R. Rinaldi, *Neues Jahrb. Mineral. Monatsh.* **1977,** 355.

10. E. K. Gordon, S. Samson, and W. B. Kamb, *Science* **154,** 1004 (1966).

11. S. Andersson and L. Fälth, *J. Solid State Chem.* **46,** 265 (1983).

12. S. Hansen, S. Andersson, and L. Fälth, *Z. Kristallogr.* **160,** 9 (1982).

13. P. H. Ribbe, *Acta Cryst.* **16,** 426 (1963).

14. K. Fischer, *Neues Jahrb. Mineral Monatsh.* **1966,** 1.

15a. J. V. Smith, *J. Chem. Soc.* **1964,** 3759.

15b. L. S. Dent and J. V. Smith, *Nature* **181,** 1794 (1958).

16. O. Jarchov, *Z. Kristallogr.* **122,** 407 (1965).

17. L. Pauling, *Z. Kristallogr.* **74,** 213 (1930).

18. J. A. Gard and J. M. Tait, *Acta Cryst. B* **28,** 825 (1972).

19. L. W. Staples and J. A. Gard, *Miner. Mag.* **32,** 261 (1959).

20. H. Nyman and B. G. Hyde, *Acta Cryst. A* **37,** 11 (1981).

21a. D. McConnell, *J. Chem. Educ.* **40,** 512 (1963).

21b. C. Cohen-Addad, P. Ducros, A. Durif, E. F. Bertaut, and A. Delapalme, *J. Phys.* **25,** 478 (1964).

21c. A. C. McLaren, R. F. Cook, S. T. Hyde, and R. C. Tobin, *Phys. Chem. Miner.* **9,** 79 (1983).

22. T. B. Reed and D. W. Breck, *J. Am. Chem. Soc.* **78,** 5972 (1956).

23. W. Nowacki and G. Bergerhoff, *Schweiz. Mineral. Petrog. Mitt.* **36,** 621 (1956).

24. W. M. Meier, *Z. Krist.* **115,** 439 (1961).

25. R. L. Withers, T. R. Welberry, G. L. Hua, and J. G. Thompson, *J. Appl. Cryst.,* in press (1988).

Subject Index

Page numbers in roman type refer to text, in italic type to data tables, and in bold font to figures. Prefix **c** to the last indicates a figure in the color section.

Formula Index

Note that the formulae frequently refer to structure types rather than chemical compounds (or minerals).

AcBr$_3$, 89
AcCl$_3$, 89
AgClO$_4$, 286, **287**, **288**, *289*
Ag$_2$F, 66
AgI, 56
Ag$_2$O, 13
AgSbO$_3$, *348*
Ag$_x$V$_2$O$_5$, 9, 35, **36**, *37*
AlB$_2$, 140, 217, *217*, **217**, 227
AlBO$_3$, 300
Al$_2$BeO$_4$, 64, *64*, **65**
AlBr$_3$, 117, **117**, *118*
Al$_4$C$_3$, 174, *174*, **c3**
Al$_4$C$_3$·2AlN, 175, **c3**
Al$_4$C$_3$·nAlN, 175
Al$_6$C$_3$N$_2$, 175, **c3**
Al$_8$FeMg$_3$Si$_6$, 88
Al$_2$MgO$_4$, 37, *37*, **38**, 64, 152, **154**, **156**, **c1**
AlN, 56, 175
AlNNi$_3$, 20
Al$_{13}$Nb$_{48}$Ni$_{39}$, **382**, 384
AlNbO$_4$, 24, 36, 360
Al$_2$O$_3$, α-, **100**, 101, 104, 113, *113*, **114**, **115**, 119, **c2**
Al$_2$O$_3$, θ-, 156
Al(OH)$_3$, 73
AlO(OH), α-, 70, *72*, **73**
AlSbO$_4$, 69
Al$_2$(Si$_2$O$_5$)(OH)$_4$, 387, 389, **390**
Al$_2$(Si$_4$O$_{10}$)(OH)$_2$, 389
AmCl$_3$, 89
As, 185

AsAg$_3$S$_3$, 230
AsCl$_3$, 266
AsF$_3$, 121, 266
AsI$_3$, 119
As$_2$S$_3$, 268, **269**, *270*
Au$_3$Sb$_4$Y$_3$, 335, *335*

BBr$_3$, 54
BCl$_3$, 54, *54*, **54**, 77
BI$_3$, 54
B$_2$O$_3$(II) (high-pressure), 59, *60*, **61**
BaAg, 236
BaBi$_2$S$_4$, 166
BaBr$_2$, 84
Ba$_2$CdS$_3$, 255
BaCl$_2$, 84, 209
BaFeO$_3$, 39
BaFe$_{12}$O$_{19}$, 39, **c1**
Ba(FeS$_2$)$_2$, 17, 307, **308**, *309*
Ba$_2$FeS$_3$, 250
Ba$_9$(Fe$_2$S$_4$)$_8$, 309
Ba$_{10}$(Fe$_2$S$_4$)$_9$, 309
Ba$_p$(Fe$_2$S$_4$)$_q$, 307
Ba$_3$Fe$_3$Se$_7$, 131
BaH$_2$, 84
BaHBr, 87
BaHg$_{11}$, 197, **199**, *199*
Ba$_2$MgWO$_6$, 41
BaMnO$_3$, 74
Ba$_2$MnSe$_3$, 255
BaNb$_2$O$_6$, 45, **46**
BaNiO$_3$, 39, 74, *74*, **75**, 258

421